AF179510

THE TRICONTINENTAL SERIES ON GLOBAL ECONOMIC ISSUES

(ISSN: 2251-2845)

Series Editor: John Whalley *(University of Western Ontario, Canada)*

Published:

Vol. 1 General Equilibrium Global Trade Models
edited by John Whalley

Vol. 2 China's Trade, Exchange Rate and Industrial Policy Structure
edited by John Whalley

Vol. 3 International Trade Agreements and Political Economy
edited by Raymond Riezman

Vol. 4 The Global Development of Policy Regimes to Combat Climate Change
edited by Nicholas Stern, Alex Bowen and John Whalley

Vol. 5 The Economics of Coercion and Conflict
by Mark Harrison

The Tricontinental Series on
Global Economic Issues

Vol. 5

THE ECONOMICS OF COERCION AND CONFLICT

Mark Harrison
University of Warwick, UK

World Scientific

NEW JERSEY • LONDON • SINGAPORE • BEIJING • SHANGHAI • HONG KONG • TAIPEI • CHENNAI

Published by

World Scientific Publishing Co. Pte. Ltd.
5 Toh Tuck Link, Singapore 596224
USA office: 27 Warren Street, Suite 401-402, Hackensack, NJ 07601
UK office: 57 Shelton Street, Covent Garden, London WC2H 9HE

Library of Congress Cataloging-in-Publication Data
Harrison, Mark, 1949–
The economics of coercion and conflict / by Mark Harrison (University of Warwick, UK).
pages cm. -- (The tricontinental series on global economic issues, 2251-2845 ; vol. 5)
ISBN 978-9814583336 (hardcover) -- ISBN 9814583332 (hardcover)
1. War--Economic aspects--History. 2. Defense industries. 3. Weapons industry. I. Title.
HC79.D4H36 2014
330--dc23

2013049595

British Library Cataloguing-in-Publication Data
A catalogue record for this book is available from the British Library.

Copyright © 2015 by World Scientific Publishing Co. Pte. Ltd.

All rights reserved. This book, or parts thereof, may not be reproduced in any form or by any means, electronic or mechanical, including photocopying, recording or any information storage and retrieval system now known or to be invented, without written permission from the publisher.

For photocopying of material in this volume, please pay a copying fee through the Copyright Clearance Center, Inc., 222 Rosewood Drive, Danvers, MA 01923, USA. In this case permission to photocopy is not required from the publisher.

In-house Editor: Philly Lim

Typeset by Stallion Press
Email: enquiries@stallionpress.com

Printed in Singapore

Dedication

To Peter Law

Contents

Acknowledgements

Many friends and colleagues have contributed to the work that I have assembled for this volume. I have been privileged in my co-authors Jari Eloranta, Paul R. Gregory, Andrei Markevich, and Nikolaus Wolf. Each of them has given me something special that I could never have aspired to on my own. I thank my fellow economic historians at Warwick, Nick Crafts and Bishnupriya Gupta, for their intellectual comradeship. Although he is no longer at Warwick it would be hard not to include Stephen Broadberry here. His insights have guided me and his restraining hand has saved me from many excesses. The University of Warwick has been my intellectual home, and its Department of Economics and its Economic and Social Research Council (ESRC) Centre on Competitive Advantage in the Global Economy have generously supported my research time and travel. At the Hoover Institution on War, Revolution, and Peace, I have found unrivalled documentary resources as well as time to think and write. At home, my loved ones have never faltered in their encouragement. I thank Anne, Sam, Jamie, and Roy for sharing their affections with me and keeping their doubts to themselves.

I am grateful to John Whalley, the series editor, whose invitation led me to make this book, and to Philly Lim, my book editor, whose eye for detail, good sense, and great patience led to a better book than I could have made on my own.

This book is dedicated to Peter Law. Peter has been my critical friend and unselfish critic for forty years; his hand is in everything I have written. No one could have been a better colleague.

Except for the Introduction and Chapter 6, which are published for the first time, all the chapters have previously appeared elsewhere. They are reproduced here without change, except for the updating of a few bibliographic references. I thank the publishers, the editors, and my co-authors for permission to reprint them.

Chapter 1, "War and Disintegration, 1914–1945", was first published in 2010 in *The Cambridge Economic History of Modern Europe* (edited by Stephen Broadberry and Kevin O'Rourke), Volume 2, pp. 133–155 (Cambridge: Cambridge University Press).

Chapter 2, "Why the Wealthy Won: Economic Mobilization and Economic Development in Two World Wars", was first published in French ("Pourquoi les riches ont gagné: Mobilisation et développement économique dans les deux guerres mondiales") in 2012 in *Deux guerres totales 1914–1918* to *1939–1945: La mobilisation de la nation* (edited by Dominique Barjot), pp. 135–179 (Paris: Economica).

Chapter 3, "Why Didn't the Soviet Economy Collapse in 1942?" was first published in 2005 in *A World at Total War: Global Conflict and the Politics of Destruction, 1939–1945* (edited by Roger Chickering, Stig Förster, and Bernd Greiner), pp. 137–156 (Cambridge: Cambridge University Press).

Chapter 4, "The Frequency of Wars", was first published in 2012 in the *Economic History Review*, 65(3), pp. 1055–1076.

Chapter 5, "Soviet Industry and the Red Army under Stalin: A Military-Industrial Complex?" was first published in 2003 in *Les Cahiers du Monde russe*, 44(2–3), pp. 323–42.

Chapter 7, "A Soviet Quasi-Market for Inventions: Jet Propulsion, 1932 to 1946", was first published in 2005 in *Research in Economic History* 23, pp. 1–59.

Chapter 8, "The Political Economy of a Soviet Military R&D Failure: Steam Power for Aviation, 1932–1939", was first published in 2003 in the *Journal of Economic History*, 63(1), pp. 178–212.

Chapter 9, "The Fundamental Problem of Command: Plan and Compliance in a Partially Centralized Economy", was first published in 2005 in *Comparative Economic Studies*, 47(2), pp. 296–314.

Chapter 10, "Accumulation and Labor Coercion Under Late Stalinism", was first published in English, but with a Russian title ("Nakoplenie i prinuzhdenie k trudu v period ≪pozdnego≫ stalinizma") in 2006 in *Rossiia v kontekste mirovogo ekonomicheskogo razvitiia vo vtoroi polovine XX veka* (edited by V. A. Vinogradov *et al.*), pp. 308–324 (Moscow: Moscow University Press).

Chapter 11, "Economic Information in the Life and Death of the Soviet Command System", was first published in 2005 in *Reinterpreting the End of the Cold War: Issues, Interpretations, Periodizations* (edited by Silvio Pons and Federico Romero), pp. 93–115 (London: Frank Cass).

Chapter 12, "Coercion, Compliance, and the Collapse of the Soviet Command Economy", was first published in 2002 in the *Economic History Review*, 55(3), pp. 397–433.

About the Author

Mark Harrison is Professor of Economics at the University of Warwick. He is a research associate of Warwick's ESRC Centre on Competitive Advantage in the Global Economy, a research fellow of the Centre for Russian and East European Studies at the University of Birmingham, and a research fellow of the Hoover Institution on War, Revolution, and Peace at Stanford University. He obtained his BA in Economics and Politics from the University of Cambridge in 1970 and his DPhil in Modern History from the University of Oxford in 1974. He was awarded the Alec Nove Prize in 1997 for his book *Accounting for War: Soviet Production, Employment, and the Defense Burden, 1940–1945* (Cambridge University Press) and shared the Russian National Prize for Applied Economics in 2012 with Andrei Markevich for their 2011 paper, "Great War, Civil War, and Recovery: Russia's National Income, 1913–1928", in the *Journal of Economic History*, 71(3), pp. 672–703.

Introduction

Mark Harrison

> **coercion**, n. "1. a. Constraint, restraint, compulsion; the application of force to control the action of a voluntary agent ... b. Forcible restraint of (action) ... 2. Government by force, as opposed to that which rests upon the will of the community governed."
>
> **conflict**, n. "1. a. An encounter with arms; a fight, battle ... b. *esp.* A prolonged struggle ... c. (without *article* or *pl.*) Fighting, contending with arms, martial strife."[1]

The study of economics begins with trade.[2] In an idealized market, sellers compete with each other for buyers. Under the rules of the market, competition is impersonal and non-violent. Each person makes their best choice, which can include staying out of the market; no one is forced to take part. As Adam Smith (1776/2005: 364) suggested, a result of the sellers' pursuit of their own private profit is that resources are "led by an invisible hand" towards their best uses. In turn, the buyers' pursuit of their own greatest satisfaction ensures that everyone gains who takes part willingly. In the outcome, the well-being of society is raised.

This model was first developed when England was still an agrarian society and is simplified in the extreme, yet it explains much of modern prosperity.

The world of coercion and conflict is, at first sight, utterly different. In this world "every man is enemy to every man ... and the life of man, solitary, poor, nasty, brutish, and short" (Hobbes, 1651/1909: 96–97). Each person's behaviour is limited not by rules but only by their own conscience

---

[1]OED Online (June 2013). Oxford University Press. http://0-www.oed.com.pugwash.lib.warwick.ac.uk/view/Entry/35725 and http://0-www.oed.com.pugwash.lib.warwick.ac.uk/view/Entry/38898 (accessed 27 August 2013).

[2]I thank David Hugh Jones and Peter Law for thoughtful comments and advice.

or, in its absence, by greed and fear. The immediate casualties are other people. In the longer term wealth is destroyed and society is impoverished. Even if a few lords or warlords retain a profit, there is never enough left over to compensate the losers.

Thus there are two worlds, one of free markets, industriousness, and enjoyment. The other is the world of tyrants, soldiers, slaves, and the lash. The gap between them is apparently unbridgeable. It turns out, however, that among the tools we have developed to analyze markets and corporations are some that can readily be turned to understand war and repression.

I am not the first to try to list these concepts and tools, and explain their application. Economists, historians, and political scientists have recently contributed excellent applied and theoretical studies of the political economy of dictatorship and coercion (Wintrobe, 2000; Gregory, 2001; Gregory and Lazarev, 2003; Dixit, 2004; Gregory, 2004; 2009). Others have done the same for interstate conflict (Offer, 1989: 7–20; Fearon, 1995; Bueno de Mesquita, 2006, Brauer and van Tuyll, 2008: 1–44, Smith, 2009: 19–53; Rockoff, 2012: 13–47; Garfinkel and Skaperdas, 2012; Findlay and O'Rourke, 2012).

If there is novelty in this chapter, it stems from examining coercion and conflict in the same framework. This is appropriate because the two are organically connected.

Adversaries, Coercion, and Conflict

To understand how coercion and conflict are connected, we need some definitions. What is conflict? There is a potential for conflict whenever two persons disagree, for example, about how to use or dispose of a resource. Resources can be of any kind — physical, financial, political, or emotional. Many such disagreements do not amount to conflict; they are fleeting, and are resolved quickly through compromise, so that they do not end in "an encounter with arms" as the Oxford English Dictionary (OED) puts it (cited above). Such readiness to compromise requires mutual recognition of the entitlements of the parties in disagreement, including the right not to agree. The fact that a particular dispute persists over a considerable period of time, however, is usually a signal that it is not of this transient and harmless nature; the parties cannot agree to each other's right to disagree. They become adversaries in a "prolonged struggle" — one of the alternative meanings that the OED gives to conflict.

If unresolved disagreement over entitlements lies at the root of conflict, how is conflict resolved? Garfinkel and Skaperdas (2012) define conflict by the presence of "adversarial investments", required for the "threat or use of coercive means." In the words of the OED conflict is inherently violent: adversaries engage in "fighting, contending with arms, martial strife." Here, economists go further: violence is a possible correlate of conflict, but it is not necessary, because expectations matter: the threat of violence can be sufficient, provided it is backed by costly investments in "coercive means" (i.e. it is more than just cheap talk). In the presence of threats, conflict among adversaries can be worked out under duress, but without violence, through negotiation or bargaining.

What is coercion? When I coerce you, I impose my choice on yours. As the OED puts it, it is "the application of force to control the action of a voluntary agent." But the agent is still voluntary, and coercion does not deprive the agent of all choice. Rather, coercion means I force you to choose between the alternatives as I have defined them, not as you would see them. Moreover, each of the alternatives I allow you is inferior to the entitlement you have or seek. A highwayman points a gun at you and demands: "Your money or your life!" That is coercion. Within it, you have a choice: you're free to choose whether to give me your wallet or die at my hand. The signal of coercion is that you cannot walk away and keep what you had. Whatever you choose, you will end up worse off than you were before. Thus, when I coerce you, it is intrinsic to the situation that you lose something to which you believe you were previously entitled.

(And, because entitlements rest ultimately on beliefs, conflicts over entitlement are always perceived asymmetrically. I stand up for my rights; you're unreasonable.)

To summarize, coercion can be distinguished from free exchange, but the difference does not lie where many would naturally assume, in the victim's absence of choice. The difference is that, when you and I trade freely, you do not suffer any loss of entitlement. Either the choices available to you through trade leave us both better off, or you can remain no worse off by choosing not to trade. Coercion differs from trade not in removal of the victim's choice, but in the restriction of choice to a set of options that is strictly inferior to those available beforehand.

Coercion is often linked to repression, but the two are not the same. When I coerce you, I leave you worse off. When I repress you, I prevent you from signaling your protest — your dissatisfaction with the outcome. Repression can be political, but it can also be economic or financial. Applied to political markets, repression means the silencing of discontent. Economic and financial repression refers to the administrative controls that conceal market stress, which would otherwise be signaled by rising prices or interest rates. Such controls include price and interest caps and allocation by rationing and licensing.

Conflict, coercion, and repression have in common the exercise of force. Force requires "adversarial investments" and so is costly to produce and use. This is what makes them all negative-sum activities — as distinct from trade, which gives rise to a positive sum. After I have forced you to do what you would not have chosen to do willingly, our joint wealth is less than it was before. My power allows me to gain, only by imposing a loss on you. If you resist, and I impose my power on you by force or threats, our joint wealth is diminished, and the smallest loss I must impose on you in order to gain myself must be larger by that amount.

Just as conflict does not have to end in violence, coercion need not be violent either, but there is generally violence when coercion is resisted. It could be that you refuse the inferior choice set that I offer you ("Your money or your life"). This suggests that my power appeared inadequate at first, and required some demonstration or reinforcement. Or, conflict arises as a prelude to coercion, if I attack you (or you attack me) preemptively, to demonstrate superior coercive means. In other words, violence expresses actual or potential resistance to coercion and is therefore linked to the scope for coercion to fail.

In what I have written so far, coercion is never productive. That perspective is perhaps too narrow, or rather it accurately reflects a bias in my research. To explain further, in the example of the highwayman, coercion was used to steal your property, and this could be thought of as destructive not only of your personal entitlement but also of the general system of property rights. A system of property rights that can be enforced within the law is generally recognized as one of the hallmarks of a well-ordered society. In that case, it would seem a good idea for the public to apply some coercion to the highwayman. With the right combination of stick and carrot, a reformed highwayman might even gain in the long run. Then, coercion would be productive of a social benefit.[3]

[3]In a similar spirit Garfinkel and Skaperdas (2012) limit their definition of conflict to exclude those investments in adversarial means that bring external benefits to third parties. They intend this to exclude from the sphere of conflict the competitive tournaments and sporting events that raise productivity or provide enjoyment. Investments in policing under the rule of law also provide external benefits to third parties and on the same logic they should also be excluded from our definition of conflict.

In a well-ordered society, coercion enforces the law, but it is also restrained by the law. Here, law enforcement and legal restraint are two sides of the same coin. One importance of legal restraint is to underpin the community's consent to the laws that are enforced. It is different in the other sort of society, the sort that is not "well-ordered." There, coercion enforces the will of the government, not the rule of law. In that case government is, as the OED puts it, "by force, as opposed to that which rests upon the will of the community governed."

Law enforcement and public coercion subject to the law are very important subjects for history and social science. On the whole, however, my research has addressed other aspects of public coercion, those that have lent support to arbitrary power including the unrestrained enforcement of arbitrary laws, when the government is relatively unconcerned about personal freedoms and private entitlements and does not answer to the courts, the electorate, or anyone but itself. In the cases that I have studied, coercion was generally associated with repression, because the government did not regard the victims of coercion as entitled to any legal redress or legitimate expression of protest.

The chapters in this book are mainly about conflict and coercion that is organized among states and groups. The state is present in every chapter, sometimes as the body that seeks to monopolize violence over its subjects, sometimes as the main source of violence within global society.

In the long run of human history the monopolization of lawful violence by the state has been one factor in the great diminution of interpersonal violence in the world since early times (Gat, 2006; Pinker, 2011; North *et al.*, 2011). But "lawful violence" must be understood carefully; it implies that alongside the state's monopolization of violence come acceptance of the rule of law, including private property rights and other rights of the citizens vis à vis the state. A problem here is that the source of these laws is the sovereign state and the source of international law is agreement among sovereigns. All sovereigns are subject to the Weingast (1995: 1) paradox: "A government strong enough to protect property rights and enforce contracts is also strong enough to confiscate the wealth of its citizens." On the same reasoning, a state that is powerful enough to monopolize violence is also powerful enough to exercise it without restraint against its own citizens and against other states.

If public coercion and conflict among states are so closely related in theory, the connection should be reflected in the facts. And it is: empirically, as discussed in Chapter 4, coercion and conflict often arise together because coercive political regimes are very often sources of conflict among states.

Before the modern era, virtually all states were authoritarian and warfare was the main function of the nation state. In European history, as Charles Tilly (1975: 42) observed, "War made the state and states made war." This relationship between the nation state and warfare began to break down only as governments became democratic, and acquired additional functions, and at the same time the rule of law took the place of rule by men.

There are many reasons why states based on the rule of law (Kant, 1795/1983) and democracy (de Tocqueville, 1835/2000) might be expected to prefer peace. These range from moral considerations, such as reluctance to spill blood except in self-defense, to the self-interest of politicians and governments under democratic arrangements. Normatively, societies that subject themselves to the rule of law are likely to extend the same protection to foreigners as to their own citizens. The structure of democracy may also impose restraints on democratic leaders; empirically, leaders that lose wars are more likely to lose office in democracies than in autocracies (Bueno de Mesquita and Siverson, 1995). Whatever the reason, it is "as close as anything we have to an empirical law in international relations" that "liberal or democratic states do not fight each other" (Levy, 1988: 662).

Given the close links between coercion and conflict, it is not surprising that the toolkits that the economist applies to these two topics are largely the same.

Rational Calculation

The single most important instrument that the economist brings to bear is the idea of rational choice. Without it, social science is nothing more than a psychological novel in which people are driven by inner forces they neither understand nor control. Are people sometimes driven by inner forces they neither understand nor control? Certainly. But there must be more to it than this, or the practice of social science becomes impossible.

When choices are rational, that is, calculated, each person is thought to weigh up the expected costs and benefits to themselves of the possible actions and to choose the one that offers the greatest surplus (or the smallest loss). As Clausewitz (1832/1982: 119) wrote:

> "War is a mere continuation of policy by other means... War is not merely a political act, but also a real political instrument, a continuation of political commerce, a carrying out of the same by other means."

Consider the idea of optimization based on the computation of costs and benefits to oneself and the balancing of first order conditions, and call it "pure" rationality. Does that fully describe human nature as we find it expressed in empirical patterns of behaviour? Clearly not. It cannot account, for example, for people who destroy themselves or the things and people they love — at least, it cannot account for this in the form I have stated it. But is it a useful starting point? Very much so, because, as I wrote in another context (Harrison, 2005):

> "If people do what they want, subject to the resource and information constraints that we can identify and if we do not understand what they do, then we are missing something important and we should not be satisfied to throw up our hands."

Consider a country threatened with overwhelming force. For the sake of argument the country is Poland and the year is 1939. Hitler wishes to acquire the Polish homeland and this is something that most Poles would prefer to avoid. But avoidance is not on offer, so their real choice is between surrender and fighting to defeat. Fighting is destructive and so must leave everyone worse off than in the case of surrender. Thinking on lines of pure rationality, a deal is available that would leave both sides better off than if war broke out, so both sides should accept it. The Poles should agree to give away the value of their homeland (which they will lose anyway), and accept from Germany in compensation a proportion of what will be saved by abstaining from resistance. For the Germans this should be an acceptable deal, since they will gain Poland and the Poles' compensation will cost them less than they would have spent on violence.

For some, it is a problem that that's not what happened. The crazy Poles did not surrender, but fought until they were defeated. Our model of pure rationality is a failure, apparently.

Actually, no. The model has told us that we are missing something, and in that sense it is a success: it has extended our interface with the unknown. When we see human agents doing something that is costly, the principle of rationality tells us that they must expect to gain some future benefit (or avoid some future loss) that is equal or greater. If we missed it, we need to work out what it was.

What were we missing? I'll consider five different ways of understanding the rationality of Polish resistance to Germany in 1939. These are all factors in the motivation of conflict that are important for economists — and others — to understand, and at the bottom they all point in the same direction. These factors are entitlement, reputation, identity, uncertainty, bounded rationality, and credibility. The first four would not automatically prevent coming to an agreement, although they would make coming to an agreement or less likely, more difficult, or more costly to Germany. The fifth, credibility, would seem to rule an agreement out, making war inevitable.[4]

Entitlement refers to the established fact that we value something more when it is already ours — when we consider that it belongs to us by right. This is one of the foundations of modern behavioural economics (e.g. Kahnemann *et al.*, 1991). This suggests a reason why the Poles were ready to fight: they considered Poland was theirs, and the value of resistance was the defense of their entitlement.

Another factor can be *reputation*. According to the historian Anna Cienciala (2011), Polish leaders of the time, such as foreign minister Józef Beck, put honor above the avoidance of conflict. If they had given away what was theirs without a fight, accepting a vassal state for Poland, they would have lost their honor. In this view the value of resistance was the safeguarding of their honor, even if they lost everything else.

The concept of *identity* fills a gap in economics (Akerlof and Kranton, 2000). Economists often describe economic behaviour as the rational pursuit of self-interest. But who are we, and how can we define our self-interest without first establishing who we are? "I could not live with myself if I did not . . .": their own self-identification is often a person's most precious possession, and occasionally one for which life itself must be sacrificed. Thus, the paradox of behaviors that lead to self-destruction including suicide, suicidal terrorism, and suicidal heroism, cannot be understood except in terms of the self that such behaviours defend (Harrison, 2006). If they had not fought for Poland, the Poles would not have been Poles. Thus, the value of resistance was to safeguard a precious identity that would otherwise have been destroyed.

Our analysis would be grossly incomplete if we did not take into account that identity has many dimensions. The Poles were not just Poles. They were also men, women, and children; elites and poor; Jews and non-Jews; soldiers and civilians; religious and secular; nationalists and

[4]Fearon (1995) lays out further grounds that may prevent agreement and lead to war within a rational-choice framework. Even when agreement is possible, each side still wishes to secure the best agreement possible. To achieve this, the government may try to exploit private information and strategic misrepresentation in such a way that agreement is impeded. These do not seem applicable to Poland in 1939, but can still be relevant elsewhere.

internationalists; and so on. They surely did not weigh up everything the same way. Poland was a young nation, and national entitlement, honor, and identity are the currency of nation-building leaders. Honor was quite possibly of more value to the elite than to the foot soldiers that were also going to have to die for it. Disputes are more likely to be resolved through violence when identities become polarized (Sen, 2006).

This reminds us, finally, that it is not nations that optimize, but persons. Each person frames the choices of others; each person in the nation must decide their own self-interest, given the choices that they expect others to make. This interaction is something else that was missing from our first attempt at "pure" rationalism.

One factor in Polish resistance may have been *uncertainty* over future costs and benefits. Our model was framed by the assumption that if it came to a conflict the Polish defeat was certain. In reality, war is a gamble: "No plan of operations can look with certainty beyond the first meeting with the major forces of the enemy," wrote the older Moltke (cited by Holborn, 1986: 289). Every Pole remembered the "miracle on the Vistula" of 1920, when the Red Army's apparently unstoppable advance on Warsaw was halted by an unexpectedly successful Polish counter-attack. It is clear that Germany's Wehrmacht in 1939 was far more of an existential threat to Poland than Soviet Russia's Red Army in 1920. There was no room for Polish optimism in any objective reading of the balance of forces. They were almost certainly going to lose.

Our *bounded rationality* restricts our capacity to make the right decision (for a recent summary, see Kahneman, 2011). The limits are cognitive as well as computational. Among other things, we overvalue small probabilities — for example, the chances of a second "miracle on the Vistula" within two decades of the first. Members of the Polish elite, with shared experience of service in the dragoons, were likely to see a cavalry charge as part of the solution to Poland's problem, even if Poland's problem had no solution at all.

Finally, any deal that Germany could have put on the table would not have been *credible*. In fact, the secrecy of Germany's war preparations, including the secret pact with Stalin that opened Hitler's way into Poland, shows that Hitler did not even try to offer the Poles a deal that could avoid war in 1939: Nor would the Poles have been wise to accept one if it was offered. Whatever agreement Hitler might have made before a Polish surrender, he would surely not have kept his word afterwards. This was the same Hitler that had recently swallowed Austria, the Czech Sudetenland,

and Czechoslovakia itself. He had a clear record of making promises when it suited him to do so and breaking them afterwards when it suited him. Any deal that Hitler offered to the Poles would not have been credible. They would not have responded to the attempt, and he did not bother to try.

The model of rationality that we began with exemplifies the saying of the statistician George Box *et al.* (1987: 74): "All models are wrong; the practical question is how wrong do they have to be to not be useful." The idea of rational choice is wrong because it is oversimplified, but it is not so wrong that it is useless. Used properly, it can still guide us to a fuller understanding of the choices that people make, even in situations characterized by ignorance, anxiety, and existential threat.

Strategic Interaction

Strategy has many meanings in the modern world, most of them only distantly related to its roots. Strategy begins with prediction of the future, and many public and private organizations claim to have "strategic plans" that are really just aspirational descriptions of the state in which they would like to be in a few years' time. There, the word "strategic" is empty, because any plan is about the future, and calling it strategic adds nothing.

The origin of the word "strategy" lies with generalship. The main task of the general is to predict and counteract the action of some adversary or rival. He must decide his best action (for example, to advance along a given front), not knowing the enemy's deployment but trying to predict it based on the fact that the enemy is simultaneously trying to predict the general's line of advance in order to decide where to deploy his defenses. "What does planning mean in war?" Answering his own question, Stalin's Marshal, Ivan Konev (1970) wrote, "We make plans alone but carry them out, if one can put it so, in company with the enemy, that is, taking into account his counteraction." It is this simultaneity and mutual engagement that captures the spirit of strategic interaction.

When we act strategically, we may or may not reveal our true preferences, or we may even conceal them. We may line the streets to cheer a hated tyrant. We may prefer peace but nonetheless plan for war.

Modern social science uses game theory to think about strategic interaction. While many particular games are studied in the contexts of conflict and coercion, one is canonical (Fearon, 1995). Suppose you have a resource that I want, that you would not give me freely, but only if I coerce you. In a dispute among states this resource might be your land; in

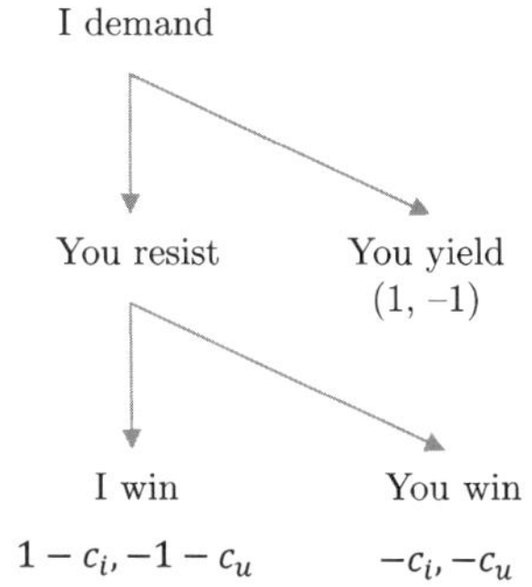

Fig. 1. Conflict Over Resources.

a domestic setting it might be your labor. Set the value of this resource equal to 1. We can resolve our conflict with or without violence. Violence costs me c_i and you c_u (think of these costs as proportions of the value at stake, which is the unit of account).[5] Then the structure of payoffs might look as in Figure 1.

The game played in the figure works like this. If I make my demand, and you yield, the resource changes hands from you to me. If you resist and I win, I gain the resource less the cost to me of violence; you lose the resource and the cost to you of violence. If you win, nothing changes hands and we both lose the costs of our own violence.

This game yields simple, intuitive results that also illustrate the principle of strategic interaction. Should I make my demand? On the face of it: Yes, if I expect you not to resist, or if I expect to gain from a fight. Both of these depend on the probability p of my winning if it comes to violence. Find our expected payoffs from violence by applying p to my winning payoff and $1 - p$ to yours. Then, you will be willing to fight me if you expect to gain from it, i.e. $1 - p - c_U > 0$; otherwise, you will yield without a fight. I will be willing to fight you if I expect to come out ahead from a conflict, i.e. $p - c_i > 0$. Unfortunately for the prospects of peaceful conflict resolution, both of these conditions can hold at the same time; they require only that the resource will not be completely consumed by the costs of violence, i.e. $c_i + c_u < 1$. Violence can be avoided only if it is so costly in prospect that a fight would destroy everything.

[5]Other assumptions are that we are risk-neutral and we have common values and full information, so there is no scope for the private information and strategic misrepresentation issues discussed by Fearon (1995).

Missing from this set-up is the option to bargain peacefully. That scope for negotiation ought to exist is shown by the fact that beforehand the value at stake is one unit whereas, after a value-destroying conflict, the most there can be to distribute between us is $1 - c_i - c_u$. By agreement we can save the value that otherwise would be destroyed $(c_i + c_u)$ and share this among ourselves. My outcome can then be better than $p - c_i$ and yours can be better than $1 - p - c_u$. As we have already discussed, however, the likelihood of a bargain is dramatically reduced or eliminated altogether by the problem that rulers who are willing to use violence to get their way, whether domestically or internationally, may also be unable to commit to agreements that avoid it.

This game is sometimes called divide-the-dollar (two players bid for a dollar; if their bids sum to less than the dollar, they receive their bids, or otherwise nothing). While international relations provide its usual setting, the divide-the-dollar game can also be used to illustrate the outcomes of domestic conflict under a coercive regime. The failure of Soviet economic reforms provides an example. The "dollar" (or ruble) here is the additional resources that would have been supplied if Soviet managers had honestly reported their assets and capabilities to the government, if their workers had given their working time fully to meeting management goals, and if the government could have committed to reward the managers and workers in return, in other words, to divide the ruble with them.

In 1929, Stalin began to transform the Soviet economy into a centralized command system. By centralizing economic controls he was able to enforce a near-complete monopoly of capital and a monopsony of wage labor and farmers' food surpluses. Between 1928 and 1940, increases in compulsory work norms and production quotas and reductions in real wages brought about a transfer of roughly 30 per cent of Soviet GNP (measured at factor costs of 1937) from household consumption to other uses, particularly investment and defense (Bergson 1961: 237).

In a market economy many workers would have responded to such a squeeze by looking elsewhere. They would have been free to find non-government buyers for their products or services or in the last resort to emigrate. Under communist rule these options were criminalized and heavily punished. In the Soviet Union most workers were limited to "inside" options which meant, for example, showing up to work but reducing effort and attention to quality. (Some important exceptions are the subject of Chapters 7 and 8.).

There was a vicious circle, summed up in the Soviet-era joke: "We pretend to work and they pretend to pay us." At intervals over the next half century, Soviet leaders made various efforts to unlock this vicious circle. The key, they believed, was to reward managers and workers for taking on more ambitious norms and quotas. Among these efforts, most prominent were the rate-busters' movement inspired in 1935 by the coal miner Aleksei Stakhanov, and the postwar management and incentive reforms of 1965 sponsored by prime minister Aleksei Kosygin.

The problem was that government promises, to divide the ruble by letting the workers keep the rewards for increased effort, were not credible. Enterprises that played the government's game tended to be rewarded only in the short term. Once they had revealed what they were truly capable of, they had given away their main bargaining chip in negotiating the next quota. In the longer term they were given higher norms and quotas, and so lost their promised share of the ruble (Harrison, 2002).

In this sense (if no other) the structure of the problem that Soviet workers faced with government promises in 1965 was identical to the one that the Poles faced with Hitler's demands in 1939: after implementing any bargain, they could anticipate that the terms would be shifted against them, and they would have given away their power to resist, so it was better not to make any deal and slug it out. As a result the equilibrium outcome of seeking to divide the ruble in the Soviet version of the game was low effort, low productivity, and endemic, unresolved conflict over norms and quotas.

In game theory terms the models shown here are highly simplified.[6] They serve mainly to illustrate the structure of strategic interaction. Although very simple, they have the virtue of suggesting how governments and their adversaries had no choice but to put themselves in each other's shoes in order to identify their own best choices, whether the adversaries were foreign leaders or their own subjects. This is the essence of strategic interaction.

Increasing and Diminishing Returns

Strategic interaction has other features that make clear its affinity to the subject of coercion and conflict. One is the importance of activities in which there are increasing returns at the margin. Most economic activities are

---

[6]In more advanced games we find that players will behave differently given opportunities to throw dice, anticipate repetition, learn about each other, form beliefs, build reputations, and purvey misinformation. For discussion of historical applications see Greif (2006).

characterized by diminishing marginal returns, and most economic models assume it.

"Diminishing returns" sounds like a gloomy sort of thing, but in practice it is what keeps society on the rails most of the time.[7] I like a glass of wine, but I enjoy the second glass less than the first. As a result I limit my consumption. Beyond a point, I stop. In other words, processes characterized by diminishing returns tend to be self-limiting and arrive sooner or later at a point of stability (which economists call equilibrium). The same processes would become explosive in the presence of increasing returns. If I enjoyed every glass more than the one before, I would quickly drink myself into unconsciousness. At the same time, if I did not enjoy the *first* glass very much, I might never start. In other words, my life would go to extremes: I would either be completely sober or there would be no stopping me.

In conflict and coercion there are often increasing returns, although in a sense that differs a little from the example of my drinking habit. In this case it is not my effort that makes returns diminish or increase, but the efforts of others. Figure 2 has two panels. Panel (a) illustrates diminishing returns.

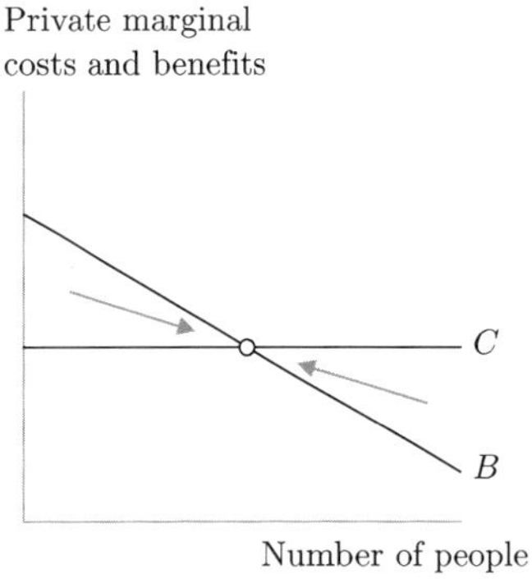

(a) Diminishing returns: private marginal benefits from taking part in an activity *decline* as the number of people engaged in it increases.

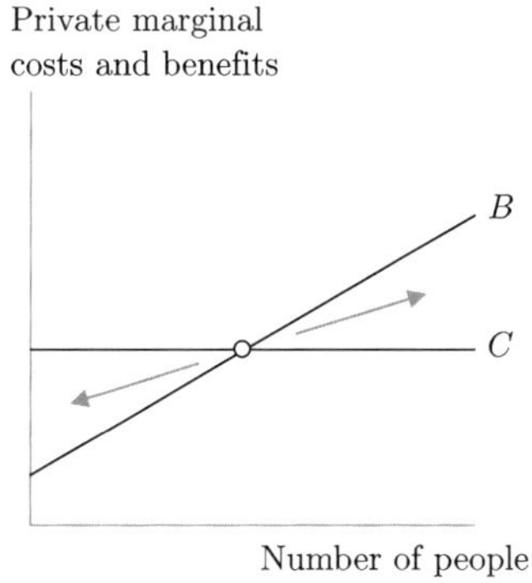

(b) Increasing returns: private marginal benefits from taking part in an activity *increase* as the number of people engaged in it increases.

Fig. 2. Increasing and Diminishing Returns.

[7]David Hugh Jones highlighted this point for me in a blog post, "Very simple thoughts about the politics of crisis" (7 April 2013), available at http://davidhughjones.blogspot.co.uk/2013/04/very-simple-thoughts-about-politics-of.html (accessed 7 April 2013).

Think of an industry that has a given market for its product. Working in that industry has certain costs, which are also taken as given. When few workers are engaged in the industry they command a high wage, so it is profitable for others to join them. Workers flow into the industry, but this also reduces the wage. Conversely, with too many workers engaged in the industry the wage is driven below the cost of working in the industry, and workers are driven out. Whichever side of the equilibrium point you begin from, the process stops in the middle, where the number of people engaged is such that the benefit to the joining worker just matches the cost of doing so. The result is a stable equilibrium. This is the normal world of everyday production and trade.

Panel (b) illustrates increasing returns. This covers situations where the private benefit to a person of doing something increases with the number of other people that are doing the same thing at the same time. Imagine a war of attrition, in which two armies grind each other down. The battle is won by the last man standing. In that battle, the more soldiers join in on my side, the more likely I am to be that last man standing. But if you, the soldier next to me, turn and run, that exposes me to the enemy and reduces my chance of survival. As each additional soldier falls out of the ranks, they reduce the benefit to me of continuing to fight. As I watch you, I ask myself continually: Should I run too? If that calculation ripples along the line, the army unravels suddenly into a fleeing mob, and the battle is lost.

Under these circumstances war becomes a game of all or nothing: we all join in together, or we all run for it. There is no balance between the two.

A similar example works for political coercion. My block leader orders me to fall into line with my neighbours and sing with them: "All hail our glorious leader!" How should I respond? My first response may not be to question whether the leader is really glorious but to find out what others will do. If my neighbours are evenly divided, I might feel free to go with my inner beliefs. If I hear they will all march and sing, my private feelings about our leader may be overwhelmed by the likely cost to me if I make a stand on my own and register dissent. If other people show resistance, however, even if I am the biggest coward on the street, I should bear in mind that it is also dangerous to be a hated tyrant's last supporter.

In other words, for anyone who is not an instinctive rebel, conforming has returns that increase with the number of other people that conform. The trick is to predict the standard to which one should conform. Hence the common phenomenon of "weathervanes" — people whose professed opinions follow the prevailing orthodoxy.

In the explosive world of increasing returns, there is no longer just one equilibrium. Rather than getting slightly inebriated, I will either be completely sober or unconscious. The army will either hold or collapse. The regime will be stable until it falls suddenly; the crowd will be fickle, and everyone will be strongly "for" until they turn "against". To make my own best choices I must continually try to work out how everyone else will behave, knowing incidentally that they are also trying to predict my own behaviour, which is more or less the definition of strategic interaction. This is, finally, a world of agonizing dilemmas: every day I must weigh integrity against survival.

The role of increasing returns should not be overstated. In wartime, many returns continue to diminish, and this accounts for important regularities. The airplane was an invention of the twentieth century, and in the 1930s most major powers gave a lot of resources to this new weapon. But beyond a point they stopped, and also modernized their armies and navies. This recognized diminishing returns to air power: sooner or later, they got a point where they figured the military value of one more plane would fall below the same value assigned to one more of the traditional guns or ships. They also limited rearmament somewhere around the point where the expected value of one more dollar given to the army fell below the value of the same given to agriculture or public works. Even in total war, this principle continued to apply: even when the threat was existential, no country gave literally everything to the war.

The returns to coercion also diminish beyond a point. An extreme case of coercion is provided by the residents of Stalin's labor camps. Behind the barbed wire of the Gulag, men and women were stripped of nearly all possessions, nearly all family contact, and nearly all control over time and movement. All their choices left them worse off. Yet, what did experience show? You can put a man in these conditions, and point a gun at his head, and it can still be hard to make him work. As Chapter 10 shows, the expansion of the Gulag saw the evolution of ever more varied and complex incentive schemes that were designed for no other purpose than to persuade the slaves to work. The stick and the carrot were not alternatives. Even when the main emphasis fell on the stick, Stalin still had to give out a few carrots. In other words, there were limits to coercion. Not even a totalitarian dictator could overcome diminishing returns.

In short, both diminishing and increasing returns are essential instruments in the toolkit that everyone needs to understand conflict and coercion, and the problem is largely to understand which one to apply in each context.

Scale and State Capacity

In conflict, scale might seem unambiguously good. Since Clausewitz, modern strategists have thought of the ideal military campaign as the application of overwhelming force to the enemy's weak points. Sometimes it worked quickly, sometimes not. When it dragged out, modern conflict often became a war of attrition in which victory would be claimed by the last man standing on the field of battle. Whether warfare took the form of a lightning blow or a bloody slog, why would less ever be worth more than more?

In fact, the wars of the twentieth century have been historically exceptional. If we go back through history it is relatively easy to find battles in which smaller armies were able to inflict overwhelming defeats on much larger forces. At Marathon (490BC), a Greek army numbered in thousands routed a Persian force of five to ten times the size. The Carthaginians achieved a similar victory over Rome at Cannae (216BC). The English did it to the French twice in a century at Crécy (1346) and Agincourt (1415).

In such cases, the smaller force defeated the larger one not so much because of an advantage of equipment or terrain but by disrupting the opposing force to a point where it suddenly changed state, from a proper army to an uncoordinated rabble that could easily be slaughtered. At times, it seems as though the smaller force was easier to hold together by ties of discipline and comradeship than a larger force where each had a smaller chance of a part in the fighting and felt less commitment and obligation to the cause of the whole.

Something that defines modern war, dating it from the time of Napoleon, is the emergence of mass armies that could be held together in defeat. Napoleon's retreat from Moscow, the Confederate retreat from Gettysburg, the Japanese retreat to the home islands from South East Asia and the South Pacific, and the German retreat from Stalingrad and the Caucasus, are all examples of vast bodies of men that somehow failed to disintegrate despite pulverizing defeats that were piled one on top of another for months and years.

If every great army of the twentieth century had disintegrated on suffering a great defeat, as the French army did in 1940, both world wars would have been over within months. The fact that they often did not, but were held together and retained their military capabilities, ensured that modern war became a war of resources, in which those armies won that could throw the greater number of men and munitions into battle and still had men standing on the battlefield when the battle was over.

In short, to have a war of resources, two great armies must meet, fight, and neither of them fall apart — something that seems to have happened quite rarely before the twentieth century. The question that follows is: When do armies fall apart, and when not? This is not an economic question, but it is one that economists have considered. Brennan and Tullock (1982) cast the soldier's motivation in the framework of a prisoner's dilemma: unless some other factor enters the soldier's calculation, whether his comrades stand and fight or turn and run, in the face of battle his best choice is always to desert. Officers build comradeship and discipline to prevent this from happening.

In history defeat has come most often, not at the point when all the soldiers are killed, but at the point when discipline and mutual obligation fail so that the army is suddenly pulled apart and is turned into a leaderless rabble. Modern armies, however, have evolved into resilient networks based on morale and commitment that can overcome the prisoner's dilemma and hold the ranks together even in quite extreme circumstances. When this factor was equally present on both sides, improved leadership and morale could no longer make the difference between victory and defeat. It was under these conditions, and only then, that scale and resources could become the decisive factor in two world wars.

Scale in modern warfare has implications for the state. To assemble an army of modest size for a season, fight a battle, and then disperse is one thing. To keep a mass army supplied and moving from battle to battle in one season after another is another. As shown in Chapters 1 and 2, the demands of twentieth century warfare have sometimes required the mobilization of half or more of a country's resources for years on end. To achieve this required immense state capacity.

State capacity has many dimensions, financial and administrative as well as coercive (Besley and Persson, 2009). The raising of war loans and taxes and the coordination and direction of human and physical resources into the supply of modern wars produced a vast derived demand for professional, non-corrupt administrative services. Autocracies of the pre-modern era had very limited capacity of this type. It was the seventeenth century revolutions, guaranteeing private property and constitutional rule in Northwestern Europe, that also gave unprecedented power to government to tax and borrow in the event of war. In military and naval power, the liberal democracies pulled ahead of their rivals, keeping the advantage through World War I. As shown in Chapters 1–4, however, by World War II some non-democracies had caught up and closed the gap, exploiting modern nationalism and repression to mobilize resources with still greater intensity, at least for short periods of time.

If pure scale has had the edge in modern warfare, does the same carry over to the exercise of coercion? That is, is it better to be the tyrant of a big country compared with a small one? To rule a large country is certainly more complex, and complexity is costly. As ruler of the world's largest territory, Stalin had to commit major resources to the control of the vast Soviet space and to border regions inhabited by ethnic minorities from the Baltic to Mongolia and China. To ensure the implementation of his orders, he had to rely on ever lengthening chains of command and multiple, overlapping agencies of control (Markevich, 2011). Compare Stalin's problem with that of Alyaksandr Lukashenka. Once a political instructor of the KGB border troops, Lukashenka rose to become the autocratic ruler of the former Soviet republic of Belarus. In Belarus, he governs no more than three per cent of the population and one per cent of the territory of the old Soviet Union. That is surely a simpler task than Stalin's.

Perhaps it is better to be a pike in a small pond than a minnow in a lake. But it is better still to rule the lake. The reasons are things we have already seen. First, dictators appear to have a propensity for war. Second, in modern war, scale counts.

Surplus Extraction

The economic analysis of coercion and conflict encourages, in some ways, a return to Marxian ideas about the economy. One of Marx's (1867/1974: 173–182) central obsessions, was his idea of surplus value: the value that a worker creates for the employer, above and beyond his own value (or maintenance cost). This idea was so important to Marx because he regarded all economies as, in essence, mechanisms whereby an elite, of one kind or another, extracted a surplus from the working poor.

Modern economics has abandoned this view as oversimplified to an extreme. The model is not only wrong; it is not useful in most contexts. In the Marxian perspective there is only one kind of surplus, called profit, one source of surplus, called labor, and one recipient, the capitalist class. In the competitive market economy of today's textbooks every transaction gives rise to a surplus, and these accrue to the producers and consumers that are party to every transaction. In other words, there are surpluses everywhere and they accrue to everyone; they are not the monopoly of one class.

At the same time, there are some economies that are undeniably extractive. In these economies there is a restricted class of proprietor, or perhaps a dictator, that controls production and distribution in a coercive way, for its own exclusive benefit. In extractive economies there is often no clear distinction between politics and economics, because the security of the proprietors rests on forcible restriction of the choices available to members of the underclass, not on universal property rights and the rule of law. In other words, the same people that want to monopolize the economy must also monopolize the state.

Manorial economies, plantation and slave economies, and communist command economies seem to fit this template better than others. Inspired by the example of Stalin's Soviet Union, Mançur Olson (1993) developed the idea of a "proprietary dictator" that defeats rivals, monopolizes a territory and extracts a surplus from it. Where competing predators would simply ruin the territory, a monopolistic ruler would rationally seek to expand the territory and even be willing to pay for public goods and infrastructure to make it more productive, because this will raise the output of the territory and so increase his revenues in the long run.

In this model economic development is not the dictator's purpose, but it is an incidental by-product of the dictator's desire to increase the surplus over time. As the example of the Soviet Union suggests, extractive regimes have sometimes been associated with prolonged growth spurts, but they have never given rise to the modern economic growth that persists unbroken for a century. One important reason may be that the monopolization of resources by an elite is antithetical to the "creative destruction" of market competition (Açemoglu and Robinson, 2012: 94). The analysis of the downfall of the Soviet model in Chapters 11 and 12, strongly suggests a link with its failure to respond to the opportunities presented by new technologies and new industries that were spurring market-economy development and globalization in the 1970s and 1980s.

Marxists used to lay great emphasis on international movements of economic surpluses. Economic historians somewhat lost interest in this subject after it became clear that international trade before the nineteenth century was simply not large enough to be a channel for significant transfers from one country to another (O'Brien, 1982).

Quantitative history suggests that this topic became much more important in twentieth century conquest (Liberman, 1996). Germany's occupation of much of Europe in World War II arose out of long-term plans for territorial expansion but the design of the occupation regime was motivated more narrowly by plans to extract a surplus of food and resources from the occupied countries by coercive means and transfer it to Germany or the German Army. A new study by Klemann and Kudryashov (2012: 104) estimates the actual contribution at around one third of Germany's war costs. (Further calculations would be required to estimate the additional surplus extracted within Germany from the slave laborers imported from occupied Europe).

The importance of the economic surplus for our subject goes beyond its role in communist and fascist states. In total war, all economies became extractive. Even in democracies, the elected government had to find resources for the army and its equipment. The resources had to come from somewhere. Beyond a point, the working poor had to be squeezed because they were the only class of society numerous enough to supply the necessary resources.

In short, to understand what overlords and dictators maximize, and how they maximize it, and to understand what all governments must maximize in total war, Marx's intuition remains useful. Although neither of the authors is a Marxist, the surplus model that we apply to Soviet labor coercion in Chapter 10 is consistent with a Marxian approach. To repeat George Box, all models are wrong — and that certainly includes the model that Marx developed. But wrong models can still yield insight, and the Marxian model is one of these. It too has a place in the economist's toolkit.

Type I Errors

One way that social science advances is by formulating and testing hypotheses against data. Evolution has programmed us all to formulate hypothesis all the time, that is, to look for patterns around us and find significance in chains of events. Our ability to do this has given us huge advantages as a species. Without it there would be no science, technology, religion, or culture: we could not understand the universe, manipulate the laws of physics, build social relationships, or peer into the future. Imposing significance upon chains of events so as to look into the future is something from which an extraordinarily wide range of people can make a living both for themselves and for others, for example politicians, priests, tipsters, entrepreneurs, novelists, journalists, economists, and historians.[8]

Our propensity to discover patterns in what we observe is not only creative. It is also a source of risk. The risk is realized when we make mistakes. The mistakes we make can be of two kinds, which statisticians call Type I and Type II. These concepts arise from statistics rather than economics, but they are of fundamental importance to the economics of coercion and conflict.

- An error of Type I is to see a pattern in the data where in reality there is none. On a dark night you hear the footsteps of an attacker behind you, but actually no one is there.
- An error of Type II is to miss a pattern in the data that is actually present. Another night, an attacker comes behind you. You're distracted and unaware until it's too late.

Which is worse? Both carry dangers. People that persistently make errors of Type II can be described as complacent. They fail continually to see the risks of terrorism and war and the environmental and health hazards around the corner. As a result, bad things are allowed to happen before we take action. Complacent people wake up in the end, only after they and others have suffered significant damage which, given better foresight, could have been avoided. They lock the stable door after the horse has bolted. Thus Type II errors can be corrected but correction is costly.

Type I errors are bad too. The people that persistently make errors of Type I can be called paranoid. They see enemies and conspiracies everywhere. The world is complicated; they think the world is a complicated gun that someone, somewhere is pointing at them. They do not recognise that many things are random; they will not accept that bad stuff often

[8]Here I draw on material that I wrote for a blog post asking "How Can We Get to See What's Coming Round the Corner?" (August 6, 2009), available at http://blogs.warwick.ac.uk/markharrison/entry/can_we_see/, and also (from the time before blogs) a short unpublished paper, "The War Against Terrorism: Type I Versus Type II Errors" (1 November 2001), available at http://warwick.ac.uk/markharrison/comment/.

happens by accident or by mistake. Nor are they able to accept that a lot of what happens is not really about them at all. Trying to forestall the bad things they anticipate, people who are gripped by Type I errors are often themselves responsible for doing things that are as bad or worse, such as mobilizing society to solve problems that do not exist or attacking people who are not enemies and would prefer to keep themselves to themselves.

Finally, there is a trade-off between the errors of Type I and Type II. If we try harder to avoid one, other things being equal, we will inevitably end up making more of the other. The only way to reduce both types of error at the same time is to have more and better data. As long as data collection is costly, we will never have complete information and we will always have to balance one kind of error against the other.

Attitudes to the consequences of Type I and Type II errors have proved to be important markers of social attitudes. Understanding the balance that is struck between Type I and Type II errors can tell you a lot about the nature of the society under study. In a court of law, the research hypothesis is that the defendant is guilty. Let's say the crime is a public order offence. Here the court can make two mistakes. It can convict the truly innocent (a Type I error) or acquit the truly guilty (a Type II error). In most liberal democracies, the courts are stacked against the prosecution, which must prove guilt beyond reasonable doubt that the defendant incited a riot. In other words, prioritizing the rights of the accused implies that we tolerate Type II errors in order to avoid errors of Type I, which we see as particularly costly.

In contrast the historical record of dictators is that, since they are not particularly concerned about the rights of citizens but very concerned to protect their own skins, they tend to operate criminal justice in a way that scoops up and penalizes many fairly innocent people in order to ensure that every guilty person is taken out of circulation. This has important implications for the study of coercion. As discussed in Chapter 10, the coercive regimes of the dictators that preside over extractive economies typically accept large numbers of Type I errors as the price of avoiding the Type II errors that might undermine their own security. In contrast, more inclusive economies that respect the rights of individual citizens tend to have arrangements, such as presuming innocence in the absence of proof of guilt, that avoid Type I errors at the price of sometimes making Type II errors.

The asymmetry in the consequences of Type I and Type II errors has strongly affected the research strategies of natural and social scientists.

Most scientists would rather not be personally responsible for precipitate action that wastes resources and destroys lives. Therefore, science is cautious, and standard criteria for statistical significance put much more weight on avoiding Type I errors than avoiding Type II errors. So the data support your hypothesis only weakly? Go back and get a bigger sample, or let somebody else try; do not prematurely announce that you have identified a cure for cancer or the root of some other evil.

Oddly enough, many historical commentators (including politicians, pundits, and some historians) do not seem to feel this self-restraint. In fact, the quality of historical writing is often judged by the ability of the historian to weave a few random threads into an interesting pattern. Most readers will be impressed by a story that is logical and is supported by some data, even though the data points are selected and not at all representative.

In relation to our subject matter, wars are especially suited to biased treatment because they comprise many great stories in which we can identify directly with the actors and we are emotionally compelled by the drama. The same is also true of the history of societies where there has been victimization on a large scale. Every victim has a story to tell and a need for acknowledgement.

Story-telling is powerful. Its intuitive appeal is much greater than models, charts, and numbers. It speaks the language of nations and politics: shared experiences, common destinies, collective rights and wrongs. It is easily voiced by leaders and heard by followers untrained in statistical thinking about trends and standard errors. As a result, while politicians may turn to economists for technical advice, they get historians to help write their speeches — Arthur Schlesinger Jr. (John F. Kennedy), Richard Pipes (Ronald Reagan), and Norman Stone (Margaret Thatcher).

What makes the power of story-telling? First, story-telling is deliberately selective. It does not try to be representative. When we scan history for stories, we look by definition for sequences of events that have a beginning, a middle, and an end. In the middle, something happens that is out of the ordinary, dramatic, and unexpected. Invariably, we rule out all those much more representative past circumstances out of which something might have come, but after which there were no surprises and nothing much happened.

Second, stories give us a way of thinking about how to handle rare and unpredictable events. Houghton (1996), for example, has shown how U.S. leaders used historical parallels to make decisions in the "novel emergency" of the 1979 Iranian hostage crisis. In this sense, a well-chosen story (in other words, one data point) can meet the needs of decision makers in a much more satisfying way than a statistician who just says: "Well, we need to collect more data."

Third, story-telling typically sounds an alarm. In history, dramatic events are rarely good news. The good news in history has generally been made up of the slow, steady progress of emancipation, literacy, and prosperity. Such good news is easily illustrated by statistics and trends, but does not make good stories. It is the bad news of tragedies and crises that makes good stories.

In fact, an entirely legitimate purpose of a good story may sometimes be to sound the alarm about the risks we face and so avert their realization. George Orwell's novel *1984*, for example, warned western societies of the dangers of totalitarian rule much more effectively than a thousand learned treatises on the subject.

There is a downside to this. Some stories can be self-fulfilling. There is a particular kind of collective story, for example, that communal identity politicians like to tell (Glaeser, 2005). These are stories of past hate crimes allegedly committed by some other ethnic or religious group against their own group: Black against White, Reds against Whites, Germans against Jews, Jews against Palestinians, Protestants against Catholics, Sunni against Shia — and, in all cases, vice versa. Such powerful stories can be extrapolated into predictions of future hate crimes yet to be committed, and then into justifications for hateful and violent action to preempt the future crimes.

When we buy a story that is untrue or unrepresentative, we make a Type I error. We accept as meaningful a coincidence between the pattern in the story and a pattern inside our brains, despite the fact that the pattern is not matched in the external world. For example, we may prefer some stories to others, not because they are true, but because they are more comfortable or more affirming of our identity. One result is that the propensity to Type I errors is strongly present in the history of conflict, where is where national identities are made and affirmed.

Consider the following popular hypotheses, which all relate to the outcome of World War II:

- Britain won the war because the British way of life inspires loyalty.
- Germany (or Japan) almost won the war because Germans (or Japanese) are disciplined and follow orders.
- France lost the war because French society was rotten.

- Italy lost the war because Italians are cowards.
- America won the war because Americans are free and equal.
- The Soviet Union won the war because Russians love their country.

These ideas all carry a high risk of Type I error, as a moment's reflection will indicate. To explain the scope for error in each case: It is doubtful that Britain could have won the war alone, in other words, without the help of the Soviet Union, the United States, and the British Empire where millions lived in ways that were quite different from those of the British at home. If Germany and Japan nearly won the war, it is likely that this was because their leaders planned it that way and it took time for others to rise to the challenge; strategy and timing were more important than the innate characteristics of the citizens. If France lost the war, its society was hardly more rotten than that of others. If Italy lost the war it was not from cowardice; quite apart from anything else the courage of Italians who resisted the dictator Mussolini must also have been a factor. If America won the war, it was with the participation of large numbers of slave descendants who were less free and less equal, and it was also with the help of the Soviet Union whose citizens were ruled by a totalitarian dictator. Finally, if the Soviet Union won the war it was despite the best efforts of millions of Russians who collaborated with the occupier or fought actively against the Red Army.

In fact, Chapter 2 will show that in the twentieth century "our nation" won wars most often by belonging to an alliance that was richer and larger than the adversary's alliance. This finding does not particularly help to affirm anyone's national identity, nor does it contribute an exciting story. Nonetheless it fits most of the patterns in the data much better than any hypothesis based on national exceptions or peculiarities.

A lesson is that the student of coercion and conflict needs to approach all good stories equipped with a strong sense of scepticism. A deep understanding and continuous awareness of the risks of the Type I error is an essential element of the economist's toolkit.

A Chapter-By-Chapter Summary

Table 1 sets out the basic tools used in this book, chapter by chapter. Not every tool is used in every chapter. Notably, no chapter could have been written without reference to the economist's idea of rationality.

Table 1. The Economist's Toolkit, Chapter by Chapter.

Chapter	Rational actors; expected costs and benefits	Strategic interaction; multiple equilibria	Diminishing and increasing returns	Scale and state capacity	Surplus extraction	Type I errors
1. War and Disintegration, 1913 to 1945	×			×		
2. Why the Wealthy Won	×			×		×
3. Why Didn't the Soviet Economy Collapse	×	×	×	×	×	
4. The Frequency of Wars	×	×		×		×
5. Soviet Industry and the Red Army under Stalin	×	×			×	
6. Contracting for Quality under a Dictator	×	×			×	
7. A Soviet Quasi-Market for Inventions	×	×			×	
8. The Political Economy of a Soviet Military R&D Failure	×	×			×	
9. The Fundamental Problem of Command	×	×	×		×	
10. Accumulation and Labor Coercion Under Late Stalinism	×		×		×	×
11. Economic Information in the Life and Death	×	×	×		×	
12. Coercion, Compliance, and the Collapse	×	×	×		×	

The Economics of Global War

The book is divided into three parts. The first part is devoted to the economics of global war. A common theme of the chapters in this part of the book is the importance of scale, resources, and state capacity in the outcomes of wars.

In Chapter 1, "War and Disintegration, 1914–1945", Jari Eloranta and I examine Europe in the first half of the twentieth century. Globalization came to an abrupt halt, replaced by protectionism, nationalism, war, and killing and destruction on an immense scale. In mid-century, globalization was resumed, and the European economies began to converge on much higher and more uniform income levels. After two world wars, three things had changed. First, European economic growth, integration, and prosperity had lost its association with empire. No doubt to the surprise of Europe's nineteenth century leaders, had they lived to see it, it proved possible to acquire wealth and wield influence without claiming imperial sway over vast stretches of faraway peoples and their lands and oceans. Second, Europe's leaders had a new sense of the importance of cooperation. They now cooperated with the United States in economic recovery, exchange rate coordination, and tariff reduction, with each other in laying new foundations for European integration, and with developing countries in decolonization and development assistance. Third, Europe's leaders had learned to use the power of the state to regulate economic life. Learning the appropriate limits of government control over the market economy in turn became a major challenge of the postwar period.

In Chapter 2, "Why the Wealthy Won", I look more closely at the roots of victory and defeat in two world wars. The chapter starts from the variation that we observe in the degrees of economic mobilization of different countries for total war in the twentieth century. Most of this variation is explained by differences in the level of economic development of each country, though not all of it and there some exceptions. There are several good reasons that help to explain why mobilization capacity should depend significantly on economic development. The empirical record is to some extent a puzzle since it seems to leave little room for other factors that would feature prominently in narrative accounts, such as national differences in war preparations, war leadership, or military organization and morale. The chapter looks at ways of solving this puzzle.

Chapter 3 asks "Why Didn't the Soviet Economy Collapse in 1942?" It appraises the economic dimensions of World War II both generally and with specific reference to the eastern front. When the Soviet war effort is examined more closely, it becomes surprising that the Soviet economy did not collapse in 1942. A rational-choice model is developed to illustrate the economic conditions under which a wartime collapse of the economy is rendered more and less likely. The possible effects of policy interventions by Stalin, Hitler, and Roosevelt on the stability of the Soviet war effort are defined.

In Chapter 4, "The Frequency of Wars", Nikolaus Wolf and I step back from particular countries and conflicts to the puzzles that arise from a quantitative overview of world history. Conflicts among states are, it seems, increasingly frequent, and the trend has been steadily upward since 1870. The main tradition of Western political and philosophical thought suggests that extensive economic globalization and democratization over this period should have reduced appetites for war far below their current level. This view is clearly incomplete: at best, confounding factors are at work. Here, we explore the capacity to wage war. Most fundamentally, the growing number of sovereign states has been closely associated with the spread of democracy and increasing commercial openness, as well as the number of bilateral conflicts. Trade and democracy are traditionally thought of as goods, both in themselves, and because they reduce the willingness to go to war, conditional on the national capacity to do so. But the same factors may also have been increasing the capacity for war, and so its frequency. The chapter concludes that we need better understanding of how to promote these goods without incurring adverse side-effects on world peace.[9]

Communism and Defense

The middle part of the book focuses on the relationship between communism and defense. In all countries where communists ruled, the mechanisms of power were designed on a similar template of coercion and repression. The two best accounts of this template are the sociological findings of Inkeles and Bauer (1959) on the Soviet Union and Anne Applebaum's (2012) new history of the imposition of communist rule on Eastern Europe.

The more we know about communist rule from the inside, the more important it seems to be that we understand how preparations for external conflict were linked to coercion at home. A common theme of the chapters in this part of the book is strategic interaction, conflict, and bargaining over the surpluses that the communist rulers were willing to share with the military and industrial interests that had the job of supplying defense needs.

Chapter 5, "Soviet Industry and the Red Army under Stalin", considers some views of the Soviet "military-industrial complex" that are current in the literature. The economic weight of the defense sector in the economic system is summarized in various aspects. The lessons of archival research are used as a basis for analysing the army–industry relationship under Stalin as a prisoners' dilemma in which, despite the potential gains from mutual cooperation, each party faced a strong incentive to cheat on the other. The chapter concludes that the idea of a Soviet military–industrial complex is not strictly applicable to the Stalin period, but there may be greater justification for the Soviet Union after Stalin.

In Chapter 6, "Contracting for Quality under a Dictator", Andrei Markevich and I examine military procurement in the Soviet economy under Stalin. This provides a novel historical context for a standard problem of market organization, that of contracting for quality. The Soviet ministry

[9] This chapter came under critical fire from leading political scientists. For their criticisms and our replies, see Gleditsch and Pickering (2013) and Harrison and Wolf (2013).

of defense was engaged in the procurement of military goods from Soviet industry. An internal market was formed and contracts were made. In the market, the contractor had power over the buyer and typically used this power to default on quality. In the background loomed a dictator, imperfectly informed, but with the power to tear up any contract and impose unlimited penalties on the contractor. The buyer's counter-action took the form of deploying agents through industry with the authority to verify quality and reject substandard goods. The final compromise restored quality at the expense of quantity. Being illicit, it had to be hidden from the dictator. Our case provides an historical illustration of the limits of dictatorship.

Chapter 7, "A Soviet Quasi-Market for Inventions", is also about the allocation of resources in the Soviet economy under Stalin's dictatorship. In this chapter the resources were designated for military research and development. The context was formed by the rapidly approaching limits of an existing aviation propulsion technology, the need to replace it with another, and profound uncertainty as to how to do so. In the process we observe the formation of an internal "market" in which rival designers proposed projects and competed for funding to carry them out. We observe rivalry and rent seeking by agents, imperfectly regulated by principals. As rent seeking spread and uncertainty was reduced, the chapter shows, the market was closed down and replaced by strict hierarchical allocation and monitoring.

Chapter 8 examines "The Political Economy of a Soviet R&D Failure", the prewar attempt to create a new aeroengine technology based on the steam turbine. From this example, we find out more about the motivations, strategies, and payoffs of principals and agents in the Soviet command economy. Alternative approaches to the evaluation of R&D failure are outlined. New archival documentation shows the scale and scope of the Soviet R&D effort in this field. The allocation of R&D resources resulted from agents' horizontal interactions within a vertical command hierarchy. Project funding was determined in a context of biased information, adverse selection, and agents' rent seeking. Funding was rationed across projects and through time. Budget constraints on individual projects were softened in the presence of sunk costs, but were hardened periodically. There is no evidence that rents were intentionally distributed through the Soviet military R&D system to win trust or reward loyalty; the termination of aviation steam power R&D in 1939 despite the sunk costs they represented was timely.

Communism and Coercion

Communist rule rested ultimately on command and coercion, and the third part of the book focuses directly on this. Again, there is a lot about strategic interaction between the state and the citizens, this time conceptualized more broadly as producers and even consumers.

Chapter 9, "The Fundamental Problem of Command", studies the problems that arose under communist rule when the dictator gave an order to an agent and advanced resources for its implementation. The temptation was for the agent to shirk or steal from the principal rather than comply; this constitutes the fundamental problem of command. Historically, partially centralized command economies enforced compliance in various ways, assisted by nesting the fundamental problem of exchange within that of command. The Soviet economy provides some relevant data. The Soviet command system combined several enforcement mechanisms in an equilibrium that shifted as agents learned and each mechanism's comparative costs and benefits changed. When the conditions for an equilibrium disappeared, the system collapsed. An extractive state had the task of aligning the incentives of its citizens to work together for the surplus that the state required to finance its own goals. The citizens faced continual temptations to shirk and steal. Keeping everything together was at least a feasible problem in the age of mass production of things. With the information revolution and the rise of the services economy this task became more and more complex and ultimately, perhaps, infeasible.

In Chapter 10, "Accumulation and Labor Coercion", Paul Gregory and I examine how the level of coercion affected Soviet wages and the surplus available for investment under Stalin. From 1940 until the death of Stalin, the conditions of Soviet labor were highly regimented under draconian legislation which prescribed imprisonment and forced labor for even minor violations. How successful was this experience? A successful use of coercion implied that fewer resources would be needed for consumption, without compromising the amount of effort supplied by the workers. In the desired outcome, accumulation would be enabled to grow. A major issue was the problem of mistakes and unintended consequences of various kinds. In the face of harsh punitive measures, Soviet workers shifted effort from production to mutual insurance through informal horizontal and vertical networks. The Stalinist leaders often interpreted attempts to protect subordinates as proof of conscious or unconscious betrayal of the party and Stalin, which then became an occasion for new repression. In this atmosphere, the likelihood of errors in the allocation of both labor and capital was only increased. In such conditions it is possible coercion did not contribute to an increase in the level of accumulation. In fact, it is quite likely that the Soviet economy's accumulation capacity improved after Stalin died and his successors reduced the level of violence in the economy.

Chapter 11, "Economic Information in the Life and Death of the Soviet Command System", shifts the focus to information problems under communism. In market economies, information adds value to transactions in three ways: it supports reputations, permits customization, and provides yardsticks. In the Soviet economy such information was frequently not produced; if produced, it was often concealed; whether concealed or not, it was often of poor quality; regardless of quality, it often suffered from low credibility outside the ruling circle. In short, the Soviet command system forced economic growth on the basis of a relatively low–value information stock. This might help explain aspects of Soviet postwar economic growth and slowdown, the collapse of the command system, and the persistence of low output after the collapse.

Chapter 12, "Coercion, Compliance, and the Collapse of the Soviet Command Economy", concludes the book. It addresses directly the problem of how and why the Soviet economy collapsed in 1991. It begins by asking whether command systems that rest on coercion are inherently unstable, and whether the Soviet economy collapsed for this reason. The problem is that, until it collapsed, the Soviet economy did not look unstable. Why did it then collapse? A game between a dictator and a producer shows that a high level of coercion may yield a stable high–output equilibrium, that stability may rest in part on the dictator's reputation, and that a collapse may be brought about by adverse trends in the dictator's regime costs and a loss of reputation. The facts of the Soviet case are consistent with a collapse that was triggered by the Russian workers' strike movement of 1989.

Conclusion

In this book, I apply the concepts and methods of analysis that I have found most useful for study of the economics of coercion and conflict. The economist's toolkit is versatile, and can be applied to many problems that are not economic at first sight. Each of the tools is useful, provided it is applied to the right context. The problem is to recognize the context and select the right tool. If you want to fix together two pieces of wood, there are times when nails will do the job and sometimes it requires screws. Hammering a screw and trying to turn a nail will get you nowhere.

Sometimes the economist's toolkit must fall short. Is economics adequate by itself for a rounded understanding of everything that is wrapped up in the subject of coercion and conflict? Clearly not. Power is about politics before it is about economics. War is a political act. Force and violence are the most multidisciplinary of human activities. They engage our emotions as well as our reason. In my own work, whether I have acknowledged it or not, I am sure that I have also borrowed many ideas drawn from psychology, political and strategic studies, international relations, and historical narrative. Nonetheless, any activity that involves the expenditure of effort for an expected return has an economic dimension, and that is the justification of the present book.

References

Açemoglu, D and Robinson, J (2012). *Why Nations Fail: The Origins of Power, Prosperity, and Poverty.* New York: Crown.

Akerlof, GA, and Kranton, RE (2000). Economics and Identity. *Quarterly Journal of Economics*, 115(3): 715–753.

Applebaum, A (2012). *Iron Curtain: The Crushing of Eastern Europe.* London: Allen Lane.

Besley, T and Persson, T (2009). The Origins of State Capacity: Property Rights, Taxation, and Politics. *American Economic Review*, 99(4): 1218–44.

Bergson, A (1961). The Real National Income of Soviet Russia since 1928. Cambridge, MA: Harvard University Press.

Box, GEP, and Draper, NR (1987). *Empirical Model Building and Response Surfaces.* New York: Wiley.

Brauer, J and van Tuyll, H (2008). *Castles, Battles, and Bombs: How Economics Explains Military History.* Chicago: University of Chicago Press.

Brennan, G and Tullock, G (1982). An Economic Theory of Military Tactics: Methodological Individualism at War. *Journal of Economic Behavior and Organization*, 3(2–3): 225–42.

Bueno de Mesquita, B and Siverson, RM (1995). War and the Survival of Political Leaders: A Comparative Study of Regime Types and Political Accountability. *American Political Science Review*, 89(4): 841–855.

Bueno de Mesquita, B (2006). Game Theory, Political Economy, and the Evolving Study of War and Peace. *American Political Science Review*, 100(4): 637–642.

Cienciala, AM (2011). The Foreign Policy of Józef Piłsudski and Józef Beck, 1926–1939: Misconceptions and Interpretations. *The Polish Review*, 56(1–2): 111–152.

Clausewitz, C von (1832/1982). *On War.* London: Penguin Books.

de Tocqueville, A (1835/2000). *Democracy in America.* Chicago: University of Chicago Press.
Dixit, AK (2004). *Lawlessness and Economics: Alternative Modes of Governance.* Princeton: Princeton University Press.
Fearon, JD (1995). Rationalist Explanations for War. *International Organization,* 49(3): 379–414.
Findlay, R and O'Rourke, KH (2012). War, Trade, and Natural Resources: A Historical Perspective. In: *The Oxford Handbook of the Economics of Peace and Conflict* by. Oxford: Oxford University Press.
Garfinkel, MR and Skaperdas, S (2012). Economic Perspectives on Peace and Conflict. In: Garfinkel MR and Skaperdas S (eds.) *The Oxford Handbook of the Economics of Peace and Conflict.* Oxford: Oxford University Press, pp. 3–19.
Gat, A (2006). *War in Human Civilization.* Oxford: Oxford University Press.
Glaeser, E (2005). The Political Economy of Hatred. *Quarterly Journal of Economics,* 120(1): 45–86.
Gleditsch, KS and Pickering, S (2013). Wars are Becoming Less Frequent: A Response to Harrison and Wolf. *Economic History Review* (forthcoming).
Gregory, PR (ed.) (2001). *Behind the Façade of Stalin's Command Economy: Evidence from the Soviet State and Party Archives.* Stanford: Hoover Institution Press.
Gregory, PR (2004). *The Political Economy of Stalinism: Evidence from the Soviet Secret Archives.* Cambridge: Cambridge University Press.
Gregory, PR (2009). *Terror by Quota: State Security from Lenin to Stalin (An Archival Study).* New Haven: Yale University Press.
Gregory, PR and Lazarev, V (eds.) (2003). *The Economics of Forced Labor: The Soviet Gulag.* Stanford: Hoover Institution Press.
Greif, A (2006). *Institutions and the Path to the Modern Economy: Lessons from Medieval Trade.* Cambridge: Cambridge University Press
Harrison, M (2002). Economic Growth and Slowdown. In: Bacon E and Sandle M (eds.) *Brezhnev Reconsidered.* London and Basingstoke: Palgrave, pp. 38–67.
Harrison, M (2005). Stalin and Our Times. In: Roberts G (ed.) *Stalin: His Times and Ours.* Dublin: Irish Association for Russian and East European Studies, pp. 67–84.
Harrison, M (2006). An Economist Looks at Suicide Terrorism. *World Economics,* 7(3): 1–15.
Harrison, M and Wolf, N (2013). The Frequency of Wars: Reply to Gleditsch and Pickering. *Economic History Review* (forthcoming).
Hobbes, T (1651/1909). *Hobbes's Leviathan.* Reprinted from the 1651 edition with an essay by the Late W. G. Pogson Smith. Oxford: Clarendon Press.
Holborn, H (1986). The Prusso-German School: Moltke and the Rise of the General Staff. In: Paret P, Craig GA and Gilbert F (eds.) *Makers of Modern Strategy from Macchiavelli to the Nuclear Age.* Oxford: Oxford University Press, pp. 281–295.

Houghton, DP (1996). The Role of Analogical Reasoning in Novel Foreign-Policy Situations. *British Journal of Political Science*, 26(4): 523–552.

Inkeles, A and Bauer, RA (1959). *The Soviet Citizen: Daily Life in a Totalitarian Society*. Cambridge, MA: Harvard University Press.

Kahneman, D, Knetsch, JL and Thaler, RH (1991). Anomalies: The Endowment Effect, Loss Aversion, and Status Quo Bias. *Journal of Economic Perspectives,* 5(1): 193–206.

Kahneman, D (2011). *Thinking Fast and Slow*. London: Allen Lane.

Kant, I (1795/1983). *Perpetual Peace and Other Essays*. Indianapolis: Hackett Publishing.

Klemann, H and Kudriashov, S (2012). *Occupied Economies: An Economic History of Nazi-Occupied Europe, 1939–1945*. London and New York: Berg.

Konev, IS (1970). *Sorok piatyi. 2/e*. Moscow: Voenizdat.

Levy, JS (1988). Domestic Politics and War. *Journal of Interdisciplinary History,* 18(4): 653–673.

Liberman, P (1996) *Does Conquest Pay? The Exploitation of Occupied Industrial Societies*. Princeton: Princeton University Press.

Markevich, A (2011). How Much Control is Enough? Monitoring and Enforcement under Stalin. *Europe-Asia Studies,* 63(8): 1449–1468.

Marx, K (1867/1974). *Capital: A Critical Analysis of Capitalist Production,* Volume 1. London: Lawrence and Wishart.

North, DC, Wallis JJ and Weingast, BR (2011). *Violence and Social Orders: A Conceptual Framework for Interpreting Recorded Human History*. Cambridge: Cambridge University Press.

O'Brien, PK (1982). European Economic Development: The Contribution of the Periphery. *Economic History Review*, 35(2): 1–18.

Offer, A (1989). *The First World War: An Agrarian Interpretation*. London: Clarendon.

Olson, M (1993). Dictatorship, Democracy, and Development. *American Political Science Review*, 87(3): 567–76

Pinker, S (2011). *The Better Angels of Our Nature: Why Violence Has Declined*. New York: Viking.

Rockoff, H (2012). *America's Economic Way of War: War and the US Economy from the Spanish-American War to the Persian Gulf War*. Cambridge: Cambridge University Press.

Sen, A (2006). *Identity and Violence: The Illusion of Destiny*. London: Allen Lane.

Smith, A (1776/2005). *An Inquiry into the Nature and Causes of the Wealth of Nations*. Edited by Jim Manis. An Electronic Classic Series publication, available at http://www2.hn.psu.edu/faculty/jmanis/adam-smith/wealth-nations.pdf (accessed 28 August 2013).

Smith, RP (2009). *Military Economics: The Interaction of Power and Money*. Basingstoke: Palgrave Macmillan.

Tilly, C (1975). Reflections on the History of European State-Making. In: Tilly, C *The Formation of National States in Western Europe*. Princeton: Princeton University Press, pp. 3–83.

Weingast, BR (1995). The Economic Role of Political Institutions: Market-Preserving Federalism and Economic Development. *Journal of Law, Economics, and Organization*, 11(1): 1–31.

Wintrobe, RP (2000). *The Political Economy of Dictatorship.* Cambridge: Cambridge University Press.

Part 1

Global Conflict

Chapter 1

War and Disintegration, 1914–1950

Jari Eloranta and Mark Harrison

Introduction

Between 1914 and 1945 Europe's economic development and integration were interrupted and set back by two world wars, and its regional patterns were brutally distorted by combat, exterminations, migrations, and the redrawing of borders.[1] World War I (the "Great War" of 1914–1918) set more than thirty countries into conflict with each other and led to ten million premature deaths. It was dwarfed only by World War II (1939–1945), in which more than sixty countries waged war and the war prematurely ended the lives of more than 55 million people (Broadberry and Harrison, 2005b). As for who fought whom, there were limited continuities: in both wars, Germany, Austria, and Hungary fought Britain, France, and Russia for much of the time. Other allegiances changed. For ease of reference, Table 1 lists the European countries that were in or out of each war and, if in, on what side.

Although punctuated by an "interwar period," the two wars can be understood as a single historical process. The process was global but the European dimension was fundamental to it. Thus, Europe was the main theatre of a vast thirty-year conflict of empires and nationalisms.

This chapter was first published by University Press in *The Cambridge Economic History of Modern Europe* (edited by Stephen Broadberry and Kevin O'Rourke), Volume 2, 133–155. © Cambridge University Press 2010.

[1]Jari Eloranta is Professor of Professor of Comparative Economic and Business History at the Appalachian State University, North Carolina, and Docent in Economic History at the University of Jyvaskyla, Finland. The authors gratefully acknowledge the advice and comments of the editors and other collaborators in *The Cambridge Economic History of Modern Europe*, in particular Joerg Baten, Stephen Broadberry, Robert Millward, Leandro Prados de la Escosura, and Kevin O'Rourke.

Table 1. Taking Sides: Belligerents and Neutrals in Two World Wars.

	World War I	World War II
Albania	Neutral	Anti-Comintern Axis
Austria (AH)	Central Powers	Annexed to Germany
Belgium	Occupied by Germany	Occupied by Germany
Bulgaria	Province of Austro-Hungarian Empire	Anti-Comintern Axis
Czechoslovakia	Province of Austro-Hungarian Empire	Occupied by Germany
Denmark	Neutral	Occupied by Germany
Estonia	Province of Russian Empire	Occupied by USSR
Finland	Province of Russian Empire	Anti-Comintern Axis
France	Entente	Occupied by Germany
Germany	Central Powers	Anti-Comintern Axis
Greece	Entente	Occupied by Italy
Hungary (AH)	Central Powers	Anti-Comintern Axis
Ireland	Province of United Kingdom	Neutral
Italy	Entente	Anti-Comintern Axis
Latvia	Province of Russian Empire	Occupied by USSR
Lithuania	Province of Russian Empire	Occupied by USSR
Netherlands	Neutral	United Nations
Norway	Neutral	Occupied by Germany
Poland	Province of Russian Empire	Occupied by Germany and USSR
Portugal	Neutral	Neutral
Romania	Province of Austro-Hungarian Empire	Anti-Comintern Axis
Russia (R)/USSR	Entente	United Nations
Serbia/Yugoslavia	Province of Austro-Hungarian Empire	Occupied by Germany
Spain	Neutral	Neutral
Sweden	Neutral	Neutral
Switzerland	Neutral	Neutral
Ottoman Empire/Turkey	Central Powers	Neutral
United Kingdom (UK)	Entente	United Nations

The first war was fought primarily by European powers in Europe; some non-European participants and colonial polities played a minor role while others intervened late in the process. It produced a fragmented continent that had spent vast amounts of physical and human capital, the countries at its centre ensnared in reparations and debts, and incapable of returning to political and economic stability. The second began in Asia but quickly spread across the world to Europe, and most participants understood that the conflict in Europe would be decisive for the outcome. The German historian Ernst Nolte (1965) was the first to define these three decades as a European "civil war". He saw it as a war launched from Russia in 1917 by communist terror and aggression, to which German National Socialism and racial genocide were a defensive response. This is not our view; if any country launched the war it was Germany, first in 1914 and again in 1939, for a variety of reasons. Undoubtedly, however, it makes sense to view the two wars and the interwar period as a continuous process — and we also find the destruction of moral limits that Thucydides first associated with civil strife.

Our period was one of economic and political instability. The average growth rates of most European states were modest compared with their performance before 1914 or after 1950 (Maddison, 1995); Angus Maddison (2001) describes these years as "a complex and dismal period, marked deeply by the shock of the two world wars and an intervening depression". Eric Hobsbawm (1996) viewed the extreme political and military outcomes of this period as characteristic of the decadence and economic failures of the twentieth century. But this is not the whole story. Bradford DeLong (2000) and Alex Field (2003; 2006) have stressed that the massive increases in living standards in the second half of the twentieth century were enabled by technological advances that continued in an unbroken stream through the most dismal episodes of the first half. In Europe as elsewhere, the rate of technological improvement remained high throughout our period and exceeded nineteenth century benchmarks in spite of military, economic, and demographic disaster (Ferguson, 2006). In short, whatever it was that came to an end in 1914, it was not the forces underlying economic growth.

Globalization, Empire, and War

In the nineteenth century globalization and empires were inextricably linked. None other than the Russian revolutionary leader Lenin, writing in 1916 (reprint, 1963), noted the association between the rise of global markets, global enterprises, and global colonial empires.

In the early twentieth century the great powers all regarded colonial empire as an entirely legitimate national pursuit. For illustration, Table 2 shows the world's colonial dependencies. In 1913, Europeans were the greatest imperialists in the history of the world, with 30 per cent of the world's population living in European colonies outside Europe that spread across more than two fifths of the world's land surface. In the same year Britain alone, which accounted by itself for one fortieth of the world's

Table 2. Colonial Dependencies in 1913 and 1938: Populations and Land Surfaces.

	1913		1938	
	Population, millions	Land Surface, million km^2	Population, millions	Land Surface, million km^2
British Dominions	19.9	19.5	30.0	19.2
British Colonies	380.1	13.5	453.8	15.0
French Colonies	48.4	10.7	70.9	12.1
Dutch Colonies	44.1	2.1	68.1	1.9
Other European Colonies	54.1	11.5	33.0	8.3
European Powers' Colonies and Dominions, total	546.5	57.2	655.9	56.5
Non-European Powers' Colonies	28.8	0.6	75.7	1.9
The World	1810.3	134.4	2168.0	134.4

Sources: Harrison (1998) and Broadberry and Harrison (2005a).

population and one five hundredth of its land surface, claimed sovereignty over 400 million people and 33 million square kilometres of Africa, Asia, America, and Australasia. Its Empire, including both self-governing dominions such as Canada and Australia, and colonial dependencies such as India and Nigeria, embraced almost one quarter of the world's population and a quarter of its land surface. Other European powers, chiefly France and the Netherlands, and after them Belgium, Germany, Italy, Portugal, and Turkey lagged far behind. Outside Europe, the United States administered some neighbouring islands, as did Japan, in addition to its Korean colony (see also Huntington, 1996).

In 1938, a quarter century later, almost nothing had changed. Germany and Turkey had lost their little empires as a result of the peace treaties that followed World War I. Italy and Japan were in the first stages of their expansion across the Mediterranean and the Pacific respectively. But these are barely apparent in the figures. Indeed, looking at the world from Berlin, Rome, and Tokyo, the lack of change in the balance of the world's colonial possessions was the problem: the old colonial powers had failed to give way to the aspirations of the new imperialists.

In fact, the pursuit of colonies by the rising powers, combined with the defense of their colonies by the established ones, contributed to repeated crises in the global equilibrium among the powers and was perceived to a large extent to be responsible for the two world wars. In the process the very idea of empire lost its legitimacy. The legitimacy of empire had never been strong among the colonial populations, and its decline became marked within the United States, the foreign policy of which became decidedly anti-colonialist. By 1950 wholesale decolonization, by consent or by force, was already under way within the British, French, and Dutch empires, the only ones of any significance that remained. This is another sense in which the two world wars formed a single historical process that began in 1914 and ended in 1945 (Modelski and Thompson, 1996).

Since Lenin it has been fashionable to link world trade and world empires with the causes of the two world wars. The truth is more complex, however. Neither globalization nor imperialism directly caused the war to break out in 1914. Norman Angell had argued in his 1909 bestseller *The Great Illusion* (reprint, 1972) that the increased economic interdependence of the great powers would make any major international conflict an impossibility, or at least bring it to a quick conclusion. Globalization had increased the interdependency of regional and global players, thereby increasing the costs of the war. Specifically, increased openness had made the European states more vulnerable to the interdiction of imported supplies of food and materials. By raising wages, economic growth had made the maintenance of standing armies more expensive. Businessmen across Europe were not pushing for war and its outbreak came as a shock (Ferguson, 1999). But there were also countervailing factors. One was the industrialization of warfare, which increased the destructive efficiency of military equipment and gave industrialized states a bigger bang for their bucks (Ferguson, 2001; McNeill, 1982). Finally, the very fact of increased vulnerability to economic disruption probably heightened the propensity of the likely participants to gamble on a rapid offensive, once war became probable (Rowe, 2005).

John Hobson (1993) has shown that both the United Kingdom and Germany spent less on their armed forces than most great powers. John Keegan (1999) describes a Europe thrust into war by communication failures and irreversible war plans based on tit-for-tat responses that were designed for the 'age-old quest for security in military superiority.' World War I was no accident, however. Historians have tended to hold Germany particularly accountable for the pre-1914 arms race and the subsequent diplomatic breakdown (e.g. Berghahn, 1973). Niall Ferguson (1999, 2001) has noted that, having started the arms race, Germany was unable to compete against its rivals, and was led therefore to gamble on a pre-emptive strike in 1914. Thus, the prewar arms race, stimulated by the competition for colonies, was a principal cause, whereas the industrialization of armaments production contributed mainly to the length and destructiveness of the war.

Table 3. Military Spending of the Great Powers in Peacetime, 1870–1913 and 1920–1938.

	Military spending, % of GDP		Military spending, % of central or federal government expenditures	
Country	1870–1913	1920–1938	1870–1913	1920–1938
Austria(-Hungary)	3.5	0.9	12.0	5.8
France	3.7	4.3	25.9	22.4
Germany	2.6	3.3	54.1	23.8
Italy	2.8	4.4	21.7	25.4
Russia (USSR)	3.9	7.1	27.9	11.9
UK	2.6	3.0	37.5	16.3
6-country mean	3.2	3.8	29.9	17.6
Japan	5.0	5.7	32.2	20.1
USA	0.7	1.2	29.4	22.4
16/17-country mean	2.7	2.8	33.3	18.0

Sources: See Eloranta (2002b: 110) for details.
Notes: The 16-country mean for 1870–1913 includes Austria(-Hungary), Belgium, Denmark, France, Germany, Italy, Japan, Netherlands, Norway, Portugal, Russia, Spain, Sweden, Switzerland, UK, and USA; the 17-country mean for 1920–1938 adds Finland.

The arms race that led towards war in 1914 is not remarkable for its economic dimension. The industrial revolution, combined with the fiscal reforms of the nineteenth century, enabled Western states to increase military spending without excessively burdening their economies. We provide two standardized measures of defense spending: the "military burden" on national resources (per cent of GDP) and the "defense share" of budgetary means (per cent of central government expenditure). In the period before 1914 most countries carried heavier military burdens than in the early nineteenth century. Within Europe, as Table 3 suggests, the great powers carried somewhat heavier burdens than others; the average military burden on the GDPs of the six great powers was 3.2 per cent, compared with 2.7 per cent for the others in the sample. Notably the United States, the emerging economic leader, devoted less than one per cent of GDP to its armed forces (Eloranta, 2003).

The arms race was fuelled by both rivalry and restraint. While some countries pushed up military spending to keep abreast of their rivals' efforts, others did the same to exploit the relative restraint of the British and American economic giants. The origins of the Great War are often seen in the consolidation of two confrontational alliances; however, when the statistical evidence is inspected for evidence of strategic interaction we find that the alliances themselves were ultimately inefficient and almost irrelevant in the spending decisions (Eloranta, 2007).

A different picture emerges when we turn to the causes of World War II. In the interwar period, military burdens were on average higher than before 1914 but, as Table 3 suggests, military spending was overtaken everywhere (except Italy) by other spending categories. As a result budgetary defense shares were almost uniformly lower, often much lower. Not reflected in the table are national variations in the timing of rearmament; the Soviet Union and Japan began to rearm at the end of the 1920s and Germany at the beginning of the 1930s. For most others intense rearmament was delayed to the mid-1930s. Hitler raised Germany's military burden from below two per cent in 1933 to nearly 20 per cent in 1938. The Japanese rearmament drive was still more impressive, with a 23 per cent military burden and more than 50 per cent defense share in 1938. Mussolini was less successful in rearmament as well as in his efforts to realize the new Roman Empire, with Italy's military burden reaching no more than five per cent in 1938. Achieving high rates of military output before that time, as in the Soviet Union, was of doubtful military value, since the rapid pace of technological change made many of the armaments produced earlier obsolete within two or three years.

Some states oscillated between policies of disarmament and rearmament. Many smaller states did not begin active rearmament until after 1935, but some had high military burdens already in the 1920s (e.g. Portugal and Finland). Sweden started the period with a high defense share, which declined noticeably by the end of the 1930s. A member of the League of Nations from the beginning, Sweden exemplifies an active pursuit of disarmament. According to Ulf Olsson (1973), the Swedish rearmament was slow to react to the worsening international security climate, and its military burden remained below two per cent until 1939. It thus resembles the slow reaction of the United States to the new arms race in the late 1930s (Eloranta, 2002b).

Relatively high military spending was not a guarantee of military success or the security of borders. Military spending determined only the increment to one of four dimensions that, according to Samuel Huntington (1996), produce military power. This was the quantitative dimension (men, arms, and resources). The other three are technological (effectiveness of the equipment); organizational (deployment and morale of the troops); and societal (ability and willingness to apply military force in various situations). Military activity itself takes place at four different levels — political, strategic, operational, and tactical — of which the political sphere contains the funding decision (Millett *et al.*, 1988). Germany, for example, put forth the quantitative resources, invested heavily in various civilian and military technologies, had the required organizational structures and training, and whipped the society into a condition ready for war in the 1930s. French policy-makers, although convinced in the 1920s that French security required high military spending, were unable to maintain military spending at a competitive level in the 1930s and this may have contributed to the technological weaknesses of the French forces in World War II. In addition, their military strategies were ill-attuned to the war of manoeuvre that was about to be unleashed.

The ascendancy of the authoritarian nations and their military spending began in the mid-1920s, and accelerated after 1933, with Germany quickly tipping the balance. It appears that the international system was destabilized by the dispersion of military power in the 1920s. The United States, failed to exercise credible military leadership commensurate with its economic power. In European democracies, domestic military spending decisions were driven more by producer lobbies than external security concerns (Eloranta, 2002b).

The outcome of the renewed rearmament competition was World War II. Where did the drive to war come from? There were both new and old elements in the mixture. Something in continuity with the period before 1914 was rivalry for colonies. The British, French, and Dutch defended their empires. The Soviet Union defended the frontiers of the former Russian Empire. The Japanese looked to create a new empire, first to the north in Manchuria and Siberia; then, when the Soviet Union proved too difficult an adversary, they turned south to the British, French, and Dutch colonies. The Italians looked to build an empire around the Mediterranean, from North Africa to the Balkans and Greece. And Germany looked towards Eastern Europe and Russia for complementary resources and markets and an ethnically restructured living space. To realise these plans the Japanese moved first (into China), then Italy moved (into North Africa), and finally Germany (into Eastern Europe). If Lenin had been alive, he would have recognized the picture. The imperial powers were redividing the world by force. Driving the imperial rivalry, however, was something new.

War, Development, and Dictatorship

Before 1913 the richest countries in Europe were becoming more democratic, while aristocratic and monarchical institutions were weakening

Table 4. Political Regime and Economic Development in Europe Across Two World Wars: Numbers of Countries.

	With GDP per head:			
	Above median		Median or below	
Polity 2 index, 1913:				
Above zero	7		5	
Zero or below		3		5
Polity 2 index, 1923:				
Above zero	9		5	
Zero or below		0		5
Polity 2 index, 1938:				
Above zero	9		2	
Zero or below		2		9
Polity 2 index, 1950:				
Above zero	11		3	
Zero or below		1		10

Sources: Polity IV Dataset from http://www.systemicpeace.org/polity/polity4.htm; GDP per head in 1990 international dollars from Maddison (2001).
Notes: Countries are Albania (1950 only), Austria (not 1938), Belgium, Bulgaria (not 1923), Czechoslovakia (not 1913 or 1938), Denmark, Finland (not 1913), France, Germany, Greece, Hungary (not 1923), Ireland (not 1913), Italy, Netherlands, Norway, Poland (not 1913), Portugal, Romania (not 1923), Russia/USSR (not 1923), Serbia/Yugoslavia, Spain, Sweden, Switzerland, Turkey, and UK. The Polity 2 index subtracts autocracy scores from democracy scores, and also fixes standardized scores, to create a composite index of the political regime suitable for time series analysis, with values ranging from +10 (strongly democratic) to −10 (strongly autocratic).

in poorer countries such as Greece, Serbia, Spain, Portugal, and Turkey. In so far as democracies rarely fight each other the prospects for peace in Europe should therefore have been improving. We measure the political regime using the Polity 2 index (based on the Polity IV data), which subtracts autocracy scores from democracy scores to create a composite index of the political regime with values ranging from +10 (strongly democratic) to −10 (strongly autocratic). Table 4 compares the degree of democracy with the level of economic development achieved. It shows that in 1913 seven of the 10 countries with GDP per head above the European median had achieved a positive Polity index — and so had half of the 10 poorer countries. World War I was launched, however, by the mobilizations of the least democratic powers in Europe: the monarchies of Germany, Austria-Hungary, and Russia.

Losing the Great War destroyed the legitimacy of the regimes that launched it. In Germany, Austria, Hungary, and Russia, the monarchies fell; Germany and Austria became markedly democratic, Hungary mildly so, and Soviet Russia not at all. Where democracy developed, however, it was fragile. The circumstances of the 1920s and early 1930s could hardly have been less favourable, and by the late 1930s the new democratic constitutions had been overridden by a new kind of dictator in Germany, Austria, Italy, Spain, Portugal, and most of Eastern Europe, while Russia had gone over from monarchical to communist absolutism with hardly a pause for breath. Table 4 shows the stark division of Europe that had developed by 1938: most rich countries continued to uphold democracy, but most poor countries had succumbed to authoritarianism. Finally, the post-World War II settlement left the continent just as polarized between rich democracies and poor dictatorships as before, with communists taking the place of fascists across eastern and south Eastern Europe.

The new ingredient in interwar imperial rivalry was totalitarianism, a term that the political theorist Hannah Arendt applied to the most oppressive regimes of the twentieth century — national socialist Germany and communist Russia. According to Juan Linz, a *totalitarian state* has: (1) centralized political power; (2) an exclusive ideology that the leader uses as an instrument both for purposes of political identification and as a guide to action; and (3) mobilization of the citizenry for collective purposes, channelled through a single party using brutal violence against real or perceived opponents (Boesche, 1996; Linz, 2000). It remains debatable where the structures of National Socialism and communism were so similar that one concept could cover them both. The consequences were similar but not identical. In the Soviet Union, Stalin created a state-owned command economy under a ruthless political dictatorship. His victims were chiefly his own subjects, whether killed by neglect in the famine that followed the forced collectivization of the countryside or killed by design during the Great Terror. Using state controls, he built up heavy industry and invested in new military technologies; in the mid-1930s the Soviet Union was possibly the biggest defense producer in the world. At the same time the secret record of Stalin's foreign policy has been shown to be 'passive-aggressive' rather than expansionist (Barber and Harrison, 2005).

Hitler's regime also exemplified the notion of totalitarian control. He added the explosive elements of anti-Semitism and racial purification to the traditional ideals of German nationalism and imperialism. His ascendancy to power in 1933, beginning with an election victory and rapid assumption of the role of a dictator, was preceded by the economic collapse of the Great Depression. German unemployment reached staggering proportions in 1932, with one third of the labor force out of work.

Hitler's appeal was based on promises to restore Germany to prosperity and imperial domination, hatred of the Jews, and fear of the communists at home and in the Soviet Union. Hitler consolidated his power by silencing his enemies and building an efficient police state. His policies reduced unemployment and aimed to make the economy self-sufficient for war. He did not aspire to total control over economic life, but was ready to squeeze living standards and real wages for the sake of rearmament (Tooze, 2007). He withdrew Germany from the League of Nations, and his decisions to remilitarise the Rhineland in March 1936 and annexe Austria were among the death blows to the League's credibility. His main target was colonial expansion (Abelshauser, 2000; Kennedy, 1989). Compared with Stalin, his policies killed more people by intention and fewer by accident. More of his victims were in other countries. But he still killed millions of his own citizens.

Compared with the totalitarianism of Hitler and Stalin, the regimes of the Italian and Iberian fascists, although bloody and repressive, do not really measure up. In Italy, Mussolini was appointed prime minister in October 1922, after the Fascists' march on Rome. He took several years to overcome parliamentarism and the other political parties and he never achieved centralized control of his polity as did Hitler and Stalin. Nor did he secure a rearmament advantage for Italy in the 1930s as he had hoped. Similarly in Portugal, where authoritarianism was on the rise from 1926, António de Oliveira Salazar only gradually acquired dictatorial authority after becoming prime minister in 1932. In Spain, Primo de Rivera's dictatorship of 1923–1930 marked the first authoritarian experiment; after a short-lived republic and a brutal civil war, Francisco Franco returned Spain to dictatorship in 1939. In Eastern Europe, the Baltic, and the Balkans, the 1930s brought the general rise of authoritarian governments, often led by prominent generals and populist leaders (Lee, 1987; Saz, 1999).

The evidence shows, however, that regime shifts were not associated with structural changes in these countries' military spending behaviour. Although authoritarian, they did not accumulate the centralized powers necessary for massive rearmament. Compared to Germany and Japan, for example, the Italian military burden in the late 1930s remained meagre (Eloranta, 2002b). Other authoritarian states failed to resolve the tension between a revisionist ideology and the requirements of survival in the international arena. Their domestic popularity rested heavily on the promise of more security and protectionist trade policies, but in the 1930s they adopted only weak foreign and commercial policy positions. In World

War II, however, especially in Eastern Europe, they tended to side with the Axis.

Waging War

The two world wars placed the economies of Europe under immense strain. What did this mean? The war demanded huge resources. In both wars, tens of millions of men and guns flung millions of tons of explosives at each other using huge machines to do so — battleships, aircraft, and tanks. The demands that this occasioned could be met chiefly by producing more, by importing more, and by consuming and investing less.

The scope for producing more in wartime was limited, and in practice most countries produced less in wartime, not more, despite determined attempts to mobilise their economies. Table 5 compares the European belligerents with three 'control' groups: the neutral countries of Europe, the former British colonies of North America and Australasia that joined the war actively from a distance, and the former Iberian colonies of South America that remained neutral or, if they joined in, did so in name only. In both wars European output fell absolutely, while the more distant regions held their ground or gained. Within Europe the countries that were occupied or defeated suffered most. In almost all regions the loss of output associated with World War II was less than that during World War I, an exception being the disastrous condition of the Soviet economy in 1944. In these terms the success of the two wars of the United States and Canadian economies, which more than doubled their GDPs from 1938 to 1944, was exceptional.[2]

Since total output was inelastic to the needs of the war, military needs imposed a tremendous squeeze on capital spending and household consumption. Table 6 shows figures for a restricted sample of belligerent countries. Setting to one side the huge methodological problems of valuing the consumer cost of the war effort statistically, it is clear that by the twentieth century the main belligerent countries had found ways of diverting between one third and two thirds of total output to warfare. To remove the necessary purchasing power from civilian markets they used a wide variety of instruments: higher taxes were important, and so were war bonds that promised repayment after victory. Also of importance were direct

[2]The Italian data report a miraculous increase in GDP in World War I; surely a statistical anomaly that is discussed by Broadberry (2005).

Table 5. Wartime GDP (Per cent of Prewar).

	Per cent of 1913:		Per cent of 1938:	
	N	1917	N	1944
Europe, total	16	90	19	96
Of which, countries that:				
Stayed out	7	97	6	112
Won	*2	100	2	99
Lost	7	75	11	91
Former British Colonies	4	105	4	207
Former Iberian Colonies	8	100	14	124

Sources: Maddison (2001), except that for GDP in 1917 we add estimates for Hungary from Schulze (2005: 86) and for Russia from Gatrell (2005: 241); for the Ottoman Empire we use a figure of two thirds of 1913 based on Pamuk (2005: 120), but the aggregate figure in the table is not sensitive to this assumption.

Notes: European countries that stayed out are, in World War I: Denmark, Netherlands, Norway, Portugal, Spain, Sweden, and Switzerland; in World War II: Ireland, Portugal, Spain, Sweden, Switzerland, and Turkey. Countries that won are in World War I: France and UK (with regard to Italy, see the note to the table); in World War II: UK and USSR. Countries that lost, meaning that their governments surrendered or their territory was entirely occupied, are in World War I: Austria, Belgium, Germany, Hungary, and the Ottoman and Russian Empires (including Finland); in World War II: Austria, Belgium, Bulgaria, Denmark, Finland, France, Germany, Greece, Italy, Netherlands, and Norway. Former British Colonies are, in both wars: Australia, New Zealand, Canada, and United States. Former Iberian Colonies are, in World War I: Argentina, Brazil, Chile, Colombia, Mexico, Peru, Uruguay, and Venezuela; in World War II: Argentina, Brazil, Chile, Colombia, Costa Rica, Cuba, El Salvador, Guatemala, Honduras, Mexico Nicaragua, Peru, Uruguay, Venezuela.
*Excludes Italy for reasons mentioned in footnote 1. If Italy is included, then 100 becomes 107.

Table 6. The Mobilization of National Resources in Two World Wars (Per cent of GDP at Current Prices).

	Government spending		Military spending	
	1913	1917	1939	1944
France	10	50	—	—
Germany	10	59	23	70
Italy	—	—	8	*21
Soviet Union	—	—	**17	53
United Kingdom	8	37	15	53

Sources: Broadberry and Harrison (2005b: 15); Harrison (1998: 21).
*1943.
**1940; figures are in constant 1937 prices.

controls on consumption and corporate activity that rationed civilian access to everything from food and textiles to machinery, fuels, and strategic materials.

The unit of mobilization was for the most part the national economy. Part of the story is also the way in which international trade was diverted to meet the national purposes of warfare. In wartime the great powers mobilized their young men for military service and their industries for war production. As a result they ceased to supply commercial exports to the world market; to the extent that they could, they sucked in food, fuel, and war goods from external sources including their respective colonial empires and their neutral trading partners. These had to accumulate credits that, they hoped, would continue to be good in the post-war period. Only one great power was comparatively rich enough to wage war and keep up an external surplus at the same time: the United States, which supplied its Allies in two world wars by doing so.

The existence of alliances created the scope for mobilization at a supranational level. Twice in this period a victorious coalition stretched around the northern hemisphere from the western seaboard of the United States through Great Britain to Russia, whose territory completed the circle across northern Europe and Asia to within a few miles of Alaska. Britain and the United States supplied financial and material aid to their allies on a large scale, particularly to France and Italy in World War I and to the Soviet Union in World War II. By combining the abundant capital of the richer economies with the abundant military manpower of the poorer ones, they maximized total fighting power from the given resource endowments of their coalition. In fact, on both occasions the winners integrated and coordinated their economic and military resources with much greater effect than the losing countries, which mobilized and fought to a considerable extent in isolation from each other. But it is also true that the winners started with endowments that were superior in both quantity and quality.

In some poorer countries the relevant scale of mobilization became sub-national. The strain of mobilization was such that efforts to integrate and coordinate a nationwide war effort failed. Instead, national economies pulled themselves apart. This tended to happen where agriculture was only partially commercialized and remained largely in the hands of small scale subsistence farmers. Under the pressure of wartime mobilization this section of the economy displayed a tendency to 'secede from the nation.' The immediate symptom was usually the emergence of urban famine. When working families in wartime Hamburg or Petrograd could no longer buy food, the main reason was not the lack of the food in the economy as a whole, but the growing reluctance of farmers to sell at any price, given the lack of industrial commodities to buy in exchange. As a result, the true scale on which mobilization could be carried out in practice was not the national economy but just the local urban economies that came under the direct control of the central government. Legally the government could claim jurisdiction over the countryside, but its economic sovereignty was de facto much more limited; it could no more command the resources of the farming households a few miles from the city limits than it could control the business decisions of manufacturers and traders in neutral countries the other side of the world (Broadberry and Harrison, 2005b; Harrison, 1998).

The economic disintegration of the Austro-Hungarian Empire during World War I illustrates this theme. The two kingdoms of Austria and Hungary were ruled by a single emperor in Vienna but had separate governments, legal systems, and currencies. More important than outward differences was an economic asymmetry: Austria was richer and more industrialized than Hungary. To mobilise its industries for war, Austria needed the surplus food products of Hungarian farmers, but was unable to supply Hungary with manufactured civilian goods on the same scale as in peacetime. As a result, trade between the two kingdoms tended to shrink; the food stayed in the Hungarian countryside while the Austrian cities were unable to scale up industrial production because of the lack of food and raw materials. At the same time two things happened. Coordination between the kingdoms of the empire deteriorated, and at the same time, within each kingdom, the farmers went their own way (Schulze, 2005). If Austria-Hungary could not coordinate itself, what chance did it have of achieving coordination with the other Central Powers?

Germany's experience of conquest in World War II provides another case. Germany's intention was to seize a colonial empire in Eastern Europe and convert the Ukraine and European Russia into a food surplus region, chiefly by killing or starving a large proportion of the inhabitants and forcing many of the survivors to flee beyond the Urals (Kay, 2006). The failure to achieve a quick victory over the Soviet Union prevented this master plan from being fully implemented; even so, the German occupation authorities made the most determined efforts to exploit the agricultural resources that fell under their control. But these resources proved exceptionally difficult to mobilise, even at gunpoint (Liberman, 1996). Almost by accident, at the same time most of Western Europe fell into German hands. During the period of occupation, Germany's import of food from rich industrialized France ran at several times the level of its food seizures in Russia (Milward, 1965). This illustrates again the difficulty of mobilizing resources when they remained in the hands of low-productivity subsistence farmers.

In the two world wars, the only low-income country to defend its economic integrity under serious attack was the Soviet Union in World War II. More than a decade earlier, Stalin and his associates had drawn the appropriate lesson from Russia's defeat in World War I: small scale peasant farming was Russia's Achilles heel in wartime (Simonov, 1996). Stalin had launched a drive to secure state control over the peasant farmers and their food surpluses by collectivizing the farms. The campaign was carried through at huge cost in lives, and the farming system that resulted was hated and inefficient (Davies and Wheatcroft, 2004). But it achieved its goal in the sense that, when war broke out again, the peasant farmers no longer had the freedom to withdraw from the market. When food was critically short, when there was absolutely not enough food in the country to keep everyone alive, and millions starved, the soldiers and war workers had enough to eat (Barber and Harrison, 1991; Harrison, 1996). Stalin had converted the peasants into the residual claimants on their own produce. As a result, when the fate of the country was on a knife edge in 1942, the economic system did not break down and the Soviet war effort sustained itself at the most critical moment of the war.

The mobilization advantage of the richer economies is equally evident in food consumption. During both world wars, for example, despite submarine blockade, the British fed themselves quite healthily and sufficiently, if monotonously, partly from their own capital-intensive commercialized agriculture, which responded speedily and flexibly to mobilization requirements, and partly from the other side of the world (Olson, 1963). In the United States in World War II, as Hugh Rockoff (1998) has observed, consumers did not have more butter, but they did have more ice cream.

The world wars took economic warfare, traditionally limited to siege and blockade, to a new level. Blockade continued to be practised, but economists from Olson (1963) to Davis and Engerman (2006) have argued that its force was easily vitiated by direct counter-measures as well as economic mobilization and substitution. Germany overcame the British blockade of its ports in World War I in part by exploiting overland trade with neutral neighbours, including Britain's trading partners. The Allies successfully used naval convoys in both wars to protect merchant vessels against Germany's submarine warfare, and so maintained the integration of the Allied war effort. Conversely, for Germany the submarine construction programme in both wars was enormously expensive, and a direct cost of the gamble to resort to unrestricted submarine warfare to in 1917 was that the United States entered the war on the side of the Allies.

In World War II heavy long-range aircraft provided the capacity to bombard the enemy's rear. Between 1940 and 1945, the Allied air forces dropped two million tons of bombs on Germany; 40 per cent of these were aimed at German industrial and transport facilities and another 30 per cent at urban areas (Zilbert, 1981). The strategic bombing of Germany was meant to dislocate the war economy and destroy the will to fight. Firebombing cities ran into diminishing returns as the war progressed; it did not break the will of the people and sometimes even had the opposite effect (Brauer and Tuyll, 2008). It also had little effect on war production since it primarily destroyed civilian resources and increased the readiness of the workers to make sacrifices. The daylight bombing of industry and transport became possible late in the war and proved more effective. Even so, German industrial facilities continued to expand because new investment more than kept pace with the war damage (Abelshauser, 2000). Strategic bombing did increase German production costs, and it forced a huge diversion of German resources from ground attack in the East to air defense in the West. The air campaign was also extremely expensive for the Allies to maintain, however, resulting in frequent disputes about its priority compared with direct support for the Normandy landings. In general, economic warfare did not determine the outcome of any conflict but may have shortened the duration (Førland, 1993).

Aftermath

The consequences of war arrived in two instalments, postmarked 1918 and 1945 respectively. There were some common features, primarily their heavy costs. Both wars resulted in large losses of life and of human and physical capital. The effects of the two wars on economic institutions and postwar performance were quite different, however. World War I cast a long shadow over interwar economic development that international institutions failed to disperse. Unresolved tensions eroded the possibilities of returning to a normal world, and Europe became fatally polarized between wealthy and poor, democracies and dictatorships. Domestic and international reforms that accompanied the ending of World War II, in contrast, helped to

Table 7. Physical Destruction in Europe in World War I: Selected Countries (Per cent of Prewar Assets).

		Physical capital			
	Human capital	Domestic assets	Overseas assets	Reparations bill	National wealth
Austria-Hungary	4.5	6.5	—	—	—
France	7.2	24.6	49.0	—	31.0
Germany	6.3	3.1	—	51.6	54.7
Italy	3.8	15.9	—	—	—
Russia	2.3	14.3	—	—	—
Turkey and Bulgaria	6.8	—	—	—	—
United Kingdom	3.6	9.9	23.9	—	14.9

Source: Broadberry and Harrison (2005b).

alleviate the economic and political problems that plagued Europe after 1918. The aftermath of World War II was quite different. The international economy recovered quickly. The Cold War began, but in spite of it Europe entered a postwar Golden Age; consumers prospered, at least for a while, on both sides of the Iron Curtain.

The human losses of World War I were terrible enough: nearly ten million soldiers died in battle and from other causes. Of 5.4 million Allied military deaths more than half were Russian and French. Of the four million on the side of the Central Powers, Germany and Austria-Hungary contributed three quarters (Broadberry and Harrison, 2005b). These losses cannot be valued in all their human dimensions. To the extent that they represent human capital that had a replacement cost, however, they can be valued financially and compared with other material losses. This is shown in Table 7. Germany, and Turkey and Bulgaria together each lost more than six per cent of their prewar human capital. The table also shows that there was a tendency for human losses to be exceeded, proportionately, by losses of physical capital. The exception was Germany, the territory of which was spared fighting. The heaviest destruction of physical capital took place in France which, being already rich and industrialized, had most to lose.

Turning to the losses associated with World War II we can give a more complete account that includes civilian as well as military deaths. Even taking this into account, the losses of World War II dwarfed those of World War I. This is most certainly the case for civilian mortality, since relatively few civilians suffered premature death during World War I. In contrast, in World War II at least 55 million people died prematurely, more

Table 8. Physical Destruction in Europe in World War II, Selected Countries (Per cent of Assets).

		Physical assets	
	Human assets	National wealth	Industry fixed assets
UK	1	5	—
USSR	18–19	25	—
Germany	9	—	*17
Italy	1	—	10
Japan	6	25	34

Source: Harrison (1998).
*West Germany only.

than half of them civilians. More than 40 million were citizens of Europe. Germany and Yugoslavia lost one in ten, the Soviet Union one in seven, and Poland one in five of its prewar population. As a proportion of the European total, three fifths were citizens of a single country, the Soviet Union, which lost 25 million (for figures see Urlanis (1971: 294–5), except for the Soviet Union itself on which see Harrison (2003)). This was not accidental; part of Germany's plan was to bring about a large reduction in the population of the Soviet Union, in order to free up food supplies for Germany.

The financial dimension of human loss may be compared with what we know about other material and financial losses in Table 8. These figures show that the rate of human losses was highly variable and in the Soviet case astonishingly heavy. Despite this, it was exceeded everywhere by the rate of physical destruction, made possible by new destructive technologies such as strategic bombardment from the air, used alongside traditional means.

Ethnic displacement and racial killing were features of both World Wars. The main episode of note in World War I was the deportation of Armenians eastward from Anatolia in 1915/16. The stated policy of the Young Turk government was to seize their property and walk them to the Syrian desert in the south. There is uncertainty over the number of deaths that resulted (from 300,000 to 1.5 million), and the precise combination of deaths arising from hunger, lack of shelter, and disease with killings in the massacres that accompanied the deportations (Simpson, 1939; Zurcher, 2000). In World War II, there is no doubt about the German intention to kill Europe's Jews, some in Germany and western Europe but most in Poland and the USSR, from 1942 onwards (but the timing, as well as the internal and external triggers of the decision are still debated). Adolf Eichmann, its chief organiser, put the number of deaths resulting at six million. In addition, the Nazis murdered hundreds of thousands of other 'undesirables' as well, including homosexuals, Gypsies, socialists, and mentally or physically disabled persons.

Still more complex and variable was the impact of the two wars on institutions and policies. World War I ended the liberal order of the nineteenth century and began a phase of de-globalization. Production and consumption for total war relied on securing vital strategic materials, raw materials, and food, and this led countries to restrict exports and expand imports when possible. The belligerents also used the denial of trade as a weapon of war, seeking to bring the enemy to its knees through blockade and starvation. International commodity markets disintegrated and the volume of world trade declined (Maddison, 1995).

More generally, in the first half of the century, mercantilist motivations and instruments became more prominent in Europe than at any time for a hundred years. The scramble for resources provoked by World War I was replaced in peacetime by a scramble for liquidity. The longer-term impact of World War I on international trade was disastrous. Globally, the share of trade in GDP fell from 22 per cent in 1913 to 15 per cent in 1929 and only nine per cent in 1938 (Estevadeordal *et al.*, 2003: 595). As for Europe, exports declined relatively and stagnated absolutely after the war; despite the initial recovery of world trade, European exports in 1929 remained below the level of 1913 (Maddison, 1989). Protectionism of both agriculture and industry gained popular support. By the mid-1920s tariffs were substantially higher than they had been in 1913 in Bulgaria, Czechoslovakia, Germany, Hungary, Italy, Romania, Spain, Switzerland, and Yugoslavia. Once the Great Depression took hold, beggar-thy-neighbor policies in the hands of increasingly nationalistic governments, and the consolidation of autarkic regional trading blocs completed the process of trade disintegration. The new restrictions on trade in the 1930s usually came in the form of quotas rather than tariffs. Trade treaties and relations became subjugated to the economic and military interests of the rearming European nations (Findlay and O'Rourke, 2007; Kennedy, 1989; Kindleberger, 1973). Europe's share of world manufacturing continued to decline, and European trade recovered only modestly by the end of the 1930s. By 1950, in contrast, all regions of the world economy, Europe included, began to share the benefits of restored trade and renewed commitments to liberalization (Maddison, 1995).

A notable exception to interwar protectionism is found in the arms trade. Military exports collapsed as World War I drew to a close, but rebounded in the early 1920s and grew substantially until the Great Depression. After the depression military exports recovered faster than world trade as a whole despite falling prices, which resulted partly from the worldwide deflation and partly from increased competition as smaller nations entered the market. Czechoslovakia and Sweden, followed by Belgium and Norway, were the rising military exporters of the 1930s; by 1935 Czechoslovakia led the way with a quarter of the market for small and medium armaments, smaller shares going to Sweden and Belgium (Eloranta, 2002a). The UK maintained its traditional domination of the trade in heavier equipment such as warships. Attempts to regulate the arms trade usually fell far short of professed goals, partly because of opposition from domestic producer lobbies (Krause, 1992; Krause and MacDonald, 1993).

Interwar economic and political instability and the rise of authoritarian regimes can be seen as extensions of World War I and the Great Depression (Feinstein *et al.*, 1997). Many institutional failures were rooted in the Versailles Peace Treaty of 1919, which created many new democracies and the appearance of a 'new Europe.' The core of the treaty, however, inflicted collective punishment on the former Central Powers and set up a cause for German nationalist resistance in the 1930s, while failing to meet the demands of many participants.

The interwar European economy was additionally stressed by war debts. The United States had provided war loans worth millions of dollars to Britain and France. In turn, the British and French had extended their own credits to Italy and Russia. As a result, the poorer countries in the coalition were able to continue fighting long after they would have run out of their own resources. Interwar economic and political relationships were bedevilled by this network of debts, most of them eliminated sooner or later by default. During World War II a similar pattern emerged in that the United States eventually made available billions of dollars of economic assistance to Britain and the Soviet Union, and Britain had its own much smaller Soviet aid programme. The difference is that, after an early pause for consideration, all the wartime assistance among the Allies was rendered free of charge, so that no substantial postwar indebtedness arose.

Institutional failures of the interwar period can perhaps best be observed in the performance of the League of Nations, established in 1920. The eighteen founder members held sway over 74 per cent of the world's population and, 63 per cent its land area, but did not include the United States, which had turned to isolationism. The record of the League was one of failure to act against various acts of aggression, Japanese, Italian, and German, over the next two decades. There were several factors to this failure, including differing disarmament and security goals of the key members, the absence of some key great powers (for example, the United States), and the failure to provide credible security guarantees for the member nations.

The fear of repeating this failure hung over the United Nations, established in 1945 by the victorious Allied coalition. The rapid emergence of the Cold War made repetition seem more than likely. For example, the first proposals for disarmament were not initiated until the 1950s, with little chance of actual success (Jolly *et al.*, 2005). There were two major differences, however. One difference is that whereas the United States had stood aloof from the interwar League, in the postwar UN the United States was the new and economically powerful hegemonic leader. Another was the additional dimensions of new institutions for multilateral coordination after 1945. The UN was complemented by other bodies that emerged in the immediate post-war period: the Bretton Woods system of exchange rate coordination under the International Monetary Fund, the International Bank for Reconstruction and Development (later the World Bank), and the Marshall Plan, which provided economic aid for rebuilding postwar Europe. Bretton Woods became the anchor of post-war stability and economic expansion until the early 1970s. (Maddison, 1995; Maddison, 2001)

Between 1948 and 1951, the United States poured \$13 billion (about \$100 billion at 2003 prices) into the economies of Western Europe. This massive aid package was named after its designer, US Secretary of State George C. Marshall. Its purpose was to help Europe recover from the devastation of the war period and at the same time promote the alternative to socialism. Most Eastern European nations had been converted into Soviet satellites or did not want to irritate the Soviet Union by accepting this aid. In Western Europe, in contrast, the Marshall plan boosted economic cooperation in the region and helped to embed the recovering economies into policies of trade liberalization and market integration (Ritschl, 2004).

In this context it is not surprising to find that postwar economic recovery after 1918 was strained and slow, whereas after 1945 it was generally rapid. Table 9 shows that Europe's neutrals and victors generally took more than three years after 1918 to return to the 1913 benchmark of GDP per head, while the losers took nearly a decade to do so. In comparison, economic recovery after 1945 took place at almost lightning speed. This was one early sign that the world after 1945 would be truly different.

Table 9. Years to Recovery of Prewar GDP per Head from the Final Year of War.

	N	World War I	*N*	World War II
European Countries that:				
Stayed out	7	3.4	6	1.3
Won	4	3.4	2	1.5
Lost	10	9.7	16	4.4

Source: Maddison (2003).
Notes: Countries that stayed out are, in World War I: Denmark, Netherlands Norway, Portugal, Spain, Sweden, and Switzerland; in World War II: Ireland, Portugal, Spain, Sweden, Switzerland, and Turkey. Countries that won are, in World War I: France, Greece, Italy, and UK (including Ireland); in World War II: UK and USSR. Countries that lost, meaning that their governments surrendered or their territory was entirely occupied, are, in World War I: Austria, Belgium, Bulgaria, Czechoslovakia, Germany, Hungary, Romania, Russia, Serbia, and Turkey; in World War II: Albania, Belgium, Bulgaria, Czechoslovakia, Denmark, Finland, France, Germany, Greece, Hungary, Italy, Netherlands, Norway, Poland, Romania, and Yugoslavia.

Conclusions

In the first half of the twentieth century European globalization came to an abrupt halt. It was replaced by protectionism, nationalism, war, and killing and destruction on an immense scale. In mid-century, globalization was resumed, and the European economies began to converge on much higher and more uniform income levels.

After two world wars, three things had changed. First, European economic growth, integration, and prosperity had lost its association with empire. No doubt to the surprise of Europe's nineteenth century leaders, had they lived to see it, it proved possible to acquire wealth and wield influence without claiming imperial sway over vast stretches of faraway peoples and their lands and oceans.

Second, Europe's leaders had a new sense of the importance of cooperation. They now cooperated with the United States in economic recovery, exchange rate coordination, and tariff reduction, with each other in laying new foundations for European integration, and with developing countries in decolonization and development assistance.

Third, Europe's leaders had learned to use the power of the state to regulate economic life. In wartime, governments had wielded immense authority over their people, production, and consumption. There is at least some reason to see the effectiveness of this power as directly linked with the level of development of the economy: as the economy became richer, the potential scope of government authority became wider and more effective. It did not follow that government ought to use this authority in peacetime as in wartime just because it could, although some thought so. One particular reason that they thought so was the unexpected success of the Soviet command economy in mobilizing to defeat Nazi Germany. Learning the appropriate limits of government control over the market economy in turn became a major challenge of the postwar period.

References

Abelshauser, W (2000). Germany: Guns, Butter, and Economic Miracles. In: Harrison, M (ed.) *The Economics of World War II: Six Great Powers in International Comparison.* Cambridge: Cambridge University Press, pp. 122–176.

Angell, N (1972). *The Great Illusion: A Study of the Relation of Military Power in Nations to their Economic and Social Advantage.* New York: Garland Publishing, Inc.

Barber, J and Harrison, M (1991). *The Soviet Home Front, 1941–1945: A Social and Economic History of the USSR in World War II.* London: Longman.

Barber, J and Harrison, M (2005). Patriotic War, 1941 to 1945. In: Sunny RG (ed.) *The Cambridge History of Russia.* Volume 3. Cambridge: Cambridge University Press.

Berghahn, VR (1973). *Germany and the Approach of War in 1914.* New York: St. Martin's Press.

Boesche, R (1996). *Theories of Tyranny, From Plato to Arendt.* University Park, Pa.: Pennsylvania State University Press.

Brauer, J and Tuyll, H von (2008). *Castles, Battles & Bombs: How Economics Explains Military History.* Chicago: Chicago University Press.

Broadberry, S (2005). Italian GDP During World War I. In: Broadberry S and Harrison M (eds.) *The Economics of World War I.* Cambridge: Cambridge University Press, pp. 305–309.

Broadberry, S and Harrison, M (eds.) (2005a). *The Economics of World War I.* Cambridge, UK: Cambridge University Press.

Broadberry, S and Harrison, M (2005b). The Economics of World War I: An Overview. In: Broadberry S and Harrison M (eds.) *The Economics of World War I.* Cambridge, UK: The Cambridge University Press, pp. 1–42.

Davies, RW and Wheatcroft, SG (2004). *The Industrialisation of Soviet Russia, Volume 5: The Years of Hunger: Soviet Agriculture, 1931–1933.* Basingstoke: Palgrave.

Davis, LE and Engerman, SL (2006). *Naval Blockades in Peace and War: An Economic History since 1750.* Cambridge; New York: Cambridge University Press.

DeLong, JB (2000). *Cornucopia: The Pace of Economic Growth in the Twentieth Century.* NBER Working Paper No. W7602. Cambridge, Mass.: National Bureau of Economic Research.

Eloranta, J (2002a). "Weak" European States in the International Arms Trade, 1920–1937: The Impact of External Threats, Market Forces, and Domestic Constraints. *Scandinavian Economic History Review*, 50(1): 44–67.

Eloranta, J (2002b). *External Security by Domestic Choices: Military Spending as an Impure Public Good among Eleven European States, 1920–1938.* Ph.D Thesis. Department of History and Civilisation. Florence, Italy: European University Institute.

Eloranta, J (2003). National Defense. In: Mokyr, J (ed.) *The Oxford Encyclopedia of Economic History.* Oxford: The Oxford University Press, pp. 30–33.

Eloranta, J (2007). From The Great Illusion to the Great War: Military Spending Behaviour of the Great Powers, 1870–1913. *European Review of Economic History* 11(2): 255–283.

Estevadeordal, A, Frantz, B and Taylor, AM (2003). The Rise and Fall of World Trade, 1870–1939. *Quarterly Journal of Economics,* 118(2): 359–407.

Feinstein, CH, Temin, P and Toniolo, G (1997). *The European Economy between the Wars.* Oxford; New York: Oxford University Press.

Ferguson, N (1999). *The Pity of War: Explaining World War I.* New York: Basic Books.

Ferguson, N (2001). *The Cash Nexus: Money and Power in the Modern World, 1700–2000.* New York: Basic Books.

Ferguson, N (2006). *The War of the World: Twentieth-century Conflict and the Descent of the West.* New York: Penguin Press.

Field, AJ (2003). The Most Technologically Progressive Decade of the Century. *American Economic Review,* 93(4): 1399–1414.

Field, AJ (2006). Technical Change and U.S. Economic Growth: The Interwar Period and the 1990s. In: Rhode P and Toniolo G (eds.) *The Global Economy in the 1990s: A Long Run Perspective.* Cambridge: Cambridge University Press, pp. 89–117.

Findlay, R and O'Rourke, KH (2007). *Power and Plenty. Trade, War, and the World Economy in the Second Millennium.* Princeton: Princeton University Press.

Førland, TE (1993). The History of Economic Warfare: International Law, Effectiveness, Strategies. *Journal of Peace Research,* 30(2): 151–162.

Gatrell, P (2005). Poor Russia, Poor Show: Mobilising a Backward Economy. In: Broadberry SN and Harrison M (eds.) *The Economics of World War I.* Cambridge: Cambridge University Press, pp. 235–275.

Harrison, M (1996). *Accounting for War: Soviet Production, Employment, and the Defence Burden, 1940–1945.* Cambridge: Cambridge University Press.

Harrison, M (1998). The Economics of World War II: An Overview. In: Harrison M (ed.) *The Economics of World War II. Six Great Powers in International Comparisons.* Cambridge, UK: Cambridge University Press, pp. 1–42.

Harrison, M (2003). Counting Soviet Deaths in the Great Patriotic War: Comment. *Europe-Asia Studies,* 55(6): 939–944.

Hobsbawm, EJ (1996). The Age of Extremes: A History of the World, 1914–1991. New York: Vintage Books.

Hobson, JM (1993). The Military-Extraction Gap and the Wary Titan: The Fiscal Sociology of British Defence Policy, 1870–1913. *Journal of European Economic History,* 22(3): 466–507.

Huntington, SP (1996). *The Clash of Civilizations and the Remaking of World Order.* New York: Simon and Schuster.

Jolly, R, Emmerij, L and Weiss, TG (2005). *The Power of UN Ideas: Lessons from the First 60 Years.* New York: United Nations Intellectual History Project.

Kay, AJ (2006). *Exploitation, Resettlement, Mass Murder: Political and Economic Planning for German Occupation Policy in the Soviet Union, 1940–1941.* New York: Berghahn Books.

Keegan, J (1999). *The First World War.* London: Hutchinson.

Kennedy, P (1989). *The Rise and Fall of the Great Powers. Economic Change and Military Conflict from 1500 to 2000.* London: Fontana.

Kindleberger, CP (1973). *The World in Depression, 1929–1939.* London: Allen Lane.

Krause, K (1992). *Arms and the State: Patterns of Military Production and Trade.* Cambridge, New York: Cambridge University Press.

Krause, K and MacDonald, MK (1993). Regulating Arms Sales through World War II. In: Burns RD (ed.) *Encyclopedia of Arms Control and Disarmament,* Volume 2. New York: Charles Scribner's Sons.

Lee, SJ (1987). *The European Dictatorships, 1918–1945.* London: Methuen.

Lenin, VI (1963). Imperialism, the Highest Stage of Capitalism. In: *Selected Works,* Volume 1. Moscow: Progress, pp. 667–766.

Liberman, P (1996). *Does Conquest Pay? The Exploitation of Occupied Industrial Societies.* Princeton, NJ: Princeton University Press.

Linz, JJ (2000). *Totalitarian and Authoritarian Regimes.* Boulder, CO: Lynne Rienner Publishers.

Maddison, A (1989). *The World Economy in the 20th Century.* Paris: OECD Publications and Information Centre Distributor.

Maddison, A (1995). *Monitoring the World Economy 1820–1992.* Paris: OECD.

Maddison, A (2001). *The World Economy: A Millennial Perspective.* Paris: OECD.

Maddison, A (2003). *The World Economy. Historical Statistics.* Paris: OECD.

McNeill, WH (1982). *The Pursuit of Power. Technology, Armed Force, and Society since A.D. 1000.* Chicago: The University of Chicago Press.

Millett, AR, Murray, W and Watman, KH (1988). The Effectiveness of Military Organizations. In: Millett AR and Murray W (eds.) *Military Effectiveness: The First World War.* Volume 1. Boston: Unwin Hyman.

Milward, AS (1965). *The German Economy at War.* London: Athlon Press.

Modelski, G and Thompson, WR (1996). *Leading Sectors and World Powers. The Coevolution of Global Politics and Economics.* Columbia, South Carolina: University of South Carolina Press.

Nolte, E (1965). *Three Faces of Fascism; Action Française, Italian Fascism, National Socialism.* London: Weidenfeld and Nicolson.

Olson, M (1963). *The Economics of the Wartime Shortage: A History of British Food Supplies in the Napoleonic War and in World Wars I and II.* Durham, NC: Duke University Press.

Olsson, U (1973). *Upprustning och verkstadsindustri i Sverige under det andra världskriget.* Göteborg.

Pamuk, S (2005). The Ottoman Economy in World War I. In: Broadberry SN and Harrison M (eds.) *The Economics of World War I.* Cambridge: Cambridge University Press, pp. 112–136.

Ritschl, A (2004). The Marshall Plan, 1948–1951. In: Whaples R (ed.) *EH.Net Encyclopedia.* EH.Net. Available online at: http://eh.net/encyclopedia/the-marshall-plan-1948-1951/ (updated 22 April 2014).

Rockoff, H (1998). The United States: From Ploughshares to Swords. In: Harrison M (ed.) *The Economics Of World War II: Six Great Powers In International Comparison.* Cambridge: Cambridge University Press.

Rowe, D (2005). The Tragedy of Liberalism How Globalization Caused the First World War. *Security Studies,* 14(3): 407–447.

Saz, I (1999). Foreign Policy under the Dictatorship Of Primo De Rivera. In: Balfour S and Preston P (eds.) *Spain and the Great Powers in the Twentieth Century.* New York: Routledge.

Schulze, MS (2005). Austria-Hungary's Economy in World War I. In: Broadberry SN and Harrison M (eds.) *The Economics of World War I.* Cambridge: Cambridge University Press, pp. 77–111.

Simonov, NS (1996). Strengthen the Defence of the Land of the Soviets: The 1927 War Alarm and Its Consequences. *Europe-Asia Studies,* 48(8): 1355–1364.

Simpson, JH (1939). *The Refugee Problem.* London: Oxford University Press.

Tooze, A (2007). *The Wages of Destruction: The Making and Breaking of the Nazi Economy.* London: Penguin.

Urlanis, B (1971). *Wars and Population.* Moscow: Progress.

Zilbert, ER (1981). *Albert Speer and the Nazi Ministry of Arms.* Rutherford, NJ: Fairleigh Dickinson University Press.

Zurcher, EJ (2000). *Turkey: A Modern History.* London: Tauris.

Chapter 2

Why the Wealthy Won: Economic Mobilization and Economic Development in Two World Wars

Mark Harrison

"It's the economy, stupid" (James Carville, managing Bill Clinton's US presidential election campaign in 1992).

The study of total war suggests two themes that might be of common interest to both economists and historians.[1] One is to evaluate the contribution of economic factors to the outcomes of wars. The other concerns the effects of wars on long-run economic development. Both topics are worthy and have attracted substantial attention in the literature (Milward, 1977; Hardach, 1977; Ránki, 1993; Overy, 1995; Harrison, 1998a; Chickering and Förster, 2000). This paper deals only with the first.

The pattern of military and economic mobilization in World War II suggests five stylized facts (Harrison, 1998a). First, victory went to the side that supplied the greatest quantity of military resources to the theatres

This chapter was first published by Editions Economica under the French title "Pourquoi les riches ont gagné: Mobilisation et développement économique dans les deux guerres mondiales" in Deux guerres totales 1914–1918 – 1939–1945: La mobilisation de la nation (edited by Dominique Barjot), 135–179. © Ed. Economica 2012.

[1]This paper was first presented to the conference on "La mobilisation de la Nation à l'ère de la guerre totale, 1914–1945: Armer, produire, innover, gérer" organized by the Département d'Histoire de l'Armement of the French Ministry of Defense and held in Paris, 26–28 October 2004. I thank the organizers and participants for comments. I have discussed the issues raised in this paper with Stephen Broadberry over many years and I have gained more than I can say from his advice.

of war. Second, superiority in military resources was based on superior wealth: the richer countries had a systematic, disproportionate advantage in their ability to supply the front with troops and military equipment. Third, are the qualifications: time and geography mattered. The richer countries needed time to make superior resources count. The countries that were closer to the front line tried harder. Fourth, the significance of other non-economic factors like leadership, organization, discipline, and morale was largely conditional on wealth, geography, and time. Given superior resources and the need and opportunity to apply them, the richer countries could solve other problems that defeated the poorer ones. Fifth, these were rules for market economies. In World War II, Stalin broke them by inventing a new kind of command economy that could produce military power out of proportion to its economic weight.

Since our project on World War II, Stephen Broadberry and I have organized a similar project on World War I which is nearly complete (Broadberry and Harrison, 2005a; 2005b). In this paper, I will pool the evidence from both wars and I will suggest that the empirical support for the predominant importance of economic factors in the first war is just as strong if not stronger than in the second.

I do not intend to narrate the story of economic mobilization in total war, but there is one aspect of the narrative that I will take for a starting point, and it is my first retreat from unbridled economic determinism. Economics would *not* have played an important role if either war had gone according to the aggressors' plan of attack. These plans were invariably for a short campaign ending in a speedy victory. The calculations made in Berlin, Rome, Vienna, and Tokyo at different times all gambled on the expectation that purely military superiority and strategic advantages would be enough to defeat the enemy long before economic factors had time to come into play. Often enough economic factors were not even considered. This mindset was not always wrong. It was almost right when Germany attacked France in 1914. It was exactly right from Japan's attack on China in 1937, and the German occupation of Czechoslovakia in 1938, through the fall of France in 1940, to the spread of German power through the Mediterranean and the Balkans in the spring of 1941. But in both world wars, a point came where it lost its relevance. It was at this point, the Battle of the Marne in 1914 and the Battle of Moscow in 1941, that economic factors began to exert their power. This is why time generally limited the role of economic factors in the two world wars: the economic factors played their part once circumstances had given them time to enter the game.

In Part 1 of the paper I will lay out the facts of the Allied superiority in military resources in two world wars. Part 2 does the same for the quantity and quality of the two sides' aggregate resources and production before the war, and also shows that each country's success in wartime production mobilization is largely explained by its pre-war starting point. Part 3 completes the triangle by showing that its pre-war starting point also largely explains each country's success in fiscal mobilization, military mobilization, and armament for capital-intensive warfare. In Part 4 the reasons underlying the strong relationship between prewar economic development and the success, or failure, of a country's wartime mobilization are considered. Part 5 concludes.

1. *Military Superiority*

In both world wars the side won that fielded the greatest quantity of men and military equipment. While this alone does not explain the outcome of either, the figures in Tables 1 and 2 certainly leave a strong impression.

In World War I, the Allied armies outnumbered those of the Central Powers by 60 per cent; the Central Powers produced more field guns and nearly as many rifles but the Allies out produced them substantially in the machine guns that dominated the infantry engagement and in the aircraft and tanks that would eventually break the defensive stalemate of the trenches. In World War II, the Allied armies outnumbered those of the Axis

Table 1. Allies vs Central Powers: Soldiers and Equipment in World War I.

	Allies (1)	Central powers (2)	Ratio, 1:2 (3)
Soldiers Mobilized, million	41.0	25.6	1.6
Weapons Produced:			
Guns, thousand	59.9	82.4	0.7
Rifles, million	13.3	12.1	1.1
Machine Guns, thousand	656	319	2.1
Aircraft, thousand	124.5	47.3	2.6
Tanks	8919	100	89.2

Source: Broadberry and Harrison (2005a: 16–17).
Note: Under Allies, soldiers mobilized cover USA, UK, France, Italy, Russia, and Serbia; the coverage of weapons produced is limited to USA, UK, France and Russia. Under Central Powers, soldiers mobilized cover Germany, Austria-Hungary, Bulgaria and Turkey; weapons produced cover only Germany and Austria-Hungary.

Table 2. Allies vs Axis: Soldiers and Equipment in World War II.

	Allies (1)	Axis (2)	Ratio, 1:2 (3)
Combatant-years, million	106.4	76.9	1.4
Weapons Produced:			
Rifles and carbines, million	25.3	13.0	1.9
Combat aircraft, thousand	370	144	2.6
Machine Guns, thousand	4827	1646	2.9
Guns, thousand	1357	462	2.9
Armoured vehicles, thousand	216	51	4.3
Mortars, thousand	516	100	5.1
Major naval vessels	8999	1734	5.2
Machine pistols, thousand	11604	1185	9.8
Ballistic missiles	0	6000	—
Atomic weapons	4	0	—

Source: Harrison (1998b: 14–16) except that numbers in the French armed forces in 1940 are corrected as noted by Harrison (2005). The number of ballistic missiles is an approximate upper limit based on Ordway and Sharpe (1979: 405–7). Of the four bombs produced by the Manhattan Project one was tested at Alamogordo, two were exploded over Japanese cities, and one remained unused.
Note: Allies are USA, UK, France, and USSR. Axis powers are Germany, Austria, Japan and, for soldiers mobilized and weapons produced other than rifles or machine pistols for which data are lacking or unreliable, Italy. Combatant-years are calculated as the cumulative sum of the strength of the armed forces of each country in each year multiplied by the proportion of that year in which the country was at war on the side of its respective coalition. For countries other than Italy, wartime supply is calculated as annual output adjusted for the number of months of wartime in each year; combatant-years are calculated similarly. For this reasons totals may differ slightly from those calculated in the source. For Italian munitions, wartime totals only are available. "Armoured vehicles" are tanks and self-propelled guns. For Germany, "major naval vessels" are submarines.

by a somewhat smaller margin, 40 per cent.[2] But in weapons and military equipment, roughly speaking, 2:1 was the *minimum* Allied advantage; the one component of military strength in which the Allied armies and navies did not dominate was in ballistic rocketry, used mainly against civilians, this was eventually offset by the American nuclear monopoly of 1945.

An objection to the weight I give these figures is that they omit the moral factor in warfare. Numbers are not the same as fighting power.

[2] I write about the "Allies" as if there was seamless continuity between the two world wars. This is convenient rather than accurate. The membership of the Allied camp in the two wars overlaps but the match is not perfect. In World War I, the United States was an Ally only by a gentlemen's agreement since no formal treaty was signed.

History has many cases when superior morale enabled a smaller army to defeat a much larger one. Not many of these come from the two world wars, however.

When we look at warfare from the point of the view of the individual we may conclude that the moral factor is the *only* thing that matters. The problem is that the collective rationality of the army differs from that of the individual. Brennan and Tullock (1982) suggest that we should think of each rival army not as a unit but as a network of individuals each bound by a moral calculus that the adversary must disrupt to win. In this calculus there are two arguments, the probabilities that my enemy will fight against me and that my comrade will fight with me. When the general has deployed his soldiers and guns his remaining problem is to convince both the enemy and each one of his men that all his men will fight, and there are various well-established mechanisms by which he can do these things. By them he strengthens the morale of his army and weakens that of the enemy.

Clearly, resources did not uniquely determine the outcome on the battlefield. It is more reasonable to claim, first, that resources decided the outcome on the battlefield when other things such as leadership, organization, and morale were equal on both sides; second, in the two world wars these other things were very often nearly equal in fact, or if they were not equal at first they tended to become roughly equal given time, so that in practice resources *did* determine the outcome on the battlefield. A well-supplied army that was losing because of deficient morale could be stiffened and defeat usually had a competitive stiffening effect; on the other hand soldiers who lacked food and ammunition would eventually be pushed back even though their morale remained high.

The evidence for the stiffening effect of defeat is that in two world wars there were so few cases of a failure of morale. Morale failed in the Russian, French, and Italian armies in 1917, 1940, and 1941 respectively; the French army also came close in 1917 and the Soviet army in 1942. The more usual case is that defeat was stiffening; the commanders on both sides proved generally successful in holding their armies together and responded to setbacks and losses with imagination and resilience. Most remarkable was the way that the German and Japanese armies of World War II were held together through years of withering losses and continuous defeats. Without the competitive stiffening effect both wars would not have lasted so long and cost so many lives.

For present purposes there is a simple implication: the Allies did not win either war because their armies were better motivated or better led or had stumbled on some clever formula for undermining the morale of the enemy. They prevailed on the battlefield because of material superiority. Our western culture has provided us with a thousand legends of individual heroism leading to victory against the odds. No doubt this happened occasionally. The prosaic norm, however, is that when British or American troops met the armies of the Axis on equal terms, man for man, and gun for gun, they often lost; when they fought on bravely despite being cornered, outnumbered, and outgunned, they were usually killed or taken prisoner.

In addition it may be objected that material superiority was not enough because it still had to be applied correctly. The choice of the *Schwerpunkt* had to be right, and this required strategic vision. But with material superiority even bad strategy could eventually prevail. Without material superiority, on the other hand, a single bad decision could lead to disaster. The Allies could afford a Gallipoli, but the Axis could not afford a Stalingrad.

A still wider objection to the sums in Tables 1 and 2 is that one should not add up the resources in different national armies without taking into account the cooperation between them. Just as international specialization and trade increase the joint value of the economic resources of different countries, in the same way military cooperation increases the fighting power of men and weapons. Just as a rabble of a thousand men is not an army whatever their uniform, half a dozen national armies without a common strategy do not make an alliance regardless of treaties and signatures. From this point of view it is probably important that in both world wars the Allies eventually pooled their economic resources and their military decision making to a greater degree than the coalition that opposed them. If so, then the ratios of Allied superiority shown in Tables 1 and 2, if anything, underestimate the true Allied advantage.

2. *Economic Superiority*

The military advantage of the Allies in two world wars was based to a much higher degree than is sometimes recognized on prewar economic advantage. A narrative account of either war necessarily begins with a detailed account of the plans and preparations of both sides. Taking a broader view, however, it appears that plans and preparations had little identifiable influence on the resources that a country actually supplied to the fighting front. By far the most important factor was its prewar size and level of economic development. To put it another way, the best way that a country could prepare for war was to arrange to be large and generally prosperous beforehand. Compared to this, nothing else mattered much.

The size of each side is measured by adding up the populations, territories, and gross domestic products of the territories at war. Populations limited the numbers of men and women available in each country for military service or war work. Territories limited the breadth and variety of natural resources available for agriculture and mining; the wider the territory, the more varied the soil types and the minerals beneath the soil tended to be. GDPs limited the volume of weapons, machinery, fuel, and rations that could be made available to arm and feed the soldiers and sailors on the fighting front. The larger the population, territory, and GDP of a country, the easier it would be for that country to overwhelm the armed forces of an adversary.

GDP was more important than either territory or population, however: A poor country might have a large population, but if most of the adults were engaged in low-productivity subsistence farming then there would be little real possibility of transferring many of them out of agriculture to the armed forces or war industry since the remaining farmers would be unable to produce enough food to keep everyone alive. Equally, a poor country might have a large territory but, without a high level of development of roads and railways, would be unable to exploit it economically or defend it militarily. Finally, a poor country typically lacked efficient government and financial services of the kind necessary to account for resources and direct them into national priorities. In short, a relatively high level of economic development was essential if territory and population were to count in war. The economic development of a country can be measured by its GDP per head of the population.

For simplicity, I will omit consideration of trade, aid, and lending between allies, and the role of trade with neutrals. These were of unquestionable importance. Economic specialization and cooperation added value to economic resources in wartime just as military cooperation increased the fighting power of military resources. In both world wars, the Allies probably maintained better economic integration than their adversaries and this increased their overall economic superiority above what the figures will show, but space is lacking to deal with this topic in any detail.

Table 3 adds up the resources on each side at the outbreak of World War I. The figures listed in the table are those reported for each territory in the year 1913. In reality, populations and outputs changed year by year during the war but for many countries and colonies we do not know by

Table 3. The Alliances in World War I: Resources of 1913.

		Territory		Gross domestic product	
Population, million		million sq. km	ha. per head	$ billion	per head, $
Allies:					
November 1914					
Allies, total	793.3	67.5	8.5	1 093.6	1 379
UK, France, and Russia only	259.0	22.6	8.7	622.8	2 405
November 1916					
Allies, total	853.3	72.5	8.5	1 210.5	1 419
UK, France, and Russia only	259.0	22.6	8.7	622.8	2 405
November 1918					
Allies, total	1 271.7	80.9	6.4	1 760.6	1 384
UK, France, and USA only	182.3	8.7	4.8	876.6	4 809
Central Powers:					
November 1914					
Central Powers, total	147.9	5.9	4.0	366.8	2 480
Germany and Austria-Hungary only	117.6	1.2	1.0	344.8	2 933
November 1915					
Central Powers, total	152.7	6.0	3.9	374.2	2 450

Source: Broadberry and Harrison (2005a: 8–10).
Notes: Figures show populations, territories, and incomes for the year 1913. Unless otherwise specified, totals include all lesser powers, colonies, and dependent territories. Territories are measured within contemporary frontiers. Currency units are international dollars at 1990 prices.

how much. The table does show how the volume of resources on each side changed purely as a result of different countries entering and leaving the war. In the first phase of the war Russia, France, and the United Kingdom were allied as the powers of the Triple Entente. They brought with them their dependencies and colonies. Other countries joined in too: Serbia and the other Yugoslav states, the British Dominions, Liberia, and Japan with her colonies. During 1915/16 a second wave of countries joined the Allies: Italy, Portugal, and Roumania. In the third wave of 1917–18 Russia dropped out but the United States joined in, bringing its own possessions, most of Central America and Brazil. Greece, Siam, and China also joined. By the end of this process governments representing 70 per cent of the world's prewar population and 64 per cent of its prewar output had declared war on the Allied side.

The bare totals on the Allied side do not give any idea of their heterogeneity. The British empire will do for illustration since it comprised some of the richest and poorest regions in the world. Britain had a prewar population of 46 million with an average income per head of nearly $5,000 (at 1990 prices). Its colonies, excluding the Dominions, had a prewar population of 380 million, mostly Indians, with an average income of less than $700. As a result, a colonial population eight times that of Britain produced a similar volume of income. However, this income was far less available than Britain's for fighting Germany for three reasons: it was hundreds or thousands of miles away from the theatre of war, the level of development of colonial government and financial services made it hard to tax, and most of it was already committed to the subsistence needs of the colonial populations. In short, the mere possession of low income territories was of little value to a great power in the war. If India helped Britain in the war it was to enable British trade and commerce rather than because Britain could mobilize Indian resources in any meaningful sense. And the trade that really mattered to the British economy in the war was with rich America and Canada, not with poor India.

The changing resources of the Central Powers, also shown in Table 3, can be described more briefly. Austria-Hungary began the war, joined immediately by Germany and soon by the Ottoman Empire. In 1915 the Central Powers were joined by Bulgaria, although not by Italy which reneged on its prewar treaty obligations. At its maximum extent, the alliance of the Central Powers comprised little more than 150 million people, but their relative lack of success in accumulating low-income colonies made them relatively well off with an average income per head of $2,450, comparable to that of Italy on the Allied side.

Table 4 compares the resources on each side at three benchmark dates: November 1914, 1916, and 1918. This table strikes a balance for each alliance as a whole, and also counting great powers only. The rationale for the latter is very simple: if low-income colonies did not count much, how do the figures look if we do not count them at all? There is some imprecision here, of course. For example, Russia is included as a great power, but much of its territory was little more developed than that of India, which is excluded; the British Dominions are also excluded although they were much richer than Russia. Still, singling out the great powers has the merit of simplicity.

Even in the first stage of the war the Allies had access to five times the population, eleven times the territory, and three times the output of the Central powers. This access was limited by relatively low average incomes across the colonial empires of Britain and France, and low incomes in Russia; we see that the average level of GDP per head on the Allied side

Table 4. **Allies versus Central Powers: Resource and Development Ratios.**

	Population	Territory	Territory per head	Gross domestic product	GDP per head
November 1914					
Total	5.4	11.5	2.1	3.0	0.6
Great Powers only	2.2	19.4	8.8	1.8	0.8
November 1916					
Total	5.8	12.3	2.1	3.3	0.6
Great Powers only	2.2	19.4	8.8	1.8	0.8
November 1918					
Total	8.6	13.7	1.6	4.8	0.6
Great Powers only	1.6	7.5	4.8	2.5	1.6

Source: Calculated from Table 3.
Note: Figures show ratios of Allies to Central Powers in populations, territories, and incomes for the year 1913. Territories are measured within contemporary frontiers. Currency units are international dollars at 1990 prices.

in 1914 was not much more than half that of the Central Powers. If we consider great powers only then the Allied advantages in population and output shrink to twice; the Allied advantage in territory actually increases, reflecting the German and Turkish propensities to colonize sandy deserts in Africa and the Middle East.

As the war continued, the Allied powers' advantage in output grew. The decisive year was 1917. When America displaced Russia, the Allied population and territory declined but its output multiplied; the average development level of the Allied powers rose above that of the Central Powers for the first time.

Table 5 covers World War II on the same lines as Table 3. It shows the resources on the territories on either side that are reported for 1938. The territories on each side changed during the war as different countries joined the war, left it, or changed sides. So too the economic potential of each alliance changed. The Allied powers were always economically more developed than the Axis powers, but again the bare totals give little idea of the heterogeneity on each side. The within-coalition variation was greater on the Allied side because it included some of the richest and poorest countries in the world: Australia and India, for example. In contrast the Axis powers were middle-countries that tended to invade other middle-income countries.

The balance of resources is made explicit in Table 6. This balance is struck twice, in 1938 as the Axis powers contemplated their options, and in 1942 when their conquests had reached their greatest extent and their global power was at its peak. It shows the tempting target presented by the prewar

Table 5. The Alliances in World War II: Resources of 1938.

Population, million		Territory, million sq. km	Territory, ha. per head	Gross domestic product, $ billion	Gross domestic product, per head, $
1938					
Allies, total	689.7	47.6	6.9	1 024	1 485
UK and France only	89.5	0.8	0.9	470	5 252
1942					
Allies, total	783.5	68.0	8.7	1 749	2 232
UK, USA, and USSR only	345.0	29.3	8.5	1 444	4 184
1938					
Axis, total	258.9	6.3	2.4	751	2 902
Germany, Austria, Italy, and Japan only	190.6	1.2	0.7	686	3 598
1942					
Axis, total	634.6	11.2	1.8	1 552	2 446
Germany, Austria, Italy, and Japan only	190.6	1.2	0.7	686	3 598

Source: Harrison (1998b: 3–9).
Notes: Figures show populations, territories, and incomes for the year 1938. Unless otherwise specified, totals include all lesser powers, colonies, and dependent territories, but China is omitted throughout. Territories are measured within contemporary frontiers. Currency units are international dollars at 1990 prices.

Table 6. Allies Versus Axis: Resource and Development Ratios.

	Population	Territory	Territory per head	Gross domestic product	GDP per head
1938					
Total	2.7	7.5	2.8	1.4	0.5
Great Powers only	0.5	0.6	1.4	0.7	1.5
1942					
Total	1.2	6.1	4.9	1.1	0.9
Great Powers only	1.8	23.5	13.0	2.1	1.2

Source: Calculated from Table 5.
Note: Figures show ratios of Allies to Axis in populations, territories, and incomes for the year 1938. Territories are measured within contemporary frontiers. Currency units are international dollars at 1990 prices.

empires of Britain and France with nearly three times the population and nearly eight times the territory of the Axis powers' sway. The temptation appears all the greater when set beside the initial inferiority of the Allied powers themselves in everything but metropolitan development level. But the success of the Axis powers that followed aroused the forces that would combine to defeat them.

By 1942, Germany and Japan appeared to stride the world. This is shown in the fact that by 1942 the overall balance of populations and GDPs on each side had become almost equal. Even the huge Allied advantage in territory had shrunk somewhat. In total war, however, the control of far-flung empires was still less important than the size and development level of metropolitan resources. Thus, Germany extracted more food from industrialized France than from the agrarian Ukraine, while Britain was fed from the United States and Canada, not India (Milward, 1977; Liberman, 1996). When it came to metropolitan resources, the decisive facts were the adhesion of the US and Soviet economies to the Allied side. The result was that even in 1942 the Allied powers out produced the Axis by 2:1.

The figures in Tables 1 to 4 are based on the assumption that in wartime, the real output of a given territory did not change. While we cannot track the changes for all countries, the figures available suggest in both wars the wartime changes in output favoured the Allies. In each case there could be an interesting national story to tell. In World War I, for example, the British and American economies expanded. Australia and New Zealand marked time. It is true that Russia began to collapse in 1916 and France in 1917, and this emphasises still more forcefully the extent to which the Allies were saved by the American entry into the war. On the side of the Central Powers, however, the dismal failure of wartime production mobilization was evident from the outset: for much of the war period the German and Austrian economies flatlined at 20 to 25 per cent below their prewar benchmarks for real output. Wartime information is unreliable for Italy and lacking for the Ottoman Empire and Bulgaria.

Figure 1 shows that wartime economic success can be largely explained on the basis of each country's prewar economic development level measured by GDP per head. Moreover, the same pattern is evident in World War II from Figure 2. Pooling the figures for twenty countries in two wars we find that three fifths of the total variation in wartime production can be explained by the prewar economic development level, leaving only two fifths of the story to be told on the basis of national peculiarities of policy, governance, and morale (regression results are reported in Appendix, Table A-1).

Finally, in economic as in military decisions the richer powers could afford mistakes. It seems likely that every country made similar mistakes in the government direction of investment. Uncontrolled mobilization led to

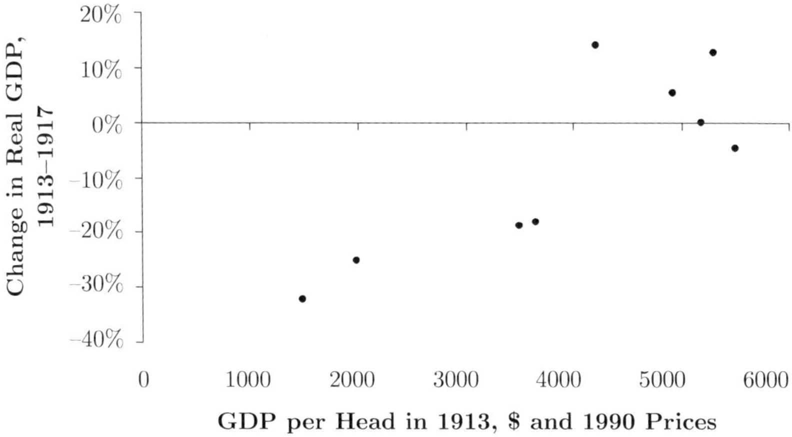

Fig. 1. Production Mobilization: Nine Countries, 1913–1917.
Source: Broadberry and Harrison (2005b).
Notes: Observations from left to right are Russia, Austria-Hungary, France, Germany, Canada, UK, New Zealand, USA, and Australia. Territories are measured within contemporary frontiers. Currency units are international dollars at 1990 prices.

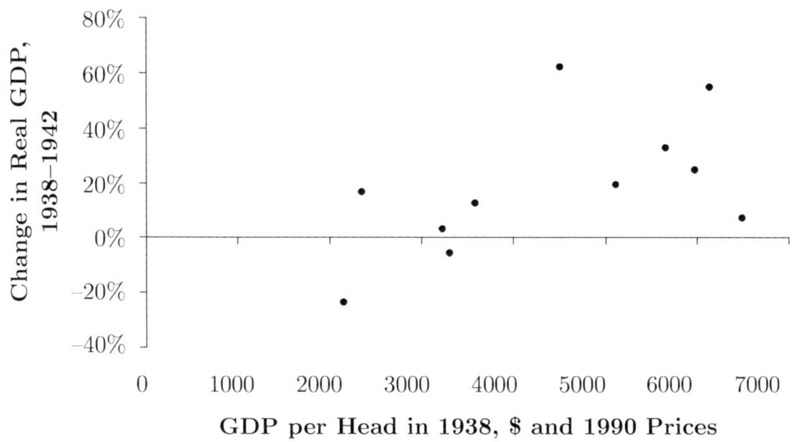

Fig. 2. Production Mobilization: Eleven Countries, 1938–1942.
Source: Harrison (1998b: 10), after correction of a spreadsheet error in the source affecting Soviet GDP as noted by Harrison (2005), and supplemented by figures from Maddison (1995: 180–3 and 194–7).
Note: Observations from left to right are the Soviet Union, Japan, Italy, Finland, Austria, Canada, Germany (excluding Austria), Australia, UK, USA, and New Zealand. Territories are measured within contemporary frontiers. Currency units are international dollars at 1990 prices.

overinvestment. The efficiency of investment was reduced by misallocation across sectors and over time, as bureaucrats misjudged the requirements of the war and its likely duration. The similarity between the pathologies of the German economy in 1917 and the United States in 1942 is striking and amounts to a syndrome of excessive mobilization that affected a number of economies at total war in the twentieth century:[3]

"The [production] programme was decreed by the military without examining whether or not it could be carried out. Today there are everywhere half-finished and finished factories that cannot produce because there is no coal and there are no workers available. Coal and iron were expended for these constructions, and the result is that munitions production would be greater today if no monster programme had been set up but rather production had been demanded according to the capacities of those factories already existing" (German Interior Minister Karl Helfferich in June 1917, cited by Feldman, 1966: 273):

"If we continue as at present, we shall have plants standing useless for lack of equipment or raw materials, or other things. Other plants will be turning scarce materials into items which cannot be used to oppose the enemy because of the lack of other things which should have been made instead. We shall have guns without gun sights, tanks without guns, planes without bomb sights, ships held up for lack of steel plates, planes which we cannot get to the field of battle because of lack of merchant bottoms" (US Army officers to the Army-Navy Munitions Board in March 1942, cited by Higgs, 2004: 507).

The consequences of these mistakes were quite different for the two countries, however. For Germany in 1917, the misallocation of investment was part of a downward economic spiral that fatally eroded the ability to maintain its armies on the eastern and western fronts. For the United States in 1942, it was a minor detriment to a spending bonanza that successfully projected its military power across two oceans at once.

To conclude, the military superiority of the Allies was matched by their economic superiority. We have measured this superiority in various

[3]On excessive mobilization in the British economy in World War II see Robinson (1951: 42–43), and in the Soviet economy Harrison (1998c, 2005: 272–4).

ways, particularly in terms of the size and development level of the great powers. On its own, this does not mean that the two were connected. The connection between a large wealthy economy in peacetime and the ability to field a large, well equipped army in war might be no more than an interesting accident. Thus, it remains to analyze the connection between the military and the economic aspect in more detail.

3. *Mobilization and the Economy*

In this section, I examine the extent to which wartime success in fielding military resources can be traced to the level of prewar economic development. The evidence will show that the comparative success of the various economies in mobilizing their resources for the war effort depended on a few factors that varied independently. The main variable was, as before, their prewar level of economic development. In the first war another factor was geography, or proximity to the front line. In the second war geography mattered less, but a new kind of economic system proved unexpectedly important.

It is convenient to start with mobilization capacity. A simple way of measuring the mobilization capacity of a country is to look at its ability to shift resources rapidly from private to public uses in time of emergency. I measure this in World War I by the shift from private to public uses of resources in each country in the first full year of warfare, and in World War II by the shift from civilian to military uses over the same period.

Figures 3 and 4 plot this shift for eight countries in World War I and six countries in World War II against their prewar development levels. In both wars, there is a group of countries among which we see a strict linear correlation, and there are some outliers. In both wars, the richer countries gained this advantage despite having tended to spend a smaller share of their national income on defense in peacetime (Eloranta 2003). Thus, their ability to transfer resources rapidly from peacetime to wartime uses was perhaps even greater than the figures imply. Finally it should be recalled that in both wars the wealthy American economy, although distant from the fighting, mobilized substantial resources for use by others, not only on its own account; it provided a further five per cent of its GDP in war loans to its Allies in World War I, and a similar proportion as military-economic aid in World War II.

The outliers in each figure are critical to establishing the sign and significance of the influence of prewar development. From Figure 3, we

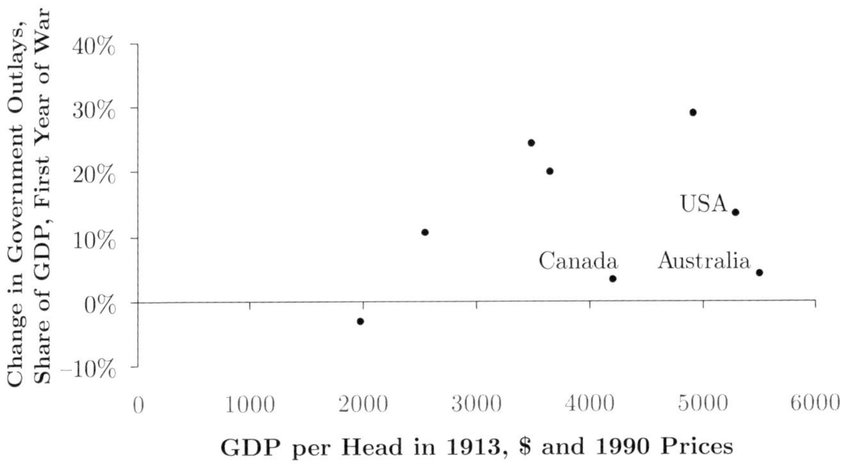

Fig. 3. Fiscal Mobilization in World War I: Eight Countries.
Source: Broadberry and Harrison
Notes: Observations not labelled within the figure are, from left to right, Austria-Hungary, Italy, France, Germany, and UK. The vertical axis measures government outlays as a share of GDP at current prices in the first full year of fighting, less the share in the previous year; for Austria-Hungary, military outlays only are counted. For France, Germany, Canada, the UK, and Australia, 1915 is compared with 1914; for Austria-Hungary, 1915/16 with 1914/15; for Italy, 1916 with 1915; for the United States, 1918 with 1917.

learn that in World War I distance mattered, so that Canada, Australia, and the United States, separated from the conflict by oceanic distances, were clearly on a different curve from the Europeans. In World War II, in contrast, the United States mobilized its economy as vigorously as others. That distance mattered in World War I, and mattered less or not at all in World War II, is not a surprise; during the twentieth century the world was shrinking continually. In Figure 4, there is a real surprise, however: although relatively poor, the Soviet Union mobilized its resources several times faster than one would predict and in fact more rapidly than any other country.

To summarize, there is a clear pattern. The prewar level of economic development powerfully influenced the capacity of economies to mobilize resources in wartime. Controlling for other variables, there was a strong positive relationship that spanned two world wars. Other variables were limited in number. Trans-oceanic distance weakened the impulse to mobilize.

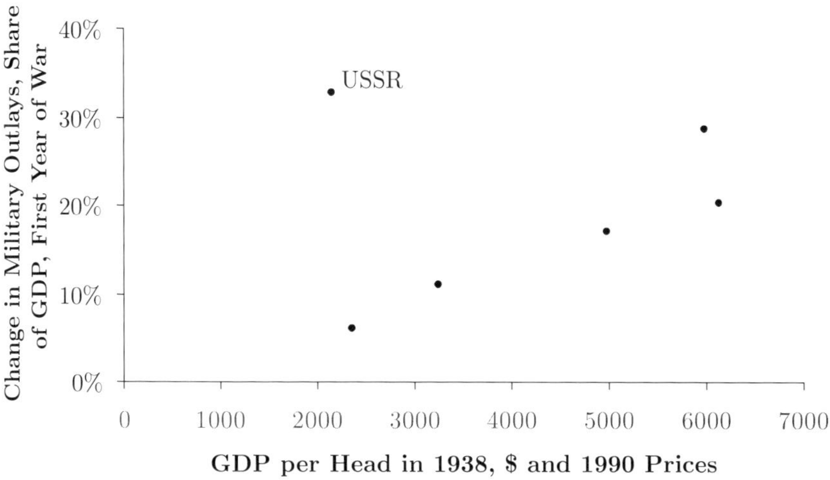

Fig. 4. Fiscal Mobilization in World War II: Six Countries.
Source: Harrison (1998b: 21).
Notes: Observations are, from left to right, the Soviet Union, Japan, Italy, Germany, the UK, and the USA. The vertical axis measures military outlays as a share of GDP or GNP in the first full year of fighting, less the share in the previous year; for the UK the net national product is the denominator; figures are at currently prevailing prices except for the USSR where constant factor costs of 1937 are used. For Germany and the UK 1940 is compared with 1939; for Italy, 1941 with 1940; and for the USA, USSR, and Japan, 1942 with 1941.

In World War II a new variable, the command system, played a big role. Controlling for these few variables we explain more than four fifths of the total variation in fiscal mobilization across fourteen countries in two wars (see Appendix, Table A-2).

These relationships persist when we turn to measure the results of mobilization in soldiers and military equipment. Figures 5 and 6 show soldiers and Figures 7 and 8 show munitions. For the first war, the widest comparisons are available on the basis of cumulative totals of soldiers mobilized during the conflict, and these are shown in proportion to the number of males aged 15–49 in each country before the war. For the second war, we have better data for the armed forces of various countries in each year than for cumulative mobilization totals, so I measure mobilization by the peak wartime number of soldiers in the armed forces as an annual average and per cent of the prewar population.

The measures in Figures 5 and 6 differ, therefore, but the patterns are similar. Figure 5 divides the countries into three distance bands. The first

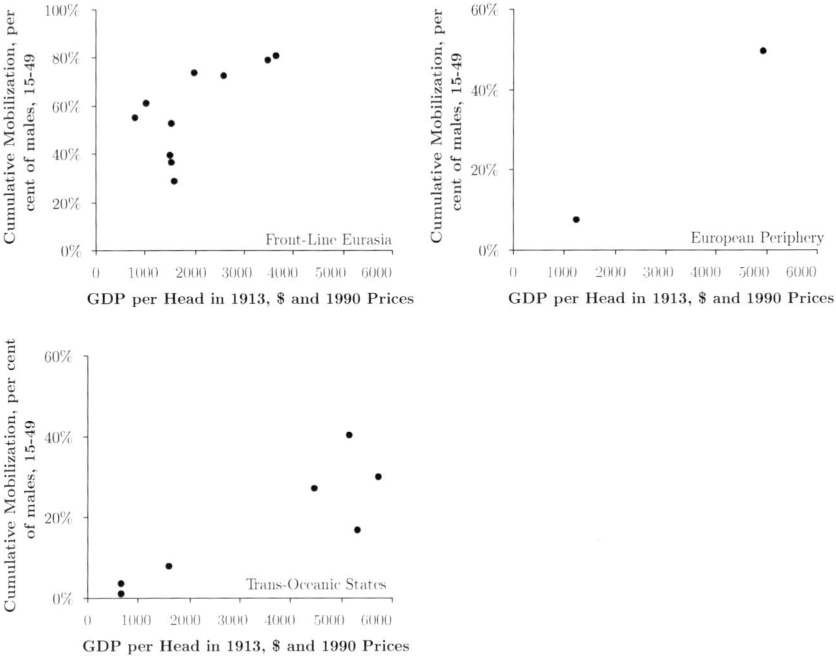

Fig. 5. Military Mobilization in World War I: Eighteen Countries and the French Colonies.

Sources: GDPs per head in 1913 from Tables 1 and 2 or, if not listed there, from Maddison (2001: 185); cumulative mobilization rates, 1914–1918, from Urlanis (1971: 209).

Note: Observations, reading from left to right in order of increasing GDP per head are as follows. Front line Eurasia: Turkey, Serbia, Russia, Bulgaria, Roumania, Greece, Austria Hungary, Italy, France, and Germany. European periphery: Portugal and UK. Non-European States: French colonies, India, South Africa, Canada, New Zealand, USA, Australia.

band comprises the front-line Eurasian states on whose territory or borders the war was fought. The second band is for the countries on the European periphery, separated from the war by land or sea, with only two members: Britain and Portugal. The third band includes countries that joined the war from oceanic distances. Within each band, i.e. controlling for distance, the figures show a strong positive dependence of the proportion mobilized in each country on its prewar income level. The distance band then controlled the height of the curve, so that dropping a band lowered the proportion substantially.

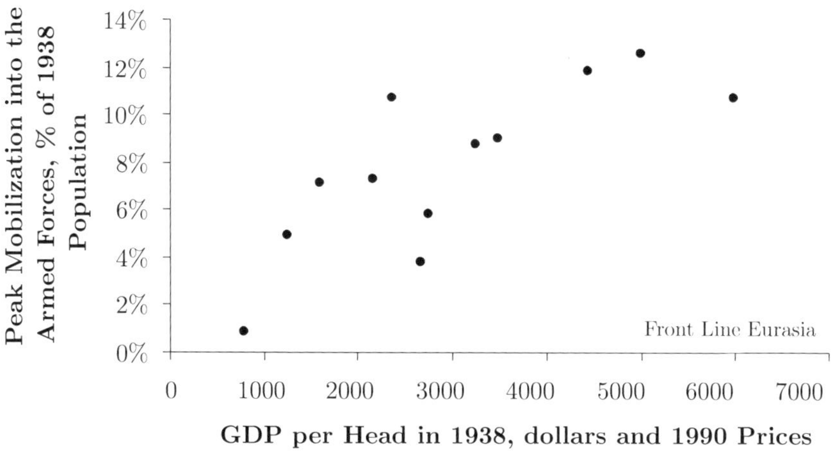

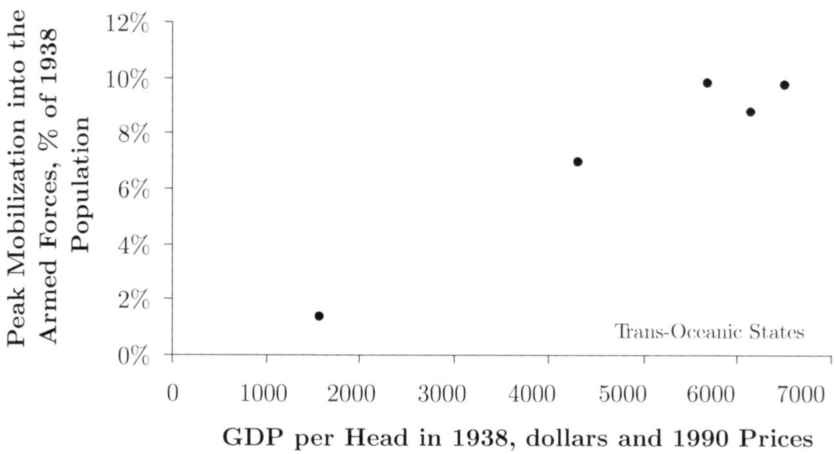

Fig. 6. Military Mobilization in World War II: Seventeen Countries.

Sources: Harrison (1998b: 3–9 and 14), supplemented by figures for wartime military personnel and pre-war populations from the Correlates of War dataset, version 2.1, at http://www.umich.edu/~cowproj. This dataset is further described by Singer (1979; 1980).

Note: The vertical axis measures the wartime maximum of the annual average level of military personnel in proportion to the 1938 population. Observations, reading from left to right in order of increasing GDP per head are as follows. Front line Eurasia: China, Roumania, Bulgaria, USSR, Japan, Hungary, Greece, Italy, Finland, France, Germany, and UK. Trans-Oceanic States: South Africa, Canada, Australia, USA, and New Zealand.

In World War II, we see the same general relationship: controlling for distance, mobilization depended strongly on prewar economic development. It is true that, when it came to mobilizing men, as distinct from resources in general, distance still mattered. Distance mattered less than in World War I because there was no longer a distinction between the European front line and periphery, an understandable result of strategic aviation. But the trans-oceanic states are still banded separately and for given development level they conscripted fewer soldiers than the front line states.

There are statistical obstacles to the pooling of results across the 36 countries represented in two world wars. Considering each war separately, we explain roughly three quarters of the total variation in military mobilization on the basis of these limited economic and geographic variables (see Appendix, Table A-3), leaving one quarter to be explained otherwise.

A notable feature of Figure 6 is the lack of Soviet exceptionalism with regard to mobilizing men (and women). It was no easier for the Soviet Union to spare workers for fighting than for any other poor or middle-income country; the idea of Russia's limitless demographic resources was just a myth. The reason was the high cost of fielding a large army on the basis of a low productivity economy that required so many workers just to feed and clothe them, let alone supply them with weapons and fuel.

Finally, the richer countries were not only able to mobilize more men. Regardless of distance, they also supplied them better. Capital-abundant economies supported capital-intensive warfare. Figures 7 and 8 plot cumulative war production in units per thousand men mobilized in wartime and per year of the war.

In World War I, we see from Figure 7 that in each case, supply rose strongly with the prewar development level of the country. The same relationship is there in World War II, but Figure 8 suggests that it is looser than before. The main reason is the reappearance of Soviet exceptionalism: during the war the Soviet economy provided equipment for its ground and air forces at the same intensity as other countries with twice or three times its income level. The same was not true of its naval shipbuilding, however. In some kinds of weapons, for example aviation, but not others, Japan also approached this performance; but then, unlike the Soviet Union, Japan was not seriously attacked until 1944.

In an alternative perspective, Figure 8 prefigures the Cold War. It shows that there were two countries that proved capable of pursuing capital

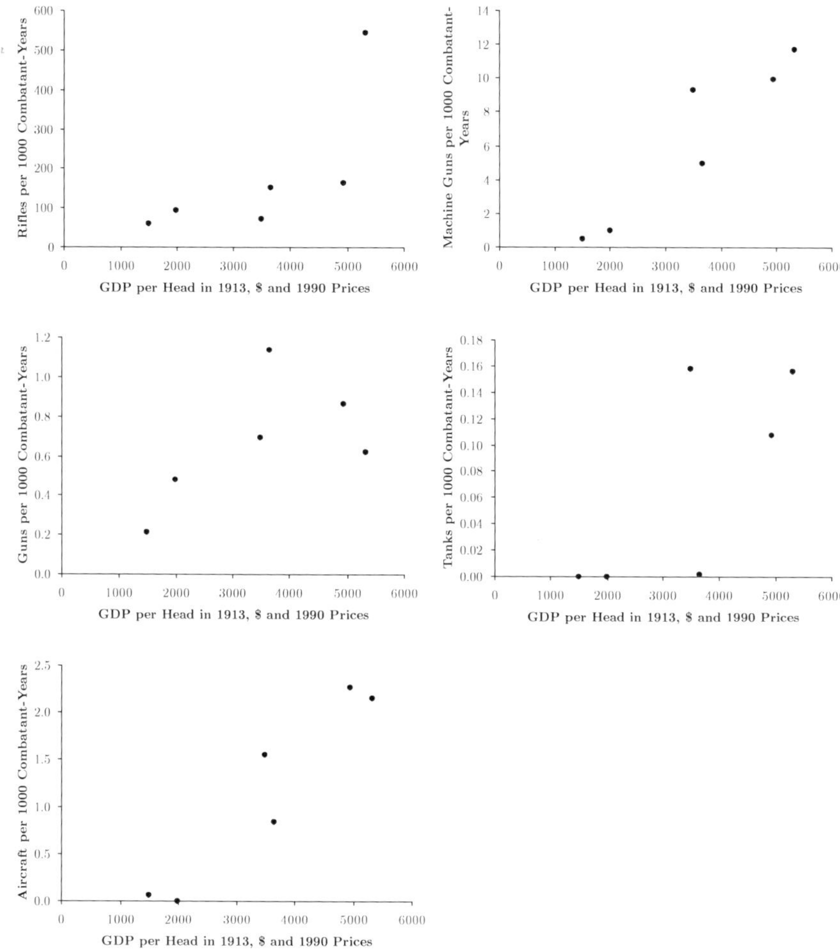

Fig. 7. The Capital Intensity of World War I: Six Countries.
Sources: As Tables 1 and 3.
Note: For each country "combatant years" are numbers mobilized multiplied by years of engagement in the war rounded to 1.5 years for the USA, 3.5 years for Russia, and 4.25 years for the others. Observations, reading from left to right in order of increasing GDP per head are Russia, Austria-Hungary, France, Germany, the United Kingdom, and the United States.

intensive warfare on a broad front in World War II: the Soviet Union and the United States. The rest were also-rans.

To summarize, the Allies fielded armies that were systematically bigger and better equipped than their adversaries in two world wars. They were

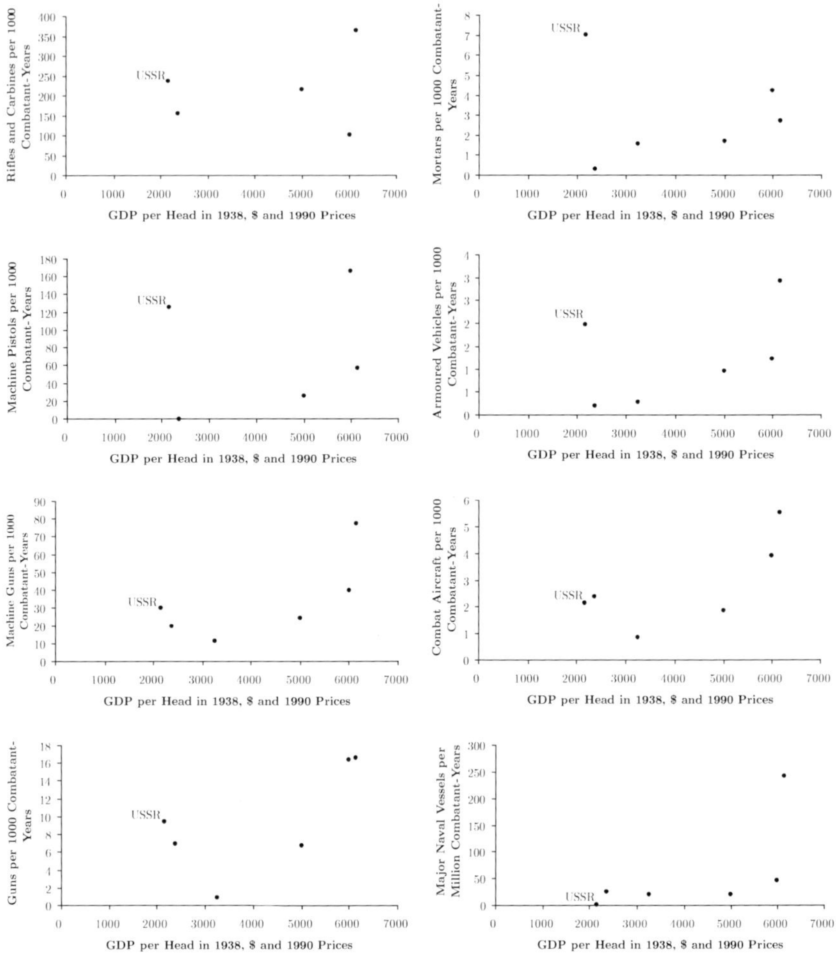

Fig. 8. The Capital Intensity of World War II: Six Countries.
Source and notes: As Tables 2 and 4. Observations, reading from left to right, are the Soviet Union, Japan, Italy, Germany, UK, and USA.

also systematically richer. The correlation of these two facts is no accident; in fact, the high prewar level of economic development of the Allied powers provides the single most powerful explanation of Allied success in wartime mobilization. It was not the only factor. Geography and the invention of the command economy also played a role; that of geography was diminishing and that of the command economy was increasing. Once these influences are taken into account, there is little left to explain in terms of national peculiarities of prewar or wartime leadership, governance, organization, or culture.

4. *Why the Poor Lost*

Countries like Russia and Austria-Hungary were large and before World War I no one doubted for a moment that they were first-rate military powers. The war showed, however, that their power was built on third-rate economic foundations. Given that they were large, why did it matter so much that they were also poor? The reason lay in agriculture: these were countries that ran short of food long before they ran out of guns and shells (Offer, 1989).

One of the most striking attributes of relative poverty was the role of subsistence farming. Contemporary observers were aware of these differences and interpreted them as follows: when war broke out, a country such as Russia would have an immediate advantage in the fact that most of its population could feed itself; moreover, the ability to divert food supplies from export to the home market would actually increase Russia's advantage. In contrast Britain would quickly starve (Gatrell and Harrison, 1993). This diagnosis could not have been more wrong. In practice the presence of a large peasantry proved to be a great disadvantage when it came to the mobilization of resources for war. Peasant agriculture behaved very much like a neutral trading partner. Why should Netherlands trade with Germany given the latter's reduced ability to pay, except under threat of invasion and confiscation? Peasant farmers made the same calculation. Thus, the Russian economy looked large, but if the observers of the time had first subtracted its peasant population and farming resources they would have seen how small and weak Russia really was. Meyendorff (cited by Gatrell, 2005: 245) described what happened in Russia as "the Russian peasant's secession from the economic fabric of the nation". And not only from Russia, for Italy, Austria-Hungary, the Ottoman Empire, and Germany all had large peasant populations that proved extremely difficult to mobilize for much the same reason.

The pattern of the peasant's secession is clearly visible from a comparison of the richer and poorer countries' experience. When war broke out British and American farmers boosted production because they were offered higher prices and responded normally to incentives. The fact that British farming had already contracted to a small part of the economy made its wartime expansion easier: there were plentiful reserves of land unused or little exploited, and the high productivity of farm labor meant that substantial increases in farm output could be achieved with relatively little extra effort (Olson, 1963).

In the poorer countries, in contrast, wartime mobilization began by taking resources away from farming, particularly young men and horses for the army. Once in the army these young men and horses still needed to be fed, of course, which implied a diversion of food supplies from rural households to government purchasers. But at the same time the motivation for farmers in the countryside to sell food was greatly reduced. These were subsistence farmers who grew food partly for their own consumption; what they sold, they took to the market mainly to buy manufactured commodities like textiles and metal goods that they needed for their families. But war dried up the supply of manufactures to the countryside. The small industrial sectors of the poorer countries were soon wholly concentrated on supplying the army with weapons and equipment, uniforms and rations. There was no capacity left to supply the countryside, which faced a steep decline in supplies. Consequently, peasant farmers retreated into subsistence activities. As the market supply of food dried up, in the towns food prices soared.

The economy began literally to disintegrate: there might still be plenty of food, but it was in the wrong place. The farmers preferred to eat it themselves than sell it for a low return. The government had to feed the army at all costs for a simple reason: hungry soldiers will not fight. Between the army and the peasantry, the urban workers were caught in a double squeeze. There was still enough food for everyone to have enough to eat; the famines that arose were localized and stemmed from the urban society's loss of entitlement (Sen, 1983; Offer, 1989), not from the decline in aggregate availability.

Aware of the unequal distribution of food, public opinion might blame unpatriotic speculators or incompetent officials, but the truth was that a poor country had few real choices. The scope for policy to improve the situation was usually more apparent than real, and government action typically made things worse: for example the Russian, Austrian, and German governments all began to ration food to the urban population, while attempting to buy up food from the countryside at purchasing prices that were fixed low for budgetary reasons. To repeat: in richer countries the government paid *more* to the food producers, and this worked, but in poorer countries the government wanted to pay *less* and this had entirely predictable results. The willingness of farmers to participate in the market was still further undermined.

Finally, the government stepped in and tried to hold prices down, creating excess demand and scope for a black market in each country. To the extent that such controls were effective, output and consumption tended to fall further. To the extent that they failed there was scope for black marketeers to step in and capture rents; as long as the rents were competed away production and consumption could both recover but popular respect for law and government would inevitably suffer in the process.

It may seem surprising to find Germany classified among the countries that lost because they were poor. Pre-1914 Germany has entered the economic history textbooks as a developed economic power, but its modernization was highly unbalanced. High levels of productivity in heavy industry co-existed with much lower productivity in light industry, and much of the service sector was also characterized by low productivity, despite Gerschenkron's (1962) focus on the modernized railways and the universal banks (Broadberry, 1998). But perhaps the most obvious sign of Germany's relative backwardness was the high share of the labor force engaged in low productivity agriculture. Germany paid a high price during the two world wars for protecting its agriculture in peacetime (Olson, 1963).

In summary, to be poor when World War I broke out was to suffer the consequences of a peasant agriculture, which was essentially a dead weight on the mobilization efforts of the country concerned. For this purpose I include Germany. The process that resulted had its inexorable conclusion in urban famine, revolutionary insurrection, and the downfall of emperors.

The story of World War II shows similarities and one difference. A similarity was that once again the poorer countries could not hold their economies together when seriously attacked. Italy and Japan remained in the war as long as the Allies were preoccupied with Germany. The Allies began to apply serious military pressure to Italy in 1943 and Japan in 1944. In each case, this pressure was quickly followed by economic disintegration and collapse. Another similarity was that peasant agriculture again proved its capacity to resist mobilization. This was particularly evident in Germany's failure to make good the deficiencies of its own low-productivity subsistence farmers at home by exploiting even lower-productivity subsistence farming in eastern Europe. I have already noted that Germany extracted more food from industrialized France than from the agrarian Ukraine, but it is also true that Britain was fed from the United States and Canada, not India.

The difference from World War I was what happened when Germany attacked Russia. Judged by its size and development level alone, the Soviet Union should have been defeated during 1942. In the two decades that separated the two conflicts, Soviet leaders had more than enough time to reflect on the disaster that had befallen Russia and its old regime in the first war. In the 1920s Stalin determined to avert a repetition. The outcome was forced industrialization based on collectivized farming which destroyed the ability of the peasants to withdraw from the market when put under pressure.

Although a disaster from the point of view of peacetime economic development, the control over agriculture that this gave him enabled Stalin to keep the economy together when war returned. In World War I, the Russian peasants had fed themselves first and fed much of the rest to their livestock or buried it in the ground while the soldiers and war workers had to fight over the scraps. In World War II, the Red Army and the war workers were fed first and the peasants became the residual claimant on available food supplies. As a result, the Soviet economy was able to mobilize itself to a degree that matched the richest of the rival powers, not the poorest. Its ability to repress consumption more generally also allowed the Soviet Union to punch militarily far above its economic weight for the rest of the twentieth century.

Conclusions

Introducing this paper, I suggested five stylized facts about military and economic mobilization in World War II. The first of these is that victory went to the side that supplied the greatest quantity of military resources to the theatres of war. Second, superiority in military resources was based on superior wealth: the richer countries had a systematic, disproportionate advantage in their ability to supply the front with troops and military equipment. Third, time and geography also mattered. Fourth, the influence of all other factors was largely conditional on wealth, geography, and time. Fifth, in World War II the patterns of mobilization in market economies were broken by an exceptional Soviet performance based on the command economy.

When we introduce the evidence from World War I, we find the first four of these patterns present in full force. When subjected to superior force, poor economies eventually crumbled. Against this historical background, the Soviet achievement in World War II appears even more remarkable.

These patterns should not be generalized too far. Broadberry and Harrison (2005a) suggest that the power of these simple ideas about the relationship between economic and military performance is confined to a relatively short historical period. The era of "total war" from 1914–1945 seems to have been unique. In both world wars, the main combatants were able to devote more than half of their national income to the war effort. This is likely to have been impossible before 1914 because until then most people were too poor to be taxed at such rates; most economies had the bulk of their resources locked up in forms of subsistence agriculture that were resistant to mobilization; before mass literacy and the telegraph, typewriter, and duplicator, commercial and government services were too inefficient to do much about it. In short, in earlier stages of global development, total war could not be staged because too many people were required to labor in the fields and workshops just to feed and clothe the population, and it cost too much for government officials to count, tax, and direct them into mass combat.

Since 1945 the economic factors in warfare may have lost significance again. This is because after the advent of nuclear weapons any rich country however small, or any large country however poor, could acquire devastating military force for a few billion dollars. Hence, the marshalling of economic resources may have played a much more vital role in the outcome of the two world wars than was likely in any period before or since.

Appendix

The regressions seek to isolate the influences of pre-war economic development, the economic system, geography, and the passage of time between the wars on the extent to which the economies in the sample could mobilize production, fiscal resources, and soldiers in wartime. In each case the regression is a strong test because it assumes for simplicity that the slope coefficients of the economic and geographic independent variables unchanged across the interwar period. For the countries included in each regression, data sources, and other remarks see the notes under the figures to which each regression relates.

Dependent Variables

Production Mobilization	The change in real GDP from 1913 to 1917 or 1938 to 1942, per cent of the initial year.
Fiscal Mobilization	The share of military or total government outlays in GDP in the first full year of warfare, less the share of the same in the preceding year.
Military Mobilization	*Cumulative Military Mobilization* (World War I) is the cumulative total of soldiers mobilized in wartime, per cent of males aged 15 to 49 in the prewar population. *Peak Military Mobilization* (World War II) is the peak value of the annual average number of military personnel, per cent of the prewar population.

Independent Variables

LnGDPC	GDP per head in 1913 or 1938, measured in dollars and 1990 prices, logarithmically transformed.
War	Equals 1 for World War II, 0 for World War I.
TransOceanic	Equals 1 for Australia, Canada, French colonies, India, New Zealand, South Africa, USA, 0 for other countries.
Peripheral	Equals 1 for the UK and Portugal, 0 for other countries.
Command	Equals 1 for the USSR in World War II, 0 for other countries including Russia in World War I.

Significance

The significance level of a statistic is denoted as follows.
* Significant at 10%
** Significant at 5%
*** Significant at 1%
**** Significant at 0.1%.

Table A-1. Dependent Variable: Production Mobilization.

	(1)	(2)	(3)
Observations	20	20	20
R-Squared	0.6940	0.7617	0.5952
F	8.5063****	8.9492****	12.4957****
Independent Variables:			
Intercept	−0.9754	−0.1898	−2.8424***
LnGDPC	0.0927	−0.0052	0.3372***
War	0.2717***	0.3165****	0.2164**
TransOceanic	0.2555**	0.2884**	—
Peripheral	0.2255	0.2601*	—
Command	—	−0.3236*	—

Sources and definitions: As Figures 1 and 2.

Explanation: On a first pass (column 1), *LnGDPC* or prewar GDP per head is not a significant influence on wartime production mobilization, but geography is. This result does not stem from failure to control for the economic system (column 2). The problem is that the countries that were further away also happened to be richer, so the distance variables *TransOceanic* and *Peripheral* are not independent of prewar GDP per head. When the distance variables are dropped (column 3) the coefficient on prewar economic development becomes positive and highly significant. The positive sign and significance of the *War* variable shows that between the two wars the mobilization capacities of all economies improved, controlling for their economic development level. The R-Squared in column 3 shows that this model explains about three fifths of the overall variation in production mobilization; this is somewhat less than in the preceding columns but its explanatory power (measured by the F of the regression) is much greater.

Table A-2. Dependent Variable: Fiscal Mobilization.

	(1)	(2)	(3)
Observations	14	14	14
R-Squared	0.4255	0.8167	0.8158
F	1.6664	7.1310***	9.9656***
Independent Variables			
Intercept	−0.7761	−2.0181**	−2.1028***
LnGDPC	0.1131	0.2680***	0.2788****
War	0.0410	−0.0249	−0.0266
TransOceanic	−0.1013	−0.1692**	−0.1770***
Peripheral	0.0700	0.0132	—
Command	—	0.3127***	0.3159***

Sources and notes: As Figures 3 and 4.

Explanation: The speed with which governments were able to mobilize resources into war spending was strongly influenced by prewar GDP per head and geography, but this effect is not apparent if the economic system is not taken into account for (column 1). Controlling for *Command* as well as distance variables (column 2), the role of *LnGDPC* emerges as strongly positive and significant. This pattern is confirmed when *Peripheral* is dropped, and it explains more than 80 per cent of the total variation in one-year fiscal mobilization. There appears to have been no significant change in fiscal mobilization capacities between the wars.

Table A-3. Dependent Variable: Military Mobilization.

	(1)	(2)
Observations	36	36
R-Squared	0.7443	0.7430
F	17.4681****	22.4092****
Independent Variables:		
Intercept	−0.6051**	−0.6113**
LnGDPC	0.1491****	0.1498****
War	−0.4106****	−0.4139****
TransOceanic	−0.2857****	−0.2837****
Peripheral	−0.2415***	−0.2398***
Command	−0.0555	—

Notes and sources: As Figures 5 and 6.
Explanation: In both wars the mobilization of men was strongly and positively associated with pre-war GDP per head. It fell significantly as we move from the frontline to the European periphery, and fell further as we move out to the states separated by oceanic distances. *Command* did not play a signficant role. The negative size and much of the significance of the coefficient attached to *War* are a statistical artefact arising from the fact that the military mobilization variable is not consistently calibrated; in World War II the numerator is a smaller concept and the denominator is a larger one than in World War I. As a result, the R-Squared may be inflated as a measure of the true variation that is explained by each model and the F of the regression may also be inflated. Given this, it is better to check for underlying patterns in each war separately.

Table A-3. continued. Dependent Variable: Military Mobilization.

	World War I	World War II	
	(3)	(4)	(5)
Observations	19	17	17
R-Squared	0.7795	0.7616	0.7473
F	17.6753****	13.8472****	20.7008****
Independent Variables:			
Intercept	−0.7121*	−0.3475****	−0.3256****
LnGDPC	0.1728***	0.0544****	0.0514****
War	—	—	—
TransOceanic	−0.4531****	−0.0344***	−0.0313**
Peripheral	−0.3529***	−0.0185	—
Command	—	—	—

Notes and sources: As Figures 5 and 6.
Explanation. Results are shown separately for World War I (column 3) and World War II (column 4 and, after dropping the insignificant *Peripheral* variable, column 5). The patterns suggested from the pooled results in columns 1 and 2 are still robustly in evidence for each war taken separately; it is indicated strongly, however, that distance played a smaller role in World War II. The variables shown continue to explain three quarters of the total variation in the dependent variable. While the R-Squareds and Fs of the regressions in columns 1 and 2 could not be relied on in theory, there is no sign that they were overstated in practice.

References

Adelman, JR (1988). *Prelude to the Cold War: The Tsarist, Soviet, and U.S. Armies in Two World Wars.* Boulder, CO, and London: Lynne Rienner.

Brennan, G and Tullock, G (1982). An Economic Theory of Military Tactics: Methodological Individualism at War. *Journal of Economic Behavior and Organization*, 3(2–3): 225–42.

Broadberry, S and Harrison, M (2005a). The Economics of World War I: An Overview. In: Broadberry S and M Harrison (eds.) *The Economics of World War I.* Cambridge: Cambridge University Press, pp. 276–309.

Broadberry, S and Harrison, M (2005b). *The Economics of World War I: A Comparative Quantitative Analysis.* Paper to the Annual Meeting of the Economic History Association, Toronto. Available at http://warwick.ac.uk/markharrison/papers/ww1toronto2.pdf (accessed 24 November 2006).

Broadberry, S (1998). How did the United States and Germany Overtake Britain? A Sectoral Analysis of Comparative Productivity Levels, 1870–1990. *Journal of Economic History*, 58(2): 375–407.

Chickering, R and Förster, S (eds.) (2000). *Great War, Total War: Combat and Mobilization on the Western Front, 1914–1918.* Cambridge: Cambridge University Press.

Eloranta, J (2003). *Responding to Threats and Opportunities: Military Spending Behavior of the Great Powers, 1870–1913.* Working Paper. University of Warwick, Department of Economics.

Feldman, GD (1966). *Army, Industry, and Labor in Germany, 1914–1918.* Princeton, NJ: Princeton University Press.

Gatrell, P (2005). Poor Russia, Poor Show: Mobilising a Backward Economy for War, 1914–1917. In: Broadberry S and Harrison M (eds.) *The Economics of World War I.* Cambridge: Cambridge University Press, pp. 235–275.

Gatrell, P and Harrison, M (1993). The Russian and Soviet Economy in Two World Wars. *Economic History Review*, 46(3), 425–452.

Gerschenkron, A (1962). *Economic Backwardness in Historical Perspective.* Cambridge MA: Harvard University Press.

Hardach, G (1977). *The First World War, 1914–1918.* Berkeley: University of California Press.

Harrison, M (ed.) (1998a). *The Economics of World War II: Six Great Powers in International Comparison.* Cambridge: Cambridge University Press.

Harrison, M (1998b). The Economics of World War II: An Overview. In: Harrison M (ed.) *The Economics of World War II: Six Great Powers in International Comparison.* Cambridge: Cambridge University Press, pp. 1–42.

Harrison, M (1998c). The Soviet Union: The Defeated Victor. In: Harrison M (ed.) *The Economics of World War II: Six Great Powers in International Comparison*, 268-301. Cambridge: Cambridge University Press, pp. 268–301.

Harrison, M (2005). The USSR and Total War: Why Didn't the Soviet Economy Collapse in 1942? In: Chickering R and Förster S (eds.) *A World at Total*

War: Global Conflict and the Politics of Destruction, 1939–1945. Cambridge: Cambridge University Press, pp. 137–156

Higgs, R (2004). Wartime Socialization of Investment: A Reassessment of U.S. Capital Formation in the 1940s. *Journal of Economic History*, 64(2): 500–520.

Liberman, P (1996). *Does Conquest Pay? The Exploitation of Occupied Industrial Societies.* Princeton, NJ: Princeton University Press.

Maddison, A (1995). *Monitoring the World Economy, 1820–1992.* Paris: OECD.

Milward, AS (1977). *War, Economy and Society, 1939–45.* London: Unwin.

Offer, A (1989). *The First World War: An Agrarian Interpretation.* Oxford: Clarendon Press.

Olson, M (1963). *The Economics of the Wartime Shortage: A History of British Food Supplies in the Napoleonic War and in World Wars I and II.* Durham, NC: Duke University Press.

Ordway, FI and Sharpe, MR (1979). *The Rocket Team.* London: Heinemann.

Overy, RJ (1995). *Why the Allies Won.* London: Pimlico.

Ránki, G (1993). *The Economics of the Second World War.* Vienna: Böhlau.

Robinson, EAG (1951). The Overall Allocation of Resources. In: Chester DN (ed.) *Lessons of the British War Economy.* Cambridge: Cambridge University Press, pp. 34–57.

Sen, AK (1983). *Poverty and Famines: An Essay on Entitlement and Deprivation.* Oxford: Oxford University Press.

Singer, JD (ed.) (1979). *The Correlates of War I: Research Origins and Rationale.* New York: Free Press.

Singer, JD (ed.) (1980). *The Correlates of War II: Testing Some Realpolitik Models.* New York: Free Press.

Chapter 3

The USSR and Total War: Why Didn't the Soviet Economy Collapse in 1942?

Mark Harrison

The aim of this chapter is to reconsider the importance of economic factors in the outcome of World War II and especially on the eastern front.[1] In a recent essay on the war (Harrison, 1998: 2), I asserted that "Ultimately, economics determined the outcome."[2] Production was decisive: the Allies outgunned the Axis because they out-produced them. Economic factors carried more weight in the Allied victory than military or political factors. For example, the Allies did not make better soldiers or provide better weapons. Nor were they better led. It is true that some of the Allies were more democratic, but being a democracy did not save the Czechs, Poles, or French, and being a dictatorship did not defeat the Soviets. The Allies won the war because their economies supported a greater volume of war production and military personnel in larger numbers. The Allied preponderance in this dimension appears so decisive that, once one has grasped it, it seems hardly necessary to pay attention to anything else.

This chapter was first published by Cambridge University Press in 2005 in *A World at Total War: Global Conflict and the Politics of Destruction, 1939–1945* (edited by Roger Chickering, Stig Förster, and Bernd Greiner), 137–156. © Cambridge University Press 2005.

[1]I thank participants in the Total War V conference, Hamburg, 29 August to 1 September 2001, and in the Department of History and Civilization seminar "Inquiries into the Age of Extremes" at the European University Institute, Florence, 23 November 2001, and also John Barber, Louis Capdeboscq, Michael Ellman, Peter Howlett, Valery Lazarev, and Arfon Rees for advice and comments.

[2]This view is directly descended from Goldsmith (1946).

The historian Richard Overy (1998) has objected that this leaves no room for "a whole series of contingent factors — moral, political, technical, and organizational — [that] worked to a greater or lesser degree on national war efforts." I agree that it was very important for each country to solve its moral, political, technical, and organizational problems of the war, and that finding solutions was always costly, sometimes prohibitively so. But on my reading of the history of the war it was always easier to solve these problems if resources were favourable. On the other hand moral, political, technical, or organizational defects could prove fatal only if resources were also lacking.

For the sake of argument, however, I will accept Overy's criticism as valid in two senses. First, he is right that economic determinism makes bad history. In fact, determinism also makes bad economics for economics is about nothing if not choices. I agree that it is desirable to understand the role of economic factors in the war in a way that does not predetermine the role of the other factors. Second, I agree with Overy that if we pay no attention to moral, political, technical, and organizational factors we will not understand the eastern front. Specifically, we will fail to grasp the reasons why the Soviet economy, no larger than Germany's before the war, industrially less developed, and seriously weakened by invasion, did not collapse and instead succeeded in supplying more soldiers and weapons to the front than Germany.

This chapter proceeds as follows. Part 1 describes the global context of the Soviet war effort. Part 2 surveys the scale and purposes of Soviet war preparations and the wartime availability and uses of resources in the Soviet Union. It is suggested that the failure of the Soviet economy to collapse in 1942 is remarkable. Part 3 presents definitions of "economic collapse" and "the point of collapse", as when it is claimed that the Soviet Union in 1942 was at or near the point of collapse, and proposes a framework for understanding the choices behind the outcomes: how did individuals decide to work with or against the national war effort? Part 4 places the strategies of the players in the Soviet economy in 1942 within this framework. Part 5 concludes.

1. *The Global Context*

The Allied economic superiority was never less than overwhelming. A few figures illustrate this point. Table 1 shows that, when the GDPs of the great powers are compared in each year of the war, the superiority of the

Table 1. Wartime GDP of the Great Powers, 1938 to 1945, in International Dollars and 1990 Prices (Billions).

	1938	1939	1940	1941	1942	1943	1944	1945
Allied Powers								
USA	800	869	943	1 094	1 235	1 399	1 499	1 474
UK	284	287	316	344	353	361	346	331
France	186	199	82	—	—	—	—	101
Italy	—	—	—	—	—	—	117	92
USSR	359	366	417	359	274	305	362	343
Allied subtotal	1 629	1 721	1 757	1 798	1 862	2 064	2 325	2 342
Axis Powers								
Germany	351	384	387	412	417	426	437	310
France	—	—	82	130	116	110	93	—
Austria	24	27	27	29	27	28	29	12
Italy	141	151	147	144	145	137	—	—
Japan	169	184	192	196	197	194	189	144
Axis subtotal	686	747	835	911	903	895	748	466
Allies-to-Axis								
Overall	2.4	2.3	2.1	2.0	2.1	2.3	3.1	5.0
USSR to Germany	1.0	1.0	1.1	0.9	0.7	0.7	0.8	1.1

Source: Harrison (1998: 10). French GDP in 1940 is allocated half to the Allies, half to the Axis. This table corrects a spreadsheet error in the source that affected figures for Soviet GDP.

Allied coalition over the Axis powers was never less than 2:1. It is true that a focus on the great powers alone might mislead since one purpose of the building of empires and influence was to expand the extra-territorial resources available to each side. Therefore, Table 2 aims to show the balance of resources on each side including those of their respective colonial empires and trading blocs in 1942, the year most favourable to the Axis. Since the world was in turmoil in 1942, we do not know exactly how many people and dollars of GDP that meant, but the estimates for 1938 get us as close as we can. Table 2 shows that even in 1942 the Allied share of the global economy dominated that of the Axis by nearly 1.9:1 in population and 1.4:1 in GDP; in resource terms, the main Allied weakness lay in the British Empire's vast, low-income African and Asian colonies. Correspondingly France was the unexpected jewel in Germany's imperial crown.

Of course wars are fought with soldiers and weapons, not GDPs. The trail of Axis victories from 1937–1942 is to be explained only by the qualitative superiority of the Japanese and German armies and the strategic advantages of their leaders. Eventually, however, the Allies translated

Table 2. World Population and GDP within 1942 Frontiers (Figures for 1938).

	Allies	Axis	Allies-to-Axis
Population (millions)			
Great Powers	345	191	1.8
Colonies	850	444	1.9
Neutral trading blocs	130	71	1.8
World	1 325	705	1.9
GDP ($ billion and 1990 prices)			
Great Powers	1 444	686	2.1
Colonies	626	866	0.7
Neutral trading blocs	259	151	1.7
World	2 329	1 703	1.4

Source: Calculated from Harrison (1998: 7, 8 and 13). Allied powers are UK, USA, and USSR; Axis powers are Germany, Italy, and Japan. Allied colonies are mainly British dominions and colonies not occupied by the enemy, plus unoccupied China; Axis colonies are occupied Europe, Africa, and Asia (including occupied China). The Allied trading bloc is Ireland plus Central and South America; the Axis trading bloc is neutral Europe and its colonies.

their economic superiority into overwhelming advantage on the battlefield. Table 3 shows that after a temporary disadvantage in 1939 and 1941, occasioned by their own late start and the fall of France, the Allies continually maintained larger armed forces than the Axis powers and this advantage reached almost 2:1 in 1943 and 1944 before the final, desperate Japanese mobilization. But nothing conveys the crushing character of the Allied advantage better than Table 4. This table shows the balance of weapons available to the two sides in 1942. It is true that in 1942 the war industries of both sides retained considerable untapped reserves. The indexes of British and Soviet war production were already close to their respective wartime peaks, but the rates of munitions output of the United States, Germany, and Japan were still accelerating. Still, 1942 was the decisive year of the war; if the Axis powers could not win in 1942 they would never get a better chance. Table 4 shows with dazzling clarity why the decision of that year was against them. In each category they were overwhelmed by Allied superiority measured at better than 2:1 (aircraft), 3:1 (rifles), 4:1 (warships), 5:1 (machine guns), 6:1 (tanks), 7:1 (guns), 13:1 (mortars), and 15:1 (machine pistols). Once we know this, what else is there to know?

Table 3. Armed Forces of the Great Powers, 1939 to 1945 (Thousands).

	1939	1940	1941	1942	1943	1944	1945
Allied Powers							
USA	—	—	1 620	3 970	9 020	11 410	11 430
UK	480	2 273	3 383	4 091	4 761	4 967	5 090
France	5 000	5 000	—	—	—	—	—
USSR	—	5 000	7 100	11 340	11 858	12 225	12 100
Allied subtotal	5 480	12 273	12 103	19 401	25 639	28 602	28 620
Axis Powers							
Germany	4 522	5 762	7 309	8 410	9 480	9 420	7 830
Italy	1 740	2 340	3 227	3 810	3 815	—	—
Japan	—	1 630	2 420	2 840	3 700	5 380	7 730
Axis subtotal	6 262	9 732	12 956	15 060	16 995	14 800	15 560
Allies-to-Axis							
Eastern front	—	—	1.1	1.5	1.4	1.9	2.3
Western and Pacific fronts	1.2	0.7	0.9	1.1	1.9	1.9	1.6

Source: Harrison (1998: 14), except that numbers in the French armed forces in 1940 are shown as held at 5 million rather than rising to 7 million on the advice of Louis Capdeboscq (personal communication, 7 January 2002) and based on figures from Crémieux-Brilhac (1990). The Allied-to-Axis ratio on the western and Pacific fronts for 1939 takes UK and France versus Germany; for 1940, the French and Italian forces are included, each with a 50 per cent weight since Italy joined the war in mid-year at the same time as the French surrendered; for 1942–3, USA and UK versus one tenth of the German armed forces, plus Italy, plus Japan, but in 1943, the Italian forces are given a weight of two-thirds corresponding to the eight months of fighting before the Italian surrender; for 1944–5, USA and UK versus one-third of the German armed forces, plus Japan. On the eastern front, USSR versus 90 per cent of the German armed forces in 1941–3, but only two-thirds in 1944–5.

Do we really need nothing else to understand the outcome of the war? This question becomes sharper when we narrow our focus from the global balance of resources to that on the eastern front. Tables 3 and 4 show that Germany was numerically just as disadvantaged in the east by the Soviet armed forces and war production as were the Axis powers facing Britain and America in the west European and Pacific theatres. Yet, Table 1 showed that the Soviet Union was not superior to Germany in overall resources. Although Soviet GDP exceeded Germany's in 1940, German wartime mobilization and the deep invasion of Soviet territory shifted the balance strongly in Germany's favour. In the most critical years of the war overall Soviet resources were only 70 per cent of Germany's, and the increment arising from Allied aid compensated only to a small extent.

Table 4. War Production of the Great Powers, 1942.

	Rifles and carbines (thou.)	Machine pistols Machine pistols	Machine guns (thou.)	Guns (thou.)	Mortars (thou.)	Tanks (thou.)	Combat aircraft (thou.)	Major naval vessels
Allied Powers								
USA	1 542	651	662	188	11.0	27.0	24.9	1 854
UK	595	1 438	284	106	29.2	8.6	17.7	239
USSR	4 049	1 506	356	127	230.0	24.4	21.7	19
Allied subtotal	6 186	3 596	1 302	421	270.2	60.0	64.3	2 112
Axis Powers								
Germany	1 370	232	117	41	9.8	6.2	11.6	244
Italy	—	—	63	5	8.5	1.5	6.7	164
Japan	440	0	71	13	1.5	1.2	6.3	68
Axis subtotal	1 810	232	251	59	19.8	8.9	24.6	476
Allies-to-Axis								
Overall	3.4	15.5	5.2	7.1	13.6	6.7	2.6	4.4
Eastern front	4.4	9.7	4.6	4.7	35.2	5.9	2.8	—
Western and Pacific fronts	2.4	27.0	5.5	9.2	3.0	7.5	2.5	—

Source: Calculated from Harrison (1998: 15–16). Half of Italian production between mid-1940 and mid-1943 is assumed to have taken place within 1942. Two-thirds of German army and air munitions produced in 1942 are assigned to the eastern front. No account is taken of the contribution of the western Allies or of Italy to the munitions supply of the eastern front.

It is true that Germany was engaged on two fronts. Taking this into account, the Soviet Union still maintained a bigger army in the field than Germany and out-produced German industry systematically in weapons other than warships.

The history of other wars and other countries suggests that the Soviet economy should have collapsed in 1942. In World War I, confronted by a small proportion of Germany's military might, Russia had struggled to mobilize itself and eventually disintegrated. The disintegration was just as much economic as military and political; indeed, it could be argued that Russia's economic disintegration had been the primary factor in both Russia's military defeat and the Russian revolution. Later, in the same war, the preponderance of the western Allies eventually brought about the economic collapse of both Austria-Hungary and Germany. Similarly, in World War II the weaker economies of Italy and Japan collapsed when these countries were seriously attacked by stronger opponents.[3] Among the poorer countries that were invaded only the Soviet Union did not undergo a complete economic disintegration. Instead, the Soviet economy mobilized its resources and the German army was overwhelmed by the scale and scope of Soviet resistance.

2. *Soviet Rearmament and Wartime Resources*

When war broke out the Soviet Union had already engaged in substantial rearmament. In 1940, the last year of less than total war (the Soviet Union had used military force only in Finland and the Baltic region), the Red Army comprised between four and five million soldiers; the military budget consumed one third of government outlays and 15 per cent of the net material product at prevailing prices. One third of the military budget was allocated to procurement of weapons, and Soviet industry produced thousands of tanks and combat aircraft, tens of thousands of guns and mortars, and millions of infantry weapons (Harrison, 1996: 68 and 284; Harrison and Davies, 1997: 372 and 394).

The strategic purposes of pre-war rearmament have been much debated. According to Lennart Samuelson's (1996; 2000a; 2000b) archival study of chief of Red Army armament Marshal M.N. Tukhachevskii, Soviet plans to build a military-industrial complex were laid down before the

[3]On Russia in World War I see Gatrell and Harrison (1993), on Italy and Japan in World War II see Zamagni (1998), and Hara (1998).

so-called war scare of 1927. These plans were not aimed at immediate armament to counter any particular military threat, since at the time none existed. They involved huge forward-looking investments in heavy and defense industry. Samuelson does not rule on their precise motivation. Nikolai Simonov (1996a; 1996b) has set these plans in the context of the Stalinist regime's basic insecurity: the Soviet leadership feared a repetition of World War I when the industrial mobilization of a poorly integrated agrarian economy in the face of an external threat resulted in economic collapse and civil war. Simonov (2000) concludes that, although the 1927 war scare was just a scare, with no real threat of immediate war, it was also a trigger for change. It reminded Soviet leaders that the government of a poor country could be undermined by events at any moment; external difficulties would immediately give rise to internal tensions between the government and the peasantry that supplied both food and conscripts. The possibility of such an outcome could only be eliminated by countering internal and external threats simultaneously, in other words by executing the Stalin package of industrialization and farm collectivization as preconditions for sustained rearmament. This has important implications for our understanding of Soviet history. It is often suggested that the Russian Civil War was an important learning experience for Bolsheviks. It is less often grasped that World War I also contributed something essential to the makeup of the communist economic development strategy.

Both Samuelson and Simonov confirm that in the mid-1930s Soviet military-economic planning was reoriented away from abstract threats to real ones emanating from Germany and Japan. As a result the pace of war production was accelerated far beyond that envisaged earlier in the decade while contingency plans for a war of the future became increasingly ambitious. In Samuelson's view, the military archives leave open the question of whether these plans were designed to support an aggressive war against Germany, rather than to counter a German attack. However, the documentation assembled by Gabriel Gorodetsky (1999) in the central political, diplomatic, and military archives has surely settled this issue: Stalin was trying to head off Hitler's colonial ambitions and had no plans to conquer Europe, although it is true that his generals sometimes entertained the idea of a pre-emptive strike, and attack as the best means of defense was the official military doctrine of the time.

It should not be forgotten that the Soviet Union remained relatively poor. The burden of pre-war rearmament on Soviet resources and incomes was much greater than that of equivalent efforts in Germany, Britain, or the United States. Moreover, the cost of what was achieved by 1940 was only a small fraction of the fresh burdens encountered when war broke out.

The outlines of the Soviet wartime mobilization of resources can be depicted briefly. Under the pressure of a deep invasion, Soviet GNP fell by one third, while the resources allocated to defense increased both relatively and absolutely. The pressure on resources was somewhat alleviated by foreign aid, which added approximately five per cent to Soviet resources in 1942 and 10 per cent in 1943 and 1944. Figure 1 compares Soviet military and civilian uses of resources with production possibilities through the war years. The bold line that wanders to the southeast before turning north marks the actual combinations of military and civilian uses of resources, or total final demand, in each year. In each year the Soviet Union's real gross national product is used to mark a boundary with a minus-45-degree

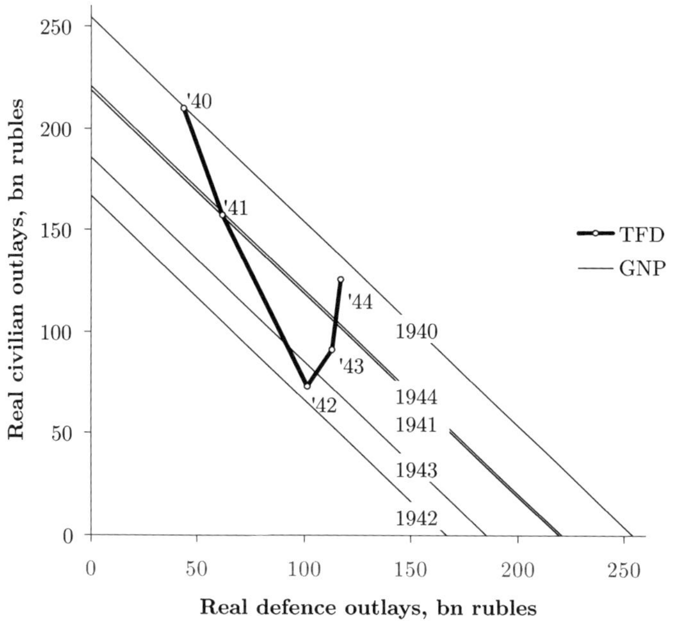

Fig. 1. Soviet Production Possibilities and Uses of Resources, 1940–1944 (Billion Rubles at 1937 Factor Costs).

Source: Harrison (1996: 104). Total final demand (TFD) is the sum of civilian and defense outlays on domestically produced and imported goods and services available for household and government consumption and investment and equals the gross national product (GNP) plus net imports.

slope that shows the alternative uses that were possible within the limits of its own production. The net import of Allied resources allowed the Soviet Union to use more resources than it produced in 1942, 1943, and 1944. The distance from the GNP line to the point representing total final demand in each year shows the difference that Allied resources made.

Wartime changes in the uses of output were accompanied by changes in both employment and productivity. Total employment fell by more than one third, while numbers engaged in military service and war production rose. The biggest shift was out of agriculture; there were smaller movements out of civilian industry, transport, construction, and services. A considerable efficiency gain in defense industry pushed output per worker far above peacetime levels. A similar process was noted (Overy, 1994: 346; Abelshauser, 1998: 155) in Germany and accounts for much of the belated surge of German war production between 1941 and 1944. There was no efficiency gain in other sectors, and labor productivity in the civilian economy declined. This raised the resource requirements of civilian output and made it more difficult to divert resources to military use.

When the war was at its most intense, the resources available to civilian producers and consumers were reduced below the minimum required to replace stocks of physical and human capital. Household consumption was already being squeezed a little by rearmament in 1940; it was squeezed ferociously in 1941–2 by the cut in overall resources and the ballooning defense budget, and squeezed even further in 1943 by the recovery of capital formation. At the low point, living standards were roughly 40 per cent below the pre-war level. Millions were overworked and malnourished, and there was substantial excess mortality amongst the civilian population.

Figure 2 illustrates a conjecture concerning the position of the Soviet economy in 1942. Again, the figure compares Soviet military and civilian uses of resources with production possibilities. Think of the vertical dashed line as showing the military replacement minimum, the minimum level of defense outlays that would maintain the Soviet armed forces and combat stocks of 1942 at a constant level while replacing their losses on and off the battlefield. The horizontal dashed line shows the civilian replacement minimum, the minimum level of civilian outlays that would maintain the Soviet population and capital stock of 1942 at a constant level while meeting subsistence and replacing wartime losses and depreciation. It is plausible that Soviet production possibilities in that year were insufficient to meet both minima at the same time, so the minus-45-degree GNP boundary

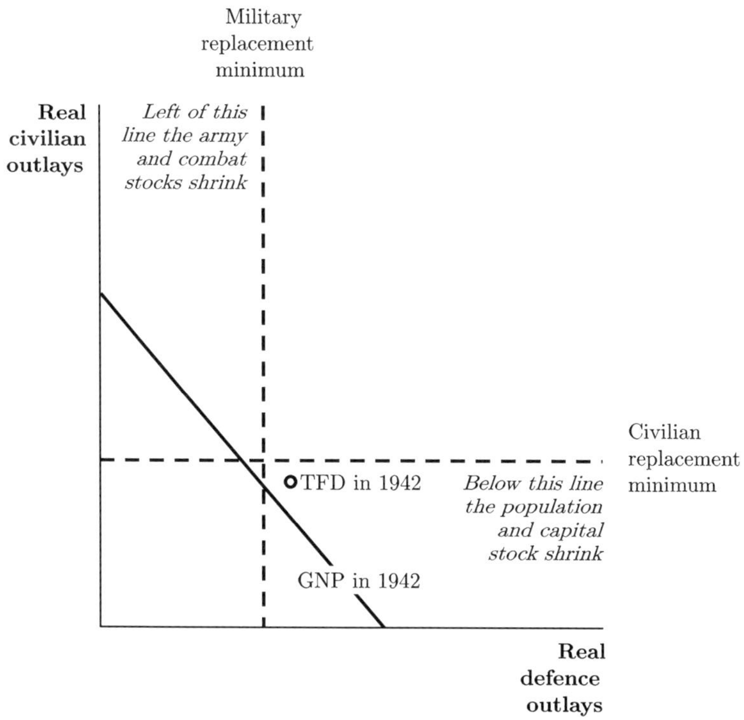

Fig. 2. Soviet Production Possibilities and Uses of Resources in 1942: A Conjecture.

falls inside the point where the dashed lines intersect. Soviet production possibilities were augmented by Allied aid, so total final demand is shown outside the GNP boundary. In 1942, the Soviet armed forces and combat stocks were rising while the population and capital stock were shrinking, so the TFD point fell to the right of the military replacement minimum although below the civilian replacement minimum.[4]

What is remarkable about the Soviet economy is that the tendency of shrinkage did *not* end in economic collapse. Despite negative net investment and millions of hunger deaths, the war effort was maintained and economic recovery followed.

[4]It should not be concluded that Allied aid was used only to augment Soviet defense outlays. In its absence both defense and civilian uses of resources would probably have been reduced but it is possible that civilian uses would have been cut by more. For discussion see Harrison (1996: 128–54).

3. *Where is the Point of Collapse?*

War production was a decisive element of the Soviet war effort. But in 1941 and 1942 its foundations were crumbling. Soviet factories could not operate without metals, machinery, power, and transportation. Their workers needed to be fed and clothed, and competed for the same means of subsistence as the soldiers on the front line and the farmers in the rear. As war production climbed, this civilian infrastructure fell away. While Soviet factories turned out columns of combat-ready vehicles and aircraft, guns and shells, civilians were starving and freezing to death. The tribulations of the other Allied economies, even Britain under submarine blockade and aerial bombardment, seem almost frivolous in comparison. Why the Soviet economy stopped short of outright collapse is therefore a proper and serious question.

How might such economies collapse? A country's war effort will collapse when citizens choose to invest effort elsewhere. In wartime the citizen may choose to allocate effort to patriotic service of the country's interest and to service of self-interest. I define self-interest broadly: it includes service of anything to the exclusion of the interest of the country. Between the country and the self are many layers of association, for example the family, the village, or ethnic group. If the latter are served in ways that conduce to the country's interest I define it as patriotic service. Otherwise self-interest is being served.

A patriotic citizen serves in whatever capacity the state directs, does his or her duty, obeys orders, accepts rations, and respects state property; call this person a mouse. Specifically, mice serve their country to prevent it from being defeated and in the hope of victory. A self-serving citizen behaves opportunistically and ignores orders or gets around regulations, goes absent without leave, jumps queues, and steals government property and the property of others; call this person a rat. Specifically, rats allocate effort between two kinds of theft. They steal nondurable goods, for example food or civilian or military materials, in the hope of surviving until victory. They also steal capital assets, for example durable goods, productive equipment, and even land titles, in anticipation of defeat and in the hope of being permitted by the enemy to establish post-war ownership rights. Thus where mice have a strategy only for victory, rats have strategies for both victory and defeat.

Citizens may choose to be rats or mice, their choice depending on the relative payoffs. In other words, mice are not better people than rats; it is not a moral choice, just a choice between payoffs. This choice

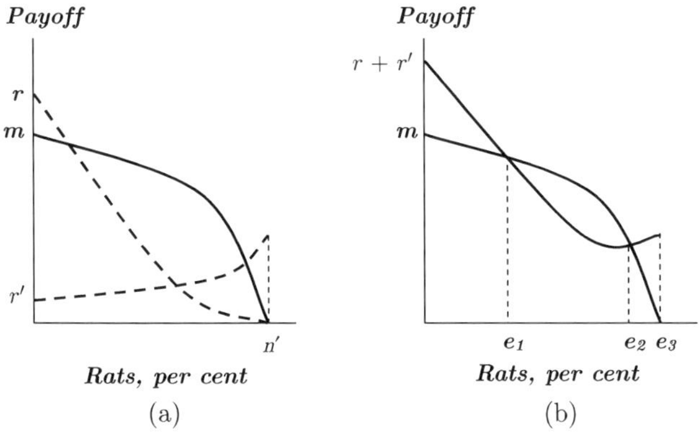

Fig. 3. The Wartime Payoffs to Serving One's Country and Serving Oneself.

is forward-looking, being based on the probability of defeat. Within the framework that I propose, the probability of defeat depends exclusively on the balance of production available to the war effort on each side. But the probability of defeat is endogenous since the level of production available to the war effort depends on people's choices.

Some possible implications are illustrated in Figure 3. Being a mouse brings a payoff. The expected return to patriotic behaviour is the citizen's share in the utility that results from defending one's country (including one's community, one's family, and oneself). This return, labelled m, will fall as the number choosing to be mice falls. At the vertical axis there are only mice, and the payoff to mice has the value m. To the right, the proportion of rats to mice increases and with the community's impoverishment the payoff to mice falls away. First, with fewer mice less output is produced. Second, the growing population of rats diverts a rising share of output away from the war effort. Both raise the probability of defeat and cut the payoff to patriotism. Third, as the probability of defeat increases rats steal a rising share of productive assets, which additionally cuts output. Eventually the payoff to mice falls to zero at a point labelled n' where defeat is certain because everyone has become a rat and output is zero.

Being a rat also brings returns. Consider panel (a). The return from stealing output is labelled r. When everyone else is being patriotic the payoff to the first rat will be substantial. The first to steal supplies will always be able to pick something of a value higher than the payoff to patriotic activity: why else do governments find it necessary to enforce wartime controls? So at the vertical axis $r > m$. But this return will fall as the number choosing to be rats increases. First, there are fewer mice producing less output for the rising population of rats to share. Second, as rats crowd in the risk of confiscation rises at first. Wartime controls are enforced by threats: crime incurs a certain probability of punishment, which confiscates the rat's payoff and reduces it to zero. Let the probability of punishment depend on the proportion of rats to mice, so that it is low when rats are few and there is little threat for mice to guard against; it rises as rats multiply, then peaks and falls again as mice become few and are overwhelmed by rats.[5] These considerations make the rats' payoff decline more rapidly at first than the returns to mice. As rats begins to outnumber mice the rats' risk of confiscation falls again, but the few remaining mice provide little output for rats to steal, so the two payoffs converge on zero at n', the point of certain defeat.

In addition to output rats also steal durable assets. The return from stealing assets is labelled r' in panel (a). Under home rule illegally held assets are always at risk of confiscation. However, rats may calculate that under enemy rule, the previous legal owner being unrepresented, possession of stolen assets will be nine points of the law.[6] As an extension, the enemy may encourage rats by offering to protect their stolen assets after victory. Then the probability that property rights over grabbed assets will become enforceable, and the incentive to grab, will rise with the probability of defeat. And the probability of defeat will rise, the more output and assets are grabbed. Of course while the country is undefeated rats still face the threat of confiscation, and this rises at first when mice are still many, but eventually the danger of confiscation will fade as rats multiply and defeat becomes more likely.

Combine the rats' expected payoff under home rule with their payoff from the enemy. In Figure 3 these give panel (b). The rats' combined payoff $r + r'$ is U-shaped; it has one maximum when rats are few and pickings are rich, and another when rats are so many that the country's defeat is

[5]In this respect the position of rats is different from that usually attributed to rent-seekers: up to a point at least, for rats there is no safety in numbers. On rent-seeking see, Murphy *et al.*, 1993: 409–14).

[6]In English law possession implies ownership unless someone with a better claim comes forward. Recently the appeal court ruled that someone who had bought a car knowing it to be stolen was entitled to keep it since the previous owner was no longer identifiable: there could be no one with a better claim. In short, "even a thief is entitled to the protection of the criminal law against the theft from him of that which he has stolen" (*The Guardian*, London, 23 March 2001).

ensured. In between there is a zone of disputed territory and unresolved conflict where the incentive to grab is weakened by impoverishment and the risks of punishment: there is less to grab, and what is grabbed cannot be held securely. But as defeat becomes more predictable the incentive to grab what's left rises again while the rats anticipate the enemy's arrival.

A result is a stable equilibrium at e_1. A few rats have invaded the community but, with the return to self-serving activity dropping away, grabbing stops at the point where the payoffs to rats and mice are equalized. Thus in any society at war a degree of rule-breaking and self-serving activity might be normal without necessarily threatening the state's survival. Further to the right, there is an unstable equilibrium at e_2 that I define as the "point of collapse". At the point of collapse the payoffs to rats and mice are equal again. To the right of this point the higher reward goes to rats and the war effort collapses unstoppably, taking the country straight to the other stable equilibrium at e_3 where the war is lost. But the existence of the "bad" equilibrium at e_3 is not a problem as long as the "good" equilibrium at e_1 is self-sustaining.

The problem presented by the point of collapse can be translated into the terms of a dictatorship of the stationary-bandit type (Olson, 1993). The dictator administers his assets through agents. Each agent will remain loyal to the dictator's interests as long as his share in the dictator's expected rents from the assets he administers exceeds the expected value of the asset if the agent stole it. If the agent was allowed to gain by stealing from the dictator he would become a roving bandit. This would reduce the expected value to all agents of serving the dictator loyally and increase the others' incentives to rove too. However, unregulated or roving banditry would also reduce the value of assets to all agents, so a rational dictator like Stalin could be expected to enforce cooperation and self-interested agents could be expected to comply. However, their incentives would change if a neighbouring bandit such as Hitler were to offer to settle on the territory, expropriate the dictator, and share the rents on his new assets with the first few agents to defect, threatening the rest with wholesale destruction.

4. *The Risks of Soviet Collapse in 1942*

When citizens chose between serving their country and serving themselves, their calculations were driven by the probability of defeat. In the framework that I propose, the probability of defeat depended exclusively on production. Thus, controlling for rats, greater initial wealth always raised the payoff per mouse relative to the payoff per rat and reduced the likelihood of a wartime collapse. A wealthier community would offer a greater private return to its defense. A poorer enemy was less likely to win and less likely in the event of victory to honor post-war claims to assets laid by rats. When Japan attacked the United States, the rewards to American mice from defending American prosperity were obviously substantial, and the Japanese ability to offer significant rewards to American rats was self-evidently limited. The size disparity of the US and Japanese GDPs ensured that the zone of stability for the US war effort was very large: even if the good equilibrium allowed for significant numbers of cheats and thieves, it remained far to the left of the point of collapse.[7] There was not the slightest chance that the US war effort would collapse into a bad equilibrium, even if a more faint-hearted or more isolationist administration might have wilfully chosen a less belligerent response to attack in the first place. This suggests that initial resource disparities can be decisive.

Think of two economies closer to each other in size, for example Germany and the USSR, engaged in a military struggle that had become too close to call. Consider the Soviet war effort in the winter of 1942. Huge Soviet wealth had already been destroyed or lost to the invader. In Figure 4, panel (a) illustrates this case. Controlling for rats, the payoff per mouse had been depressed by capital losses. Controlling for mice, the anarchy in the civilian economy and the dangers of outright defeat had raised the payoff per rat. The net effect was to shift the good equilibrium dangerously close to the point of collapse. Stalin could rationally fear that with only a small additional capital loss the good equilibrium and the point of collapse would converge and then disappear, making a disintegration of the Soviet war effort inevitable. This case is illustrated in panel (b): there is only one equilibrium where collapse has already occurred.

Under the circumstances shown in panel (a) of Figure 4, the exact positions and slopes of the various schedules became critically important, and the contingent "moral, political, technical, and organizational" factors came fully into play. Fearing destabilization, and with few means available to raise the payoff to mice, the Soviet regime did everything it could to depress the payoff to rats. It is true that the latter was fixed in part by the expected policies of a victorious enemy, and Stalin was helped

[7] On food rationing violations in the United States in World War II see Mills Rockoff (1987).

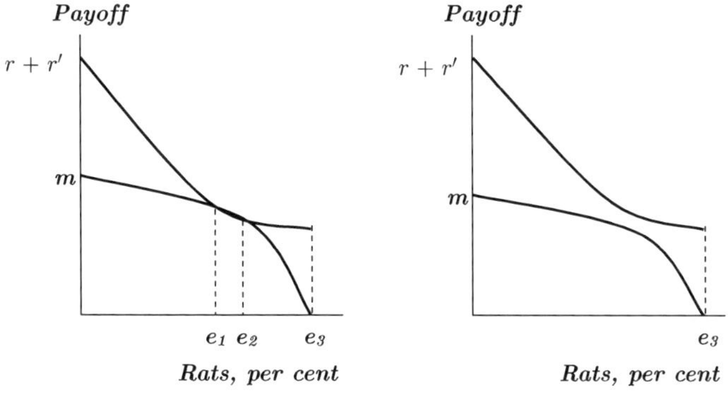

(a) Risk of Collapse is High (b) Collapse is Inevitable

Fig. 4. Two More Cases.

by the fact that Hitler promised little or nothing to ethnic Russians. The Soviet authorities also downshifted the expected payoff from German occupation by threatening potential collaboration with death: even if the enemy prevailed, collaborators would not live to receive any benefit.

Various experiences of 1941–2 testify both to the risks of destabilization and the importance of the Soviet countermeasures taken to strengthen the stable equilibrium.[8] For example, in 1941 expectations were widespread that Soviet resistance to German invasion would follow the same course of unravelling and collapse as that already followed by Poland, Netherlands, Belgium, France, Norway, Greece, and Yugoslavia. These forecasts were reinforced by the ease with which the Wehrmacht moved into the Baltic and the western Ukraine and the warmth of its reception there. Such expectations raised the expected value to individuals of pursuing a strategy for defeat and threatened the existence of the "good" equilibrium. No single episode illustrates this more clearly than the Moscow "panic" of mid-October 1941: with the enemy a few kilometres distant, wrongly believing Stalin had left the city, crowds rioted and looted public property. The authorities took determined steps to counter such perceptions of likely defeat. Stalin suppressed information about Red Army setbacks and casualties. Many were executed for spreading defeatist rumours about events on the front line that might simply have been the truth. Moscow and Leningrad were closed to refugees from the occupied areas in the

[8]Unless otherwise noted, all cases are taken from Barber and Harrison (1991).

autumn of 1941 to prevent the spread of information about Soviet defeats. Evacuation of civilians from both Leningrad and Stalingrad was delayed by the authorities' desire to conceal the real military situation.

Despite this millions implemented or contemplated strategies for defeat. Huge numbers of Red Army soldiers rejected orders that prohibited surrender or retreat. Against orders, millions of encircled soldiers surrendered to the invader in the autumn and winter of 1941 and the spring of 1942. Some prisoners who survived the winter of 1942 subsequently went over to the German side and fought alongside the Wehrmacht, for example General A.A. Vlasov's "Russian Liberation Army", and the Germans also recruited national "legions" from ethnic groups in the occupied areas. At the end of July 1942 when the Germans' summer offensive reached Rostov on Don, significant numbers of Red Army troops ran away from the front line. The risks arising from such behaviours led Stalin to impose the most severe penalties. His Order no. 270 of 16 August 1941 stigmatized the behaviour of Soviet soldiers who allowed themselves to fall into captivity as "betrayal of the Motherland" and inflicted social and financial penalties on the families of the prisoners of war. His Order no. 227 of 28 July 1942 ("Not a step back") combated defeatism in the retreating Red Army by deploying military police behind the lines to shoot stragglers and men retreating without orders and officers who allowed their units to disintegrate. While the war continued Stalin singled out several national minorities suspected of collaboration, for example the Chechens, for mass deportation to Siberia. After the war the *Vlasovtsy* were mercilessly pursued, and Vlasov himself was horribly executed.

Against the same background civilians made similar calculations; this led them to withdraw their human capital from the war effort and steal or conspire to steal productive assets including land titles. In the countryside in the summer of 1941 defeatism stimulated speculative talk about sharing out state grain stocks and collective livestock. In 1941–2 there were widespread reports of collective farmers secretly agreeing the re-division of the collective-farm fields into private property in anticipation of the arrival of German troops. They did not know that Hitler was determined to offer no concessions to Russian peasants, but the Germans permitted some decollectivisation in the north Caucasus and this stimulated local collaborationism. Some of the trains evacuating the plant and equipment of the Soviet defense industries from the southern and western regions to the remote interior in the autumn and winter of 1941 were looted as they moved eastward. In the urban economy, although labor discipline became highly militarized, lateness, absenteeism, and illegal quitting remained widespread. Wartime "deserters" from war work on the industrial front were doggedly pursued and hundreds of thousands were sentenced to terms in prisons and labor camps while the war continued (Filtzer, 2000).

Regardless of the prospects of defeat or victory food crimes became widespread. People stole food from the state and stole from each other. Military and civilian food administrators stole rations for own consumption and for sideline trade. Civilians forged and traded ration cards. In the winter of 1942, Red Army units in the Caucasus began helping themselves to local food supplies (Zolotarev, 1998). Food crimes reached the extreme of cannibalism in Leningrad in the winter of 1941.[9] But when millions lived on the edge even quite trivial violations of food regulations could have lethal consequences for individuals who suffered losses as a result, and food crimes in general were harshly punished, not infrequently by shooting.

In short, it is apparent that the stability of the Soviet war effort was seriously at risk in 1941 and 1942. Millions of Soviet citizens faced desperately hard choices between serving the state and serving their own interests and the interests of those around them with whom they identified. Strategies for victory and defeat diverged. However, beyond a certain point the danger that citizens might choose defeat in ever increasing numbers was not realized. Both Stalin and Hitler played their part in stabilizing the Soviet war effort by closing off the options of honorable surrender and the restoration of private property under German occupation.

Roosevelt also contributed to Soviet stabilization. The first instalment of wartime Allied aid that reached the Soviet Union in 1942, although small by later standards, amounted to some five per cent of Soviet GNP in that year. Although Allied aid was used directly to supply the armed forces with both durable goods and consumables, indirectly it probably released resources to households. By improving the balance of overall resources, it brought about a *ceteris paribus* increase in the payoff to patriotic citizens. In other words, Lend-Lease was stabilizing. We cannot measure the distance of the Soviet economy from the point of collapse in 1942, but it seems beyond doubt that collapse was near. Without Lend-Lease it would have been nearer.

Stalin himself recognized this, although he expressed himself more directly. He told Khrushchev several times that the Soviet Union had

[9] In addition to Barber and Harrison (1991), see Moskoff (1990); and, on Leningrad, Dzeniskevich (2001) and Barber (2001).

suffered such heavy losses that without Allied aid it would have lost the war.[10]

Conclusion

The outcome of the war was decided by production, and production rested on overall resources and their mobilization into the war effort. Taking a global view we can see that Allied superiority measured by overall resources was never in question. Moreover, once the Axis powers had exhausted their purely military advantages, it was not particularly difficult for the Allies to translate economic superiority into overwhelming superiority on the battlefield.

But every generalization has its limits. The limits of this one are to be found on the eastern front, where the war was most bitterly contested. On the eastern front, the Red Army soon outnumbered and outgunned the Wehrmacht. Yet the Soviet Union did not have an overall economic advantage over Germany. It should be considered surprising that, under the pressure of deep invasion and devastating military setbacks the Soviet war effort did not completely unravel in 1942.

The failure of the Soviet economy to collapse in 1942 demands explanation. In that year the Soviet war effort rested on a knife-edge. A battle of motivations took place in which a hundred million people made individual choices based on the information and incentives available. The decisions that individuals made were aimed either at victory or at defeat. The battle of motivations took place in the context of a balance of resources between the two sides that was indecisive. Within this context policy interventions by Stalin, Hitler, and Roosevelt could make a difference. Thus, where the balance of overall resources was indecisive, "moral, political, technical, and organizational factors" decided the outcome.

References

Abelshauser, W (1998). Germany: Guns, Butter, and Economic Miracles. In: Harrison M (ed.) *The Economics of World War II: Six Great Powers in International Comparison.* Cambridge: Cambridge University Press, pp. 122–174.

[10]N.S. Khrushchev (1999: 598–9 and 638) I thank Michael Ellman for this reference.

Barber, J and Harrison, M (1991). Soviet Home Front: 1941–1945. London: Longman.

Barber, J (2001). *Zhizn' i smert' v blokadnom Leningrade. Istoriko-meditsinskii aspekt.* St Petersburg: Dmitrii Bulanin.

Crémieux-Brilhac, J-L (1990). *Les Français de l'an 40 (Volume 2): Ouvriers et soldats.* Paris: Gallimard.

Filtzer, D (2000). Labour Discipline and Criminal Law in Soviet Industry, 1945–1953. PERSA Working Papers no. 8, University of Warwick, Department of Economics.

Dzeniskevich, AR (2001). Banditizm (osobaia kategoriia) v blokirovannom Leningrade. *Istoriia Peterburga*, 1: 47–51.

Gatrell, P and Harrison, M (1993). The Russian and Soviet Economy in Two World Wars. *Economic History Review*, 46(3), 425–452.

Goldsmith, R (1946). The Power of Victory: Munitions Output in World War II. *Military Affairs*, 10: 69–80.

Gorodetsky, G (1999). *Grand Delusion: Stalin and the German Invasion of Russia.* New Haven, CT: Yale University Press.

Hara, A (1998). Japan: Guns Before Rice. In: Harrison M (ed.) *Economics of World War II: Six Great Powers in International Comparison.* Cambridge: Cambridge University Press, pp. 224–267.

Harrison, M (1996). *Accounting for War: Soviet Production, Employment, and the Defence Burden, 1940–1945.* Cambridge University Press: Cambridge.

Harrison, M (1998). The Economics of World War II: An Overview. In: Harrison M (ed.) *The Economics of World War II: Six Great Powers in International Comparison.* Cambridge: Cambridge University Press, pp. 1–42.

Harrison, M and Davies, RW (1997). The Soviet Military-Economic Effort under the Second Five-Year Plan (1933–1937). *Europe-Asia Studies*, 49(3): 369–406.

Khrushchev, NS (1999). *Vremia, liudi, vlast'.* Volume 1. Moscow: Moskovskie novosti.

Mills, G and Rockoff, H (1987). Compliance with Price Controls in the United States and the United Kingdom During World War II. *Journal of Economic History*, 47(1): 191–213.

Moskoff, W (1990). *The Bread of Affliction: the Food Supply in the USSR during World War II.* Cambridge: Cambridge University Press.

Murphy, KM, Shleifer, A and Vyshny RW (1993). Why is Rent-Seeking so Costly to Growth? *American Economic Review Papers and Proceedings*, 83(2): 409–14.

Olson, M (1993). Dictatorship, Democracy, and Development. *American Political Science Review*, 87(3): 567–76.

Overy, RJ (1994). *War and Economy in the Third Reich.* Oxford: Oxford University Press.

Overy, RJ (1998). Who Really Won the Arms Race? *The Times Literary Supplement*, 13 November 1998, 4–5.

Samuelson, L (1996). *Soviet Defence Industry Planning: Tukhachevskii and Military-Industrial Mobilization.* Stockholm: Stockholm School of Economics.

Samuelson, L (2000). The Red Army's Economic Objectives and Involvement in Economic Planning, 1925–1940. In: Barber J and Harrison M (eds.) *The Soviet Defence-Industry Complex from Stalin to Khrushchev.* London and Basingstoke: Macmillan, pp. 47–69.

Samuelson, L (2000). *Plans for Stalin's War Machine: Tukhachevskii and Military-Economic Planning, 1925–1941.* London and Basingstoke: Macmillan.

Simonov, NS (1996a). *Voenno-promyshlennyi kompleks SSSR v 1920–1950-e gody: tempy ekonomicheskogo rosta, struktura, organizatsiia proizvodstva i upravlenie.* Moscow: ROSSPEN.

Simonov, NS (1996b). 'Strengthen the Defence of the Land of the Soviets': The 1927 'War Alarm' and its Consequences. *Europe-Asia Studies*, 48(8): 1355–64.

Simonov, NS (2000). The 'War Scare' of 1927 and the Birth of the Soviet Defence-Industry Complex. In: Barber J and Harrison M (eds.) *The Soviet Defence-Industry Complex from Stalin to Khrushchev.* London and Basingstoke: Macmillan, pp. 33–46.

Zamagni, V (1998). Italy: How to Lose the War and Win the Peace". In Harrison (ed.) *Economics of World War II: Six Great Powers in International Comparison.* Cambridge: Cambridge University Press, pp. 177–223.

Zolotarev, VA (1998). Velikaia Otechestvennaia. Tyl Krasnoi Armii v Velikoi Otechestvennoi voiny 1941–1945 gg.. Dokumenty i materialy, *Russkii Arkhiv*, 25(14): 304–5.

Chapter 4

The Frequency of Wars

Mark Harrison and Nikolaus Wolf

"War made the state and states made war" (Tilly 1975: 42).

Wars are becoming more frequent.[1] More precisely, the frequency of bilateral militarized conflicts among independent states has risen steadily over 131 years from 1870 to 2001. In this paper we consider how to evaluate this as a fact, how to explain it, and how to respond to it.

Part I of the paper reviews the data, and finds cause for concern. Part II outlines some reasons for puzzlement. The puzzle is that the world has become more globalized and more democratic; on both counts it should have got more peaceful, not less. In Part III we go back to the data. We find that the rising frequency of conflict is everywhere; it is not explained by a few bellicose powers. We show the answer to the puzzle will be related to the changing number of states. In Part IV, we discuss the historical relationship between war and state formation. In Part V, we raise some issues about

This chapter was first published in 2012 in the *Economic History Review*, 65(3), 1055–1076. © The Economic History Society 2011.

[1]Nikolaus Wolf is Professor of Economic History at Humboldt University, Berlin. He is also a research affiliate of CEPR, a research fellow of CESifo, and a research associate of the ESRC Centre on Competitive Advantage in the Global Economy at the University of Warwick. Earlier versions of this paper were presented to the annual Defense Economics conference held at the Institute for Defense Analyses of the U.S. Department of Defense, Washington, DC, on 17 November, 2008, the University of Birmingham Centre for Russian and East European Studies current affairs seminar, 25 February, 2009, and the German Historical Institute conference on War in Transnational and Long Run Perspective, Moscow, 2 July, 2010. We thank Alex Apostolides, Stephen Broadberry, Nick Crafts, Jari Eloranta, Bishnupriya Gupta, James Harrison, Les Hannah, Alex Klein, Bas van Leeuwen, Anandi Mani, Philippe Martin, Sharun Mukand, Dennis Novy, Eugenio Proto, Jeremy P. Smith, and the editor and referees, for advice and comments.

how the factors conducive to peace and war have been analysed in the literature and suggest lines of further investigation, in particular underlying determinants of state capacity for war. Technological, fiscal, and commercial aspects of the capacity for war are discussed in Part VI; the issue here is that these capacities are promoted by the same forces of democratization and globalization that are supposed to discourage conflict. Part VII concludes that, if the frequency of conflict has been increasing, it may be not because we want it; more likely, it is "because we can".

PART I

Many indicators of interstate conflict have been flat or declining for decades or longer. This includes the number of wars in each year since 1816 (Gleditsch, 2004: 243), the number of military fatalities in each year since 1946 (Hewitt, 2008: 114), and the annual probability of bilateral interstate conflict since 1950 (Martin *et al.*, 2008: 866). In the most recent years, despite conflicts associated with the breakup of the Soviet and Yugoslav states in the early 1990s, the downward trends have continued (Gleditsch, 2008: 693–694).

One indicator has moved persistently in the wrong direction. How many countries are at war at any given time? Exploiting the Uppsala dataset on armed conflicts (described by Gleditsch *et al.*, 2002), backdated to 1946 and updated to 2005, Joseph Hewitt (2008) has noted upward trends in the annual percentage shares of all countries in the world that are at war, and of all possible country-days at war, over the postwar period. Nils Gleditsch (2008: 694) has dismissed these observations as statistical artifacts of a trend to coalition wars in which countries participate symbolically, at increasing distance, without ever exchanging fire with the adversary. This comforting inference is undermined, however, by an observation from the Correlates of War dataset (Martin *et al.*, 2008: 867): between the 1950s and the 1990s, the average distance separating country pairs at war fell by one half (from more than 5,000 kilometres to less than 2,500).

Using the Correlates of War dataset, updated in 2007, we trace the origin of the upward trend in the frequency of bilateral conflicts as far back as 1870.[2] We show that it has proceeded with surprisingly little interruption

[2]This is the Militarized Interstate Disputes dataset, described by Ghosn *et al.* (2004); we refer to version 3.1 (2007). The time series we use are reported in an unpublished Appendix, Table A1, available from http://warwick.ac.uk/markharrison/data/frequency/.

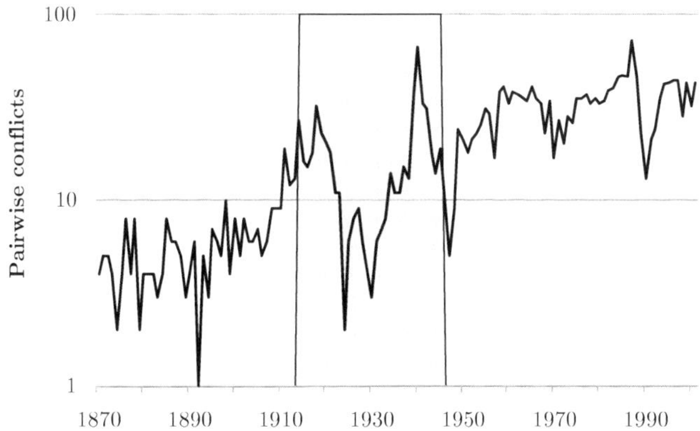

Fig. 1. Militarized Disputes between Pairs of Countries Since 1870.

Notes: Disputes are coded from level (1 no action) through 2 (threat of force), 3 (display of force), 4 (use of force), and 5 (war). We use all disputes of level 3 (the closing of a border or the dispatch of ships or troops) and above. For the full dataset, see unpublished Appendix Table A1. The shaded area delimits the 1914 to 1945 period.

Source. The Militarized Inter-State Disputes dataset, version 3.1, at http://www.correlatesofwar.org, described by Ghosn *et al.* (2004).

through two World Wars nearly to the present day. Befitting a phenomenon that is older than the oldest person alive today, we suggest that deep causes are at work.

Figure 1 charts the number of pairs of countries that have disputed with each other in each year from 1870 to 2001. This is a greater number than the number of wars for two reasons: first, it accounts for the number of countries involved in each conflict, rather than the number of conflicts; second, it has wider coverage than formal states of war, because it includes displays as well as uses of military force. The chart measures the number of pairwise disputes on a logarithmic scale, partly to give a clear picture of what has happened at the lower frequencies.

Viewed in this way, the chart demonstrates a clear log-linear trend; the frequency of bilateral conflicts has been rising steadily for over a century at two per cent per year.[3] To be sure, there was a good deal of

[3]We can reject a unit root at the five per cent level, although not at the one per cent level. Hence the series is quite close to stationary. In empirical political science, war is generally acknowledged to be auto-regressive in the sense that conflict in one period makes conflict more likely in the next. If our series of conflict frequency is regressed on time and itself with a one-year lag, there is around 50 per cent year-on-year persistence. Controlling for that, the underlying trend growth rate is reduced to about 1 per cent year, but this trend is very significant. For details, see the Appendix, Tables A2 and A3.

disturbance around the two world wars. But the surprising character of this disturbance is as follows: between 1914 and 1945, the conflicts that would normally have been distributed across the three decades arrived a little early and were squeezed into World War I, or were delayed until World War II, with an unexpected lull in between. After 1945, however, the frequency of conflict snapped back to the same trend it had followed up to 1914 and continued on that path through to the end of the millennium.

In principle, the number of pairwise conflicts in a time period, or the *absolute frequency*, is the product of two underlying variables into which it can therefore be decomposed. One component is the number of country pairs, which has increased enormously since the nineteenth century. In 1870, the world contained fewer than 50 independent states. By the end of the twentieth century, there were more than 180. This was associated with the breakup of empires (Austro-Hungarian, German, Ottoman, Russian, French, British, Dutch, Belgian, Portuguese, and Soviet) and federations (Czechoslovak and Yugoslav). As a result the total number of possible country pairs in the world between whom relations of peace or war could exist grew from around one thousand to over 17,000.

After the increase in the number of possible pairs is stripped out of the data, we are left with the other component, the *relative frequency* of conflicts, that is, the absolute frequency of pairwise conflict normalized for the number of pairs. The number of countries since 1870 and the relative frequency of conflicts among them are shown together in Figure 2. As the chart shows, in the first 80 years the number of countries did not change much; the relative frequency of disputes fluctuated wildly and tended to rise. Then, over the next half century, the relative frequency of disputes fell back to the level of the 1870s and below, but the number of countries increased dramatically and this took over as the main driver that kept the absolute number of conflicts on its upward trend. Thus, the steady increase in the absolute frequency of conflicts was driven, statistically speaking, by quite different forces in different periods.

Historically it is very interesting. From a present-day standpoint, should we feel concerned? Normalized for the number of country pairs, the relative frequency of war does not show a trend and is lower today than in the 1870s. This might seem to reassure, but should not do so. Normalized for the number of planets that all countries must share — that is one, exactly — the

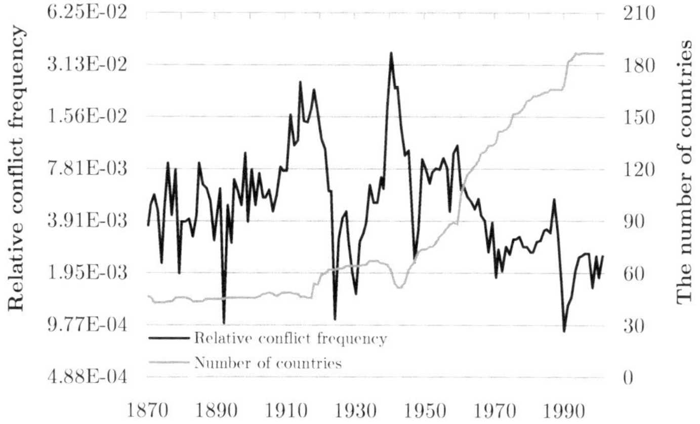

Fig. 2. The Relative Frequency of Pairwise Militarized Disputes and the Number of Independent States Since 1870.

Sources: Conflict data as Figure 1. Number of countries from Martin *et al.* (2008).

absolute frequency of conflict today is similar to what it was during World War I. (The intensity of conflict measured by forces and expenditures is much lower, admittedly; we do not face death and destruction on the scale of the Great War.)

We have more conflicts now than then, apparently, because we have more states. But the number of states is not an exogenous or random variable. When new states come into being, what motivates them is the demand for sovereignty (Alesina and Spolaore, 2003: 2). And sovereignty includes decisions over peace — and war — with neighbors, including former compatriots. In fact it is not at all uncommon for new states to plunge into war, like Serbia or Georgia, or be born out of war, like Kosovo. If state formation and war frequency are indeed linked, and the linkage has a clear historical dimension, this should motivate historians to enquire more deeply into the nature of the connection.

PART II

The data are a surprise, given the longstanding traditions of western political and philosophical thinking on the future of war. According to these traditions, the global trends towards democracy and globalization should make war increasingly a minority sport. In fact, war *is* a minority sport. The problem is that the minority has been growing.

The expected relationship between war and globalization is, on the face of it, clear cut. For many reasons, modern states ought to prefer trade to war. On the eve of World War I, Norman Angell (1911: 62–63) wrote:

> "Men are fundamentally just as disposed as they were at any time to take wealth that does not belong to them... But their relative interest in the matter has changed. In very primitive conditions robbery is a moderately profitable enterprise... But to the man whose wealth so largely depends upon his credit... dishonesty has become as precarious and profitless as honest toil was in more primitive times."

In more contemporary terms, trade is a positive-sum interaction; war is negative-sum. Trade costs have fallen (Jacks *et al.*, 2008). War costs are high and rising (Edelstein, 2000; Stiglitz and Bilmes, 2008; Glick and Taylor, 2010). Victory in war can bring one-sided gains but the gains are reversible if conflict is renewed. And, in wars of choice, victory is not only uncertain but unlikely. On the record of all wars since 1700, to start one attracts a 56 per cent probability of defeat (Eckhardt, 1989: 93).

More or less the same tradition affirms that the spread of democracy should crowd war out of the global community. Whoever else they fight, the evidence is compelling that "Liberal or democratic states do not fight each other" (Levy, 1988: 661; and also Russett, 1995).

The reasons have been much debated. According to one interpretation, democratic norms make leaders more likely to exercise self-restraint. Possibly, moral constraints are weakened when the foreign state has an authoritarian ruler (Doyle, 1986; Maoz and Russett, 1993; Russett, 1995). In another view, democratic institutions may constrain leaders to resolve conflicts non-violently, abroad as at home, or punish them more severely or more certainly if they embark on violence, or if they resort to war and lose. If democracies are thereby more selective about the wars they fight, and so more likely to win them, other states (or more specifically other democracies) may be less likely to attack them. Autocrats, in contrast, can steal the benefits of war while shifting the costs onto their subjects and avoiding punishment (Maoz and Russett, 1993; Bueno de Mesquita and Siverson, 1995; Bueno de Mesquita *et al.*, 1999; Bueno de Mesquita *et al.*, 2003; Bueno de Mesquita, 2006; Jackson and Morelli, 2007).

The democratic peace continues to be debated. Where skepticism persists, it is often based on counter-examples, such as the wars undertaken by the United States, the world's largest rich democracy. According to Sebastian Rosato (2003; 2005), the democratic peace was "an imperial peace based on American power"; the United States enforced peace among the democracies after 1945 because the Cold War allowed it to, and made war against some dictators for the same reason. In this view the democratic peace was a temporary accident. A more popular and radical criticism sees the democratic peace as a sham; it detects the hand of aggressive American imperialism intervening with rising frequency in resource conflicts around the globe (e.g. Pelletière, 2004).

In themselves, counter-examples may be of questionable significance because they can always be interpreted otherwise than as grounds for refutation: they may reflect randomness, or selection bias, or the influence of confounding factors (Doyle, 2005; Kinsella, 2005; Slantchev *et al.*, 2005). In the next section we will consider specifically whether the rising frequency of wars might be attributable to one rich country or a small group of them.

Some other recent qualifications to the democratic peace are of interest. Here are a few. First, according to George Downs and David Rocke (1994), elected leaders that face punishment by the electorate because their policies are failing have an incentive to gamble on resurrection, for example by starting wars or by persisting with them, in the hope that something will turn up. Joseph Stiglitz and Linda Bilmes (2008) have applied a similar argument to America's war in Iraq.

Second, observing the record of the former Soviet and Yugoslav states, Edward Mansfield and Jack Snyder (1995; 2005) have proposed that new or incompletely established democracies are particularly vulnerable to risky adventures in nation-building.[4] Georgia seems to have supplied recent out-of-sample confirmation. Along similar lines, Baliga *et al.* (2011) suggest that limited democracies might on balance be still more aggressive than dictatorships, if leaders are uncertain about the aggression of their neighbors and face a challenge from a hawkish minority in their own country.

Michael Doyle (2005) has suggested that democracy is a dynamic process qualified by values and institutions at the same time. The democratic peace rests, he argues, on a tripod of republican representation, commitment to human rights, and transnational interdependence that falls when any one leg is missing.

Finally, Conconi *et al.* (2008) have looked in the data to find that democracies where leaders are subject to term limits are as likely to make war as autocratic states — and term limits are increasingly widespread. It

[4]For discussion, see Narang and Nelson (2009), and Mansfield and Snyder (2009).

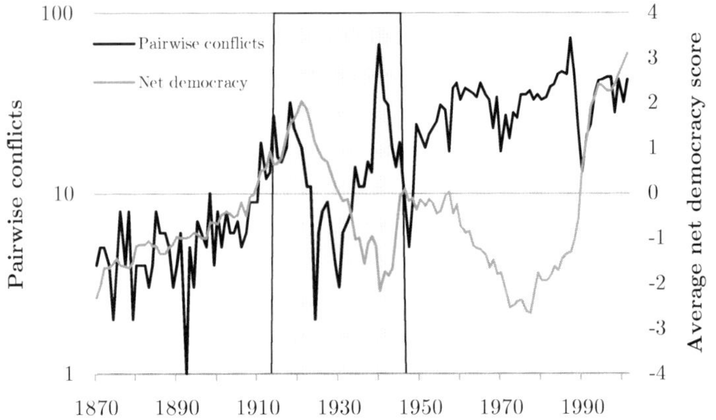

Fig. 3. Democratization Since 1870.

Note: The shaded area delimits the 1914 to 1945 period.

Sources: Conflict data as Figure 1. Net democracy: the *Polity2* variable from the Polity IV dataset at http://www.systemicpeace.org, described by Marshall and Jaggers (2007).

is the democracies without term limits, where established leaders retain the option of continuing to compete for office, that account for the democratic peace.

In short, the idea that democratization necessarily spreads peace has been qualified in various ways. We question whether these qualifications are adequate to the task of explaining a trend towards the rising frequency of war that has persisted for 130 years. The full difficulty is illustrated in Figures 3 and 4 which deal, respectively, with the spread of democracy and trade. In Figure 3, we report the average balance of democratic versus authoritarian attributes of political systems across all countries in the world in each year. This measure, from the Polity IV dataset, shows clearly the spread of democracy in the late nineteenth century, reversed by the rise of fascist, communist, and military dictatorships that began in the 1920s and continued through the 1970s. Beginning in the 1980s, democracy swept around the world once more, but it was not until the new millennium that average scores exceeded historical records. Figure 4 shows the average of a standard measure of trade openness over the same period. It reflects a relatively open global economy in the late nineteenth century, the profound setback to globalization of the interwar period, and a recovery that began in the 1950s, recovered to the level of the previous century in the 1970s, and went on to far exceed it.

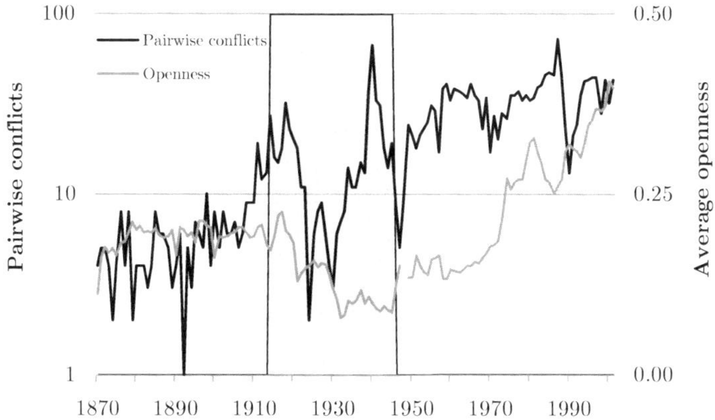

Fig. 4. Trade Openness Since 1870.

Note: The shaded area delimits the 1914–1945 period.

Sources: Conflict data as Figure 1. Openness data from Martin *et al.* (2008).

These charts show clearly that the world in the year 2001 was more open and more democratic than the world of 1870 — and also more conflict ridden. It is true that trade and democratization grow together. From study of the endogenous relationship between trade and democratization since 1870, Ernesto López Córdova and Christopher Meissner (2008) confirm that more open countries have been consistently more democratic. Most likely trade has tended to drive democracy, but with long lags and through uncertain and varying channels. But based on our own figures, as trade and democracy have spread, so have wars. Over significant sub-periods, for example from 1870 to 1913 and from the mid-1970s to 2001, the positive associations of openness and democratization with the absolute frequency of wars have been particularly close. Thus, if we have not got the general relationship between economic and political progress and war completely and utterly wrong, to say the least, we have missed some important confounding factors.

PART III

In this section, we explore the data in cross section and time series. Initially, we ask whether the rising frequency of wars is attributable to a single country or small group of countries. Consider the United States. We are all aware of America's wars, but the result is a selection bias. The share

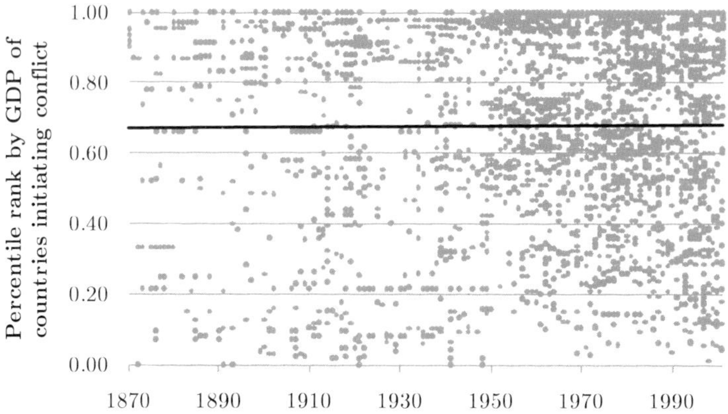

Fig. 5. Countries Originating Conflicts, 1870–2001, Ranked by GDP.

Note: The solid line is a linear time trend. The slope of the trend is not significantly different from zero; see unpublished Appendix Table A5.

Sources: As Table 1.

of the United States in the global count of pairwise conflict-years was 9.3 per cent from 1870 to 1945, 11.2 per cent from 1946 to 1991, and 10.8 per cent from 1992 to 2001. Subtracting the conflicts of the United States from the global series makes little difference to its level and no difference to its upward trend.[5]

More generally, Figures 5 and 6 show that the propensity to intervene militarily is everywhere, in all parts of the global income distribution. Countries with larger GDPs (Figure 5) have a tendency to throw their weight around somewhat more often than others. Table 1 makes this more precise. The upper block of bold figures in the table covers the whole period; all countries are ranked by GDP in every year. We find that, of all those originating a dispute, the median country was ranked at 0.73 in the global

[5]The influence of the United States in our data is possibly our most frequently asked question when we have presented this work. Using our full dataset with *FW* for the annual pairwise frequency of conflict, and t for time in years with 1871 as year zero, we get:

$$\mathrm{Ln}(FW) = 0.6187 + 0.5344\ \mathrm{AR}(1) + 0.009387\,t\ (N = 130, R^2 = 0.77)$$

and excluding conflicts involving the United States, we get:

$$\mathrm{Ln}(FW\ no\ US) = 0.5552 + 0.5419\ \mathrm{AR}(1) + 0.009245\,year\ (N = 130, R^2 = 0.77)$$

All coefficients in both equations are significant at the 0.0001 per cent level. For more detail see the Appendix, Tables A3 and A4.

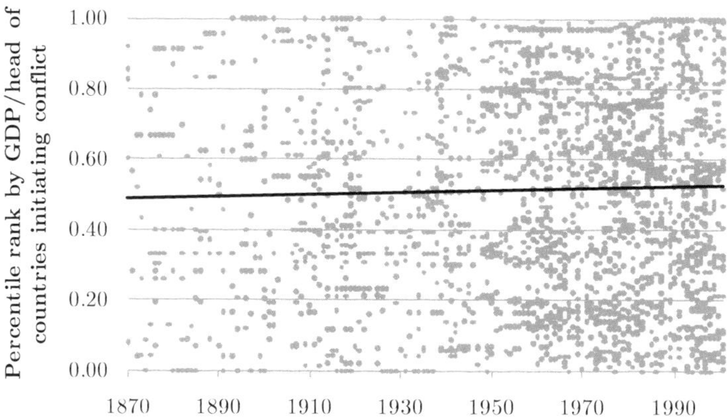

Fig. 6. Countries Originating Conflicts, 1870–2001, Ranked by GDP Per Head.

Note: The solid line is a linear time trend. The slope of the trend is upward at 3.1 percentiles per century (significant at 10 per cent); see unpublished Appendix Table A6.

Sources: As Table 1.

GDP distribution at the time. It is possible to say that the top 27 per cent of countries by GDP accounted for one half of all military interventions on average, as long as it is understood that the population of the "top 27 per cent" differed from year to year as the total number of countries changed and countries exchanged ranks. We also measure the quartile points of the distribution. Subject to the same qualification, the "top seven per cent" of countries by GDP originated one quarter of conflicts, while the bottom half also originated one quarter. There is no trend in these shares over time. Countries with smaller economies have contributed as much as larger ones, therefore, to the increase in the frequency of wars.

Bellicosity was an attribute of size, but not of development level. Figure 6 and the lower block of bold figures in Table 1 show that, over the period as a whole, countries that were richer in GDP per head were not disproportionately responsible for the instigation of disputes. The "top 24 per cent" of countries originated 25 per cent of the conflicts, and so on down the distribution. Before 1914, the economically more developed countries were slightly *less* likely than others to embark on military interventions. As the note to Figure 6 makes clear, the average rank of countries that originated conflict tended to rise over time, but at a glacial pace measured in percentiles per century. Thus, poorer countries have started as many

Table 1. Pairwise Conflicts, 1870–2001, and Countries Originating them, Distributed by GDP and GDP Per Head.

	1870 to 2001	1870 to 1913	1914 to 1945	1946 to 1990	1991 to 2001
Conflicts	3168	334	606	1724	504
Conflicts per year	24.2	7.8	18.9	38.3	45.8
Percentile rank of countries that originated disputes,					
By GDP:					
Quartile 3	**0.93**	0.94	0.94	0.91	0.91
Quartile 2 (median)	**0.73**	0.80	0.85	0.71	0.72
Quartile 1	**0.50**	0.33	0.45	0.53	0.42
By GDP per head:					
Quartile 3	**0.76**	0.69	0.79	0.77	0.71
Quartile 2 (median)	**0.54**	0.45	0.49	0.56	0.53
Quartile 1	**0.27**	0.26	0.30	0.27	0.26

Note: The upper block of bold figures in this table can be read as follows. When all countries are ranked by the size of their GDP in every year, we find that, of all those originating a dispute, the median country was ranked at 0.73 in the global percentile GDP distribution at the time. The country at the upper quartile was ranked at 0.93 and the country at the lower quartile at 0.50. Thus, one quarter of countries originating conflicts had GDPs larger than 93 per cent of all countries at the time, while one quarter had GDPs smaller than half of all countries.

Sources: For this table, we merge two datasets, the Militarized Inter-State Disputes dataset, version 3.1, at http://www.correlatesofwar.org, described by Ghosn *et al.* (2004), and the national income and population dataset of Angus Maddison at http://www.ggdc.net/maddison. In the Militarized Inter-State Disputes dataset, we identify countries that originated disputes of level 3 or higher, as defined in the notes to Figure 1. We match 162 countries by name; many are straightforward, but we match Austria with Austria-Hungary, and Russia with both Russian Federation and USSR. Within the period, Ethiopia sometimes includes Eritrea and sometimes not, and similarly for Germany (the German Democratic Republic), Indonesia (Timor), Korea (North Korea), and the UK (the Irish Republic). In this way we buy additional observations at the expense of some measurement error, which matters more for size (GDP) than for development level (GDP per head). In each year we attach a percentile rank to all matched countries by GDP and GDP per head. There are many missing annual observations in the Maddison dataset; particularly before 1950, some countries are represented by infrequent benchmarks. To lessen the risk of selection bias (since countries that are poorer in data tend also to be poorer in income) we give each country with missing data its percentile rank from the year when it is next observed. Again, we increase coverage at the cost of some mismeasurement, but the latter is limited to the extent that country rankings change only slowly. In this way, we create 3,168 matches out of 3,224 conflicts that were originated between 1870–2001. We then find the percentile ranks, year by year, of the countries originating disputes, by GDP and GDP per head in each year over the whole period and in each sub-period shown. Unpublished data files are available at http://warwick.ac.uk/markharrison/data/frequency/.

conflicts as richer ones and have contributed very nearly as much as richer ones to the increase in the frequency of war.[6]

Closer analysis of time series can sharpen our focus on this puzzle. First, the relationship between militarized disputes on one hand and democracy and trade openness on the other will emerge as statistically weak. Second, the changing number (and hence size) of states should be a crucial factor in any explanation for the frequency of wars. Third, the statistics will tell us that the relationship between these factors and the frequency of militarized disputes is highly non-linear.

If we simply regress the number of militarized disputes on the degree of trade openness and on the average degree of democratization (all in logs), and control for serial correlation (Table 2, specification 1), we find that both openness and democracy significantly reduce the frequency of war. Once we control for the number of states, however (specification 2), the positive influence of democracy vanishes.

An econometric perspective allows us to explore whether the relationship among variables is sufficiently well approximated as linear with all explanatory variables entering separately, or more complex than this. Using Ramsey's (1969) Regression Equation Specification Error Test (RESET) on specification 2, we can test (specification 3) whether any transformation of the set of dependent variables — including cross-products between them, such as effects of democracy and trade on country formation — can improve the fit. While the linear specification initially appeared to fit the data rather well with an adjusted R^2 of 0.77, the RESET test very clearly rejects this simple model. It points us toward possible interactions between democracy, trade, and the number of countries as well as potential nonlinearities in their effect on war over time. In short, we should rethink how democracy and openness matter.

Before we leave econometrics, we carry out some robustness checks, specification 2 (controlling for the number of states) as our baseline. In Table 3, specification 4 examines the effect of splitting our democracy measure into levels of political competition versus executive restraint. Based on the powerful reasoning of Paul Collier (2009), we look for evidence that

[6]Which countries have accounted for most conflicts? Over the entire period and the 3,168 conflicts in Table 1, the United States comes in second place (originating 161 conflicts) behind Russia/USSR (219). The United Kingdom (119) is fourth, following China (151). Germany (102) is sixth, after Iran (112). France is tenth, after Israel, Turkey, and Iraq. Thus middle- and low-income countries are well represented and even small countries are present.

Table 2. Conflict, Democracy, Trade Openness, and the Number of Countries.

Dep. Variable = Log(FW)	Specification 1		Specification 2		Specification 3: Unrestricted RESET equation[a]	
	Coefficient	t-stat[b]	Coefficient	t-stat[b]	Coefficient	t-stat[b]
Constant	2.522	2.471	−5.424	−4.272	1.942	2.022
Log(Openness)	−0.794	−1.925	−0.737	−2.620	0.017	0.111
Log(Democracy)[d]	−0.676	−1.624	−0.096	−0.262	0.184	1.408
Log(# of Countries)	...	...	1.602	8.205	−0.299	−1.404
AR(1)	0.900	25.529	0.651	7.241	0.229	10.033
Fitted Value^2	...	...	...	...	−0.143	−1.281
Adj. R^2	0.727		0.762		0.770	
Prob (RESET F-test)	...		...		0.024	
DW-stat	2.666		2.333		1.950	
# of obs	130		130		130	

Notes:
[a]Ramsey's RESET Test Statistics: F-Statistic: 4.251 (Probability: 0.04), Log likelihood ratio: 4.382 (Probability: 0.036). The RESET test considers whether the inclusion of further variables or non-linear combinations of the regressors makes a significant contribution to explain the variation of the dependent variable.
[b]Based on Newey-West HAC Standard-Errors and Covariance.
Sources: FW is the conflict measure shown in Figure 1. Openness and # of countries: Martin et al. (2008). Democracy: the Polity2 (or net democracy), variable from the Polity IV dataset at http://www.systemicpeace.org, described by Marshall and Jaggers (2007), averaged over all sample countries for a given year.

stable democracy is founded on executive restraint, and the spread of the latter might favour peaceful conflict resolution. But neither is significant.

We also check for the effects of eliminating pairwise conflicts involving the United States from the dependent variable. This exclusion does not change much. Specification 5 reports the effect of democracy as somewhat more positive, but still insignificant. Splitting democracy into political competition and executive restraint also remains ineffective.

PART IV

As illustrated by Figure 2 and Table 2, the upward trend in the frequency of wars is tightly related to the increase in the number of sovereign states. Given the dramatic change in the number of states since 1870, and especially after the two world wars, an understanding of this relationship seems to be

Table 3. Robustness Tests.

	Specification 4 Log(FW)		Specification 5 Log(FW no US)		Specification 6 Log(FW no US)	
Dep. variable	Coefficient	t-stat	Coefficient	t-stat	Coefficient	t-stat
Constant	−6.059	−4.014	−5.488	−2.396	−6.526	−3.878
Log(Openness)	−0.773	−2.598	−0.803	−2.628	−0.851	−2.622
Log(Democracy)	...	...	−0.207	−0.265	...	...
Log(PolComp)	−0.099	−0.070	...	...	−0.239	−0.166
Log(ExConst)	0.443	0.192	...	...	0.701	0.304
Log(# of Countries)	1.601	7.095	1.638	6.723	1.623	6.818
AR(1)	0.647	7.331	0.655	9.464	0.652	8.122
Adj. R^2	0.760		0.764		0.763	
DW-stat	2.323		2.180		2.170	
# of obs	130		130		130	

Note: Specification 4 tests specification 2 (from Table 2), splitting the democracy variable into separate components for political competition and executive constraint (as Figure 5). Specification 5 tests specification 2, eliminating pairwise conflicts involving the United States from the dependent variable. Specification 6 tests specification 4, again eliminating U.S. pairwise conflicts from the dependent variable.
Sources: As Table 2.

crucial for any explanation of the absolute frequency of wars in general, and of the role of democratization and trade in particular.

Empirical studies on the issue usually treat the changing number of countries as exogenously given, and either use it as an additional control variable, or focus on the relative frequency of wars standardized by the number of country-pairs (as do Martin *et al.*, 2008: 866). This approach can mislead, however, insofar as the incidence of warfare has been at the heart of the process of state-formation and wars have served to create, consolidate or destroy states. We take one lesson from Charles Tilly (1975: 42) who, considering the consolidation of the European state system, in the quotation that heads our paper, proposed that "war made the state and states made war". Another lesson is available from Douglas Gibler (2007) who suggests that peace and democracy are joint symptoms of stable borders, not the other way around.

If we limit our attention to the period after 1871, many wars in Europe and elsewhere began in attempts to revise existing political borders, either as a struggle for independence from empires — often supported by external powers — or as an effort to expand existing empires. Tensions within the British Empire (for example, the two Boer Wars) and within the Austro-Hungarian and Ottoman Empires in the build-up to World War I come to mind as conflicts over the very number and size of independent political entities. The increase in the number of states after 1918, largely due to the partitioning of the Austro-Hungarian and Ottoman Empires, was not only a direct outcome of World War I; aspirations for the creation of new independent states lay equally at the origins of the war (Henig, 2002: 42–43).

Similarly, World War II arose from Nazi Germany's efforts to build an autarkic continental empire, beginning with the subjugation and liquidation of independent states such as Poland and Czechoslovakia (Overy, 1987: 31–37).

After 1945, the larger colonial empires were dismantled in a largely (if not entirely) peaceful way. Many of the newly independent states were only weakly integrated by European standards, however. They inherited fragmented populations and fragile economic structures that were designed for an imperial periphery. In consequence, many of these states were prone to militarized conflicts over boundaries, motivated either by ethnic tensions or economic pressures. Comparing the challenges faced by African states to the European experience of state formation, for example, Jeffrey Herbst (1990: 136) argued: "It should be obvious that the incentives that African leaders have to incite wars for the purpose of state-making are significant and may become much stronger in the future". Against this background, there is a clear case for treating the number and size of countries as endogenous to the frequency of wars.

As we argued earlier, the data suggest that changes in the number of states not only affect the absolute frequency of wars but also interact with the effects of democratization and trade, and hence most probably affect the relative frequency of wars. Consider the theoretical framework provided by Alberto Alesina and Enrico Spolaore (2003), who have highlighted interactions between the formations of states on the one hand side and democracy, trade openness, and the development of international institutions on the other. In a nutshell, they argue, given all countries are composed of heterogeneous populations, global economic integration should strengthen the formation of smaller independent states; so should democratization, via tendencies of further decentralization and eventual separatism. The global spread of democracy and declining trade costs after 1945, together with the observed increase in the number of states, seems to lend empirical support to these ideas. But it also challenges our understanding of the frequency of wars.

Alesina and Spolaore (2003: 221) conclude by conceding that they have not explored "how a configuration of countries might affect the level of conflict" except for the impact of an "exogenous" change in the likelihood of international conflict on state formation. Yet the likelihood of international conflict is clearly *not* exogenous; it is what we would like to explain.

PART V

How much do we really know? Less than we should, apparently. There is a vast and long-standing international relations literature on war and peace. The literature was once based on intuitive inference from narratives and comparisons, but has been transformed over the last thirty years by new data and the application of quantitative methods. Large-scale open-access cross-country panel datasets have been created that deal with war and peace, political regimes, and historical macroeconomic and trade variables.[7] We should know more than ever before about the correlates of war and peace. Yet, what do we know?

As might be expected, the literature that has resulted, being voluminous, is of variable quality. Not all of the data now available have been well used; among thousands of regressions that have been reported (Green *et al.*, 2001) are many with potentially biased or otherwise dubious estimates, for example because of the neglect of fixed effects in pooled regressions.

In some ways the present state of the field is reminiscent of the literature on global economic growth and divergence a decade or more ago. Abhijit Banerjee (2007) has described how economists strayed into thinking of global development as a machine that produced growth using levers labeled "investment", "education", and "trade". In much the same way, estimation strategies have typically modeled global relations as a machine with big push-buttons marked "democracy" and "trade". Economists have learned, however, that, while the big buttons have some power as statistical drivers of global development in the aggregate, their power has intrinsic limits. The buttons become particularly unreliable when applied in the context of any given country. One likely reason (Rodrik, 2007: 22–25) is that

[7] See the Correlates of War project at http://www.correlatesofwar.org, the Polity IV project at http://www.systemicpeace.org, the UCDP/PRIO (Uppsala Conflict Data Program at the Department of Peace and Conflict Research, Uppsala University, and International Peace Research Institute, Oslo) dataset at http://www.prio.no/CSCW/Datasets/Armed-Conflict/UCDP-PRIO/, the Penn World Tables at http://pwt.econ.upenn.edu, and the national income and population dataset of Angus Maddison at http://www.ggdc.net/maddison.

their operation is at least partly confounded by unobserved cross-country variation in institutions.

Where next for the study of peace and war? Experience suggests three possible correctives. One is to look inside the regressors: democracy and trade are complex phenomena that may have multiple or non-linear effects. An example of work in this spirit would be the investigation of term limits in democracies (Conconi *et al.*, 2008), but other aspects are also likely to be deserving of closer study. Paul Collier (2009: 15) has argued that electoral competition may impede effective governance for development unless accompanied by checks on executive power. Intuitively, electoral competition without executive restraint might be as damaging for international relations as it can be for domestic development. As we have seen, however, our aggregate data do not confirm this.

Another desirable corrective is to seek reconciliation of cross-section results with what time series and narratives tell us. The virtue of cross sections is that they enlarge the data; but the fact is that we live our lives through time. When we ask what may happen next year, it is not always helpful to be told what would happen if Argentina became Britain in a timeless way, since countries (and country pairs) are likely to be otherwise different in ways that we cannot control. Narratives of democratization in particular countries, for example, have shed light on the hypothesis of democratic peace where pooled cross-section studies have failed to do so or may even have misled. In principle fixed effects are intended to exclude the variation across units, exploiting only the variation within units over time. However, this works only under some rather restrictive assumptions, for example that the variation across units remains unchanged over time.

A third corrective is to rethink the units of analysis themselves, because it is not always clear what the unit should be: the country or the pair, for example. We should not treat the number of sovereign states and their capacity to wage war as exogenous. The nature of "state and legal capacity" generally, and its relationship to propensities for peace and war, are the subject of recent work by Timothy Besley and Torsten Persson (2008; 2009). Following their lead, further research in this field should also incorporate issues of state formation, institutional change and openness.

Such an agenda faces two obvious challenges. The first challenge is that empirical studies into these issues must find a way to capture the process of state formation as an endogenous variable. But the data are intrinsically unsuited to this. The data currently used in empirical studies are defined on given lines of national state boundaries (for example, data on trade between states, state institutions, or conflicts between states). The state made statistics, and statistics defined the state (e.g. Tooze, 2001: 1–39). Therefore, we face great difficulty in treating boundaries — that is, the geographical reach of institutions — as varying endogenously over time, and this is one factor that tends to limit our focus to variation in the cross-section.

One solution might employ narratives and case studies that explore both developments over time and interactions between regions. Another solution would follow the lead of those pioneering the use of data on a grid of equally sized regions that are defined strictly by geographical position (Ramankutty *et al.*, 2002; Buhaug and Rød, 2006; Wimmer and Min, 2006; Michalopoulos, 2012). Their data would require extension to cover institutional characteristics including political independence, variables reflecting trade costs, and the prevalence of conflict. This is a feasible but still enormous task.

A significant gain from this approach might be to weaken the intellectual barriers that arise when statistics are based on state borders. In reality, violence runs seamlessly from unorganized and organized crime through civil war to militarized interstate conflict. But social science struggles to recognize this as a continuum of interconnected phenomena. Instead, our data and models slice it up into artificial segments. Scholars specialize in one segment or another. The possibility of integrating insights is diminished.

The second challenge arises directly from our earlier results and the broad trends visible in the data. An understanding of the frequency of wars apparently needs to consider not only the relationship of war to state formation, institutional change, and trade, but should crucially consider all these factors as interrelated. For example, democratization may impose constraints on political leaders that reduce the probability of war and enhance trade integration. Simultaneously, democratization might transform public finance and hence as a by-product increase the capacity to wage war. Trade integration, by enabling countries to consume outside their production possibilities, may also increase the capacity for war. Hence, the second challenge is to open the "black boxes" of institutions, boundaries, and trade and inspect the multiple interactions through which each affects the frequency of war.

Put in a simpler way, a focus on the appetite or "demand" for war is reasonable, justifiable, and convenient if the number of states can be treated as exogenous; but this may have led us to neglect "supply-side" or capacity-for-war factors that are also relevant. We will consider technological, fiscal, and commercial aspects of this at greater length. Globalization and democratization both ought to have diminished the appetite for war — and may well do so in cross section. But they may also have promoted the formation of states and the capacity for war over long periods of time, and this may explain some of what we see in the historical time series.

We frame the rest of our contribution as narrative, rather than as quantitative analysis. This is for a variety of reasons, including the tentative nature of our investigation, but the most important reason is that we see our existing statistical categories, within the frontiers of given states, as too crude and artificially bounded to bear the weight of statistical inference.

PART VI

We consider the technological, fiscal, and commercial capacities for war. First, the relative costs of destructive power have been declining for centuries. From the Middle Ages, they fell more rapidly in western Europe than elsewhere. The result, described by Philip Hoffman (2010), was a European comparative advantage in the "gunpowder technology", reflected first in trade, then in conquest. The growth of technological capacities for war has continued to the present day. A preoccupation with public finances is likely to perceive only the rising unit cost of major weapon systems; in place of horses and sailing vessels, we now have billion dollar planes and hundred billion dollar warships. But the destructive power of these systems has risen even faster than unit costs. The advent of atomic weaponry meant that the same destruction that previously required the repeated application of mass bomber formations in multi-year campaigns could be achieved by one plane and one bomb. In terms of "bangs per buck", writes Niall Ferguson (2001: 40), "military technology has never been cheaper".

Competitive arms races followed inevitably. How could such weaponry become affordable on a rising scale? There was fiscal as well as industrial revolution. In the Middle Ages most citizens were poor, but that was not all. Tax compliance was low and sovereign debt was unattractive to lenders. Often, rulers raised military forces in kind: local overlords supplied the king with armed men and food. As a result, the ruler could wage external war only by consensus of the nobility. Or the king raised taxes to pay the army; conditional on having done so, he gained freedom of military action, but he could raise the taxes in the first place only through the overlords, and this again required their consent. Nor could rulers borrow to any great extent because, at this stage, there was no real distinction between public finance and the personal finance of the king; lenders were reluctant, not knowing if the king would be bound by his word, or if his debts would die with him.

Comparative data (Karaman and Pamuk, 2010) suggest that no sixteenth-century ruler could extract more than five per cent of GNP in central revenues from the territory of the kingdom. The local burden on peasants might well be higher, but much of what could be raised locally was dissipated locally in paying off overlords or tax farmers. Only adding to the size of the kingdom could add to central revenues, but this risked diminishing returns as tax-raising authority was delegated across a wider territory.

The seventeenth century saw a fiscal revolution in northwestern Europe. Dutch and English fiscal ratios climbed to 10 and then 20 per cent of national income. Patrick O'Brien (2001; 2005) has charted the progress of this revolution in England between 1500–1800. In the middle of these three centuries, fall the English Civil War of 1642–1651 and the Glorious Revolution of 1688. Before 1642, English revenues were only once, briefly, more than five per cent of national income; after 1688, they were never less than that, and increasingly much more.

What drove the transformation of public finance? The Civil War and the Glorious Revolution destroyed absolutism and set new restraints on the executive — at least, the executive was now restrained in everything but the making of war (O'Brien, 2001; 2005; Açemoglu *et al.*, 2005). Abroad, the government aggressively promoted the Atlantic trade by extending naval power, a policy that won taxpayers' support and built tax compliance. At home, credible guarantees against default widened the market for sovereign debt (Tilly, 1975; 1990; Hoffman and Rosenthal, 1997; Bonney, 1999; Ferguson, 2001).

Western Europe and its colonial offshoots saw a revolution in public finance. Where other regions of Europe and the Near East did not follow, there was fiscal divergence. Nomadic empires such as that of the Mongols failed because they could not mobilize sufficient resources to compete with China or Russia (Perdue, 1996). At the end of the eighteenth century the fiscal ratio of an agrarian state like the Ottoman Empire remained as it was. Through the nineteenth century, the gap widened (Cardoso and Lains, 2010). In fact, by the early twentieth century, the liberal democracies could put half or more of national income into war. In World War I Germany, not a liberal democracy, did this too, but only by exhausting its economy in the attempt to compete. In contrast, the agrarian empires of the Ottomans, Romanovs, and Habsburgs struggled to mobilize their resources at all (Broadberry and Harrison, 2005: 14–15).

Later in the century, the non-democracies caught up and eliminated the gap. The extraordinary 60-to-70 per cent fiscal ratios of Nazi Germany, militarist Japan, and the Soviet Union in World War II stand out (Harrison, 1998: 20–21). Behind this lay the fact that, by the 1940s, dictators of varying hues had learned to substitute the instruments of modern nationalism and modern repression for their adversaries' advantages of fiscal transparency and voluntary tax compliance.

What the dictators could not match was the capitalist democracies' commercial capacities for war. This aspect of state capacity is illustrated by a twentieth-century paradox. Since the Napoleonic era, European governments have worried about food security. Britain relied overwhelmingly on imported calories. Despite this, in two world wars Britain had little difficulty in feeding its people (Olson, 1963). In contrast, those countries that believed themselves secure were the first to run short of food. In the last quarter of the nineteenth century, Germany's leaders worked hard to limit their exposure to international trade and to protect agriculture. In 1914 Russia went to war, assured of the availability of a large peacetime surplus of exportable grains. Yet it was Russian and German cities that were stalked by hunger (Offer, 1989; Broadberry and Harrison, 2005; Collingham, 2011).

It was easier for Britain to feed itself from the other side of the world than for Berlin, Vienna, St Petersburg, or Constantinople to induce farmers thirty miles distant from the capital to feed their own people. Why was this? Britain had invested not in agriculture but in something more important: the gains from international trade. These were not only direct gains in the Ricardian sense of returns to specialization, but also indirect gains from the establishment of an overseas trading network that would robustly survive the disruptions of continental war. The Russian, German, Austro-Hungarian, and Ottoman Empires had inferior external networks, although Russia was helped by peripheral membership of the Anglo-French network. But there was more: these countries, with their large peasant populations, could not maintain the integration of their own domestic markets under the pressure of wartime mobilization. Unable to trade with the cities on peacetime terms, their peasant farmers seceded from the war effort, retreating into subsistence activities, leaving the soldiers and war workers without food.

To varying degrees, these countries had a commercial capacity for war that was greatly inferior to Britain's. They thought they were safe; they perceived the British to be at risk. When war broke out, they expected Britain to starve. Using commerce rather than agriculture, however, the

British fed themselves to standards little short of peacetime through two world wars. In both world wars, moreover, the Allies were able to multiply the military value of coalition resources through long-distance economic cooperation that the Central and Axis Powers could not match.

The lesson of this narrative is straightforward: war and trade are not exclusive. The same conclusion can also be reached in other ways. Using panel data from 1950 to 2000, Martin *et al.* (2008) have shown that trade had a double effect on the relative frequency of pairwise conflict. More bilateral trade reduced this frequency, but more multilateral trade raised it. Over time both multilateral and bilateral openness increased on average, but the net effect was positive. For any country pair separated by less than 1,000 kilometres, globalization from 1970 to 2000 raised the probability of conflict by one fifth (from 3.7 to 4.5 per cent). On the interpretation of Martin and his co-authors, the same forces that widened the scope of multilateral trade made bilateral war less costly. As long-distance trade costs fell, open economies could increasingly wage war against some (most likely close by), while continuing to reap the gains from trade with others (at a distance). This phenomenon might also offer a key to the weaknesses of economic sanctions identified by Lance Davis and Stanley Engerman (2003; 2006: 417–426).

From various angles, therefore, it is possible to identify something that it is convenient to call the commercial capacity for war; this capacity is increasing in trade liberalization, and also in the information, communication, transportation, and transaction technologies that account for much of modern economic growth.

PART VII

The evidence suggests that, normalized by the number of countries in the world, the risk of war is lower today than at the end of the nineteenth century. Normalized by the number of planets we have to share, however, it is of the same frequency (if not intensity) as during World War I. There has been a steady upward trend in the number of bilateral conflicts over 130 years.

Existing explanations of the resort to war in terms of the political incentives facing rulers subject to varying moral and cultural norms and constitutional arrangements, widespread in modern political science and political economy, are clearly both necessary and productive. We argue that an emphasis on preferences and incentives, which we call the demand side of the decision for war, cannot fully or convincingly explain the aggregate picture. It is necessary also to consider the supply side — the capacity for war. In this sense, we conclude, if the frequency of conflict has been increasing, it may be not because we want it; more likely, it is "because we can".

The rising frequency of bilateral conflicts is reflected right across the global distributions of countries by size and wealth. Wealthier countries have not been responsible for more than their share of military interventions. If their share has risen over time, it is at a rate that is all but imperceptible. Countries that are economically above the median of the economic size distribution have contributed more than their share, but the upward trend of the overall frequency is also present among countries that are smaller (and poorer) than the median.

The upward trend may turn out to have been driven by things we would otherwise welcome as global improvements. For example, the hunger for political participation and national self-determination has been satisfied in many troubled regions, and this has led to the formation of new states. The growing number of states is an important explanatory factor in the rising frequency of wars, but this does not make the trend a statistical artefact because the number of states is not exogenous.

In modern times just as much as in the Middle Ages, new states have been born amid conflict. The demand for statehood is also a demand for the capacity to engage in national self-determination by force, and each new state has added a focus for potential conflict. With the downfall of empires, moreover, democracy has become more typical — and, with democracy comes improved fiscal capacity. As a result, countries that adopt democracy are likely to be able to raise taxes or borrow more in order to promote national adventures without recourse to domestic repression.

With more borders there is more cross-border trade. Beyond this, moreover, falling trade costs are another modern boon that has allowed many countries to benefit from specialization and increased economic interdependence. Wider markets have in turn increased the scope for smaller countries to self-insure against asymmetric shocks. A moral hazard that we associate with insurance, however, is that the insured can then engage in risky behaviour at lower cost. In the same way, small states that reduce risks through multilateral exchange may become more inclined to risky action in bilateral relations. To complete the picture, continuously rising global productivity has lowered the costs of production and consumption — and destruction, too.

We see lessons for policy and history. In policy terms, democracy is good, but without nation there is no democracy, and nation-building is a double-edged process. Similarly, falling trade costs and wider multilateral exchange have been powerful promoters of economic growth and development, but may also have cheapened war. How can we encourage democracy to spread in ways that don't offer gains to nation-building adventurists? How can we lock countries into regional or global trade without freeing their hands for confrontational foreign adventures at shorter range? These questions may hold some of the keys to a peaceful twenty-first century.

For history, we have identified some unsolved problems in the relationship between economic progress and organized violence, and we have proposed some answers. An underlying issue is that our historical categories and statistics have been limited by the existence of states and their borders. In historical reality, there is a continuum of violence from organized crime through civil conflict to inter-state warfare. As violence flows from one category to another, it drops out of one specialist field and one dataset and pops up in others. There is a unified process to which the formation and destruction of states and state borders is endogenous. This process challenges historians and social scientists to work together to understand it.

References

Açemoglu, D, Johnson, S and Robinson, JA (2005). The Rise of Europe: Atlantic Trade, Institutional Change, and Economic Growth. *American Economic Review*, 95(3): 546–79.

Alesina, A and Spolaore, E (2003). *The Size of Nations*. Cambridge, MA: MIT Press.

Angell, N (1911). *The Great Illusion: a Study of the Relation of Military Power to National Advantage* (3rd ed.). London: W. Heinemann.

Baliga, S, Lucca, D and Sjöström, T (2011). Domestic Political Survival and International Conflict: is Democracy Good for Peace? *Review of Economic Studies*, 78(2): 458–486.

Banerjee, A (2007). Inside the Machine: Toward a New Development Economics. *Boston Review*, 32(2): 12–18.

Besley, T and Persson, T (2008). Wars and State Capacity. *Journal of the European Economic Association*, 6(2–3): 522–530.

Besley, T and Persson, T (2009). The Origins of State Capacity: Property Rights, Taxation, and Politics. *American Economic Review*, 99(4): 1218–1244.

Bonney, R (ed.) (1999). *The Rise of the Fiscal State in Europe, 1200–1815*. Oxford: Oxford University Press.

Broadberry, SN and Harrison, M (2005). The Economics of World War I: an Overview. In: Broadberry SN and Harrison M (eds.) *The Economics of World War I*. Cambridge: Cambridge University Press, pp. 3–40.

Bueno de Mesquita, B (2006). Game Theory, Political Economy, and the Evolving Study of War and Peace. *American Political Science Review*, 100(4): 637–642.

Bueno de Mesquita, B and Siverson, RM (1995). War and the Survival of Political Leaders: A Comparative Study of Regime Types and Political Accountability. *American Political Science Review*, 89(4): 841–855.

Bueno de Mesquita, B, Morrow J, Siverson, RM and Smith, A (1999). An Institutional Explanation of the Democratic Peace. *American Political Science Review*, 93(4): 791–807.

Bueno de Mesquita, B, Smith A, Siverson RM and Morrow J (2003). *The Logic of Political Survival*. Cambridge, MA: MIT Press.

Buhaug, H and Rød, JK. (2006). Local Determinants of African civil wars, 1970–2001. *Political Geography*, 25(3): 315–335.

Cardoso, JL and Lains, P (eds.) (2010). *Paying for the Liberal State: the Rise of Public Finance in Nineteenth-Century Europe*. Cambridge: Cambridge University Press.

Collier, P (2009). *Wars, Guns, and Votes: Democracy in Dangerous Places*. New York: Harper Perennial.

Collingham, L (2011). *The Taste of War: World War Two and the Battle for Food*. London: Penguin Press.

Conconi, P, Sahuguet N, and Zanardi M (2008). *Democratic Peace and Electoral Accountability*. ECARES Working Paper no. 2008/25, Université Libre de Bruxelles.

Davis, LE and Engerman, SL (2003). Sanctions: Neither War nor Peace. *Journal of Economic Perspectives*, 17(3): 187–197.

Davis, LE and Engerman, SL (2006). *Naval Blockades in Peace and War: an Economic History since 1750*. Cambridge: Cambridge University Press.

Downs, GW and Rocke, DM (1994). Conflict, Agency, and Gambling for Resurrection: the Principal-Agent Problem goes to War. *American Journal of Political Science*, 38(2): 362–380.

Doyle, MW (1986). Liberalism and World Politics. *American Political Science Review*, 80(4): 1151–1169.

Doyle, MW (2005). Three Pillars of the Liberal Peace. *American Political Science Review*, 99(3): 463–466.

Eckhardt, W (1989). Civilian Deaths in Wartime. *Bulletin of Peace Proposals*, 20(1): 89–98.

Edelstein, M (2000). War and the American Economy in the Twentieth Century. In: Engerman SL and Gallman RE (eds.) *The Cambridge Economic History of the United States: The Twentieth Century*. Volume 3. Cambridge: Cambridge University Press, pp. 329–405.

Ferguson, N (2001). *The Cash Nexus: Money and Power in the Modern World, 1700–2000*. New York: Basic Books.

Ghosn, F, Palmer, G and Bremer, S (2004). The MID3 Data Set, 1993–2001: Procedures, Coding Rules, and Description. *Conflict Management and Peace Science*, (21): 133–154.

Gibler, DM (2007). Bordering on Peace: Democracy, Territorial Issues, and Conflict. *International Studies Quarterly*, 51(3): 509–532.

Gleditsch, KS (2004). A Revised List of Wars Between and Within Independent States, 1816–2002. *International Interactions*, 30(3): 231–262.

Gleditsch, NP (2008). The Liberal Moment Fifteen Years On. *International Studies Quarterly*, 52(4): 691–712.

Gleditsch, NP, Wallensteen, P, Eriksson, M, Sollenberg, M and Strand, H (2002). Armed Conflict 1946–2001: a New Dataset. *Journal of Peace Research*, 39(5): 615–637.

Glick, R and Taylor, AM (2010). Collateral Damage: Trade Disruption and the Economic Impact of War. *Review of Economics and Statistics*, 92(1): 102–127.

Green, DP, Kim, SY and Yoon, DH (2001). Dirty Pool. *International Organization*, 55(2): 441–468.

Harrison, M (1998). The Economics of World War II: an Overview. In: Harrison M (ed.) *The Economics of World War II: Six Great Powers in International Comparison*. Cambridge: Cambridge University Press, pp. 1–41.

Henig, R (2002). *The Origins of the First World War* (3rd ed.). London: Routledge.

Herbst, J (1990). War and the State in Africa. *International Security*, 14(4): 117–139.

Hewitt, JJ (2008). Unpacking Global Trends in Violent Conflict, 1946–2005. In: Hewitt JJ, Wilkenfeld J and Gurr TR (eds.) *Peace and Conflict 2008*. Boulder, CO: Paradigm Press, pp. 107–118.

Hoffman, PT (2010). Prices, the Military Revolution, and Western Europe's Comparative Advantage in Violence. *Economic History Review*, 64(S1): 39–59.

Hoffman, PT and Rosenthal J-L (1997). The Political Economy of Warfare and Taxation in Early Modern Europe: Historical Lessons for Economic Development. In: Drobak J and Nye JV (eds.) *The Frontiers of New Institutional Economics*. San Diego: Academic Press, pp. 31–55.

Jacks, DS, Meissner, CM and Novy, D (2008). Trade Costs, 1870–2000. *American Economic Review*, 98(2): 529–534.

Jackson, MO and Morelli, M (2007). Political Bias and War. *American Economic Review*, 97(4): 1353–1373.

Karaman, K and Pamuk, Ş (2010). Ottoman State Finances in European Perspective, 1500–1914. *Journal of Economic History*, 70(3): 593–629.

Kinsella, D (2005). No Rest for the Democratic Peace. *American Political Science Review*, 99(3): 453–457.

Levy, JS (1988). Domestic Politics and War. *Journal of Interdisciplinary History*, 18(4): 653–673.

López Córdova, JE and Meissner, CM (2008). The Globalization of Trade and Democracy. *World Politics*, 60(4): 539–575.
Majumdar, S and Mukand, SW (2004). Policy Gambles. *American Economic Review*, 94(4): 1207–1222.
Mansfield, ED and Snyder, J (1995). Democratization and War. *Foreign Affairs*, 74(3): 79–97.
Mansfield, ED and Snyder, J (2005). *Electing to Fight: Why Emerging Democracies go to War.* Cambridge, MA: MIT Press.
Mansfield, ED and Snyder, J (2009). Pathways to War in Democratic Transitions. *International Organization*, 63(2): 381–390.
Maoz, Z and Russett, B (1993). Normative and Structural Causes of Democratic Peace, 1946–1986. *American Political Science Review*, 87(3): 624–638.
Marshall, MG, and Jaggers, K (2007). *Political Regime Characteristics and Transitions, 1800–2006.* Center for Systemic Peace, Polity IV Project at http://www.systemicpeace.org (accessed on 29 October, 2008).
Martin, P, Mayer, T and Thoenig, M (2008). Make Trade not War? *Review of Economic Studies*, 75(3): 865–900.
Michalopoulos, S (2012). The Origins of Ethnolinguistic Diversity. *American Economic Review*, 102(4): 1508–39.
Narang, V and Nelson, RM (2009). Who are These Belligerent Democratizers? Reassessing the Impact of Democratization on War. *International Organization*, 63(2): 357–379.
O'Brien, PK (2001). Fiscal Exceptionalism: Great Britain and its European Rivals: from Civil War to Triumph at Trafalgar and Waterloo. Working Paper No. 65, London School of Economics, Department of Economic History.
O'Brien, PK (2005). Fiscal and Financial Preconditions for the Rise of British Naval Hegemony, 1485–1815. Working Paper No. 91/05, London School of Economics, Department of Economic History.
Offer, A (1989). *The First World War: An Agrarian Interpretation.* Oxford: Oxford University Press.
Olson, M (1963). *The Economics of the Wartime Shortage: A History of British Food Supplies in the Napoleonic War and in World Wars I and II.* Durham, NC: Duke University Press.
Overy, RJ (1987). *The Origins of the Second World War.* New York: Simon & Schuster.
Pelletière, SC (2004). *America's Oil Wars.* New York: Praeger.
Perdue, PC (1996). Military Mobilization in Seventeenth and Eighteenth-Century China, Russia, and Mongolia. *Modern Asian Studies*, 30(4): 757–793.
Ramankutty, N, Foley, J, Norman, J and McSweeney, K (2002). The Global Distribution of Cultivable Lands: Current Patterns and Sensitivity to Possible Climate Change. *Global Ecology and Biogeography*, 11(5): 377–392.
Ramsey, JB (1969). Tests for Specification Errors in Classical Linear Least Squares Regression Analysis. *Journal of the Royal Statistical Society, Series B*, 31(2): 350–371.
Rodrik, D (2007). *One Economics, Many Recipes: Globalization, Institutions, and Economic Growth.* Princeton: Princeton University Press.

Rosato, S (2003). The Flawed Logic of Democratic Peace Theory. *American Political Science Review*, 97(4): 585–602.

Rosato, S (2005). Explaining the Democratic Peace. *American Political Science Review*, 99(3): 467–472.

Russett, B (1995). And Yet it Moves. *International Security*, 19(4): 164–175.

Slantchev, BL, Alexandrova, A and Gartzke, E (2005). Probabilistic Causality, Selection Bias, and the Logic of the Democratic Peace. *American Political Science Review*, 99(3): 459–462.

Stiglitz, JE and Bilmes, LJ (2008). *The Three Trillion Dollar War: the True Cost of the Iraq Conflict*. New York: W. W. Norton & Company.

Tilly, C (1975). Reflections on the History of European State-Making. In: Tilly C (ed.) *The Formation Of National States in Western Europe*. Princeton: Princeton University Press, pp. 3–83.

Tilly, C (1990). *Coercion, Capital, and European States, AD 990–1992*. Oxford: Oxford University Press.

Tooze, JA (2001). *Statistics and the German State, 1900–1945: The Making of Modern Economic Knowledge*. Cambridge: Cambridge University Press.

Wimmer, A and Min, B (2006). From Empire to Nation-State: Explaining Wars in the Modern World, 1816–2001. *American Sociological Review*, 71(6): 867–897.

Supplementary Data

An unpublished Appendix with Tables A1 to A4 is available with other data files at http://warwick.ac.uk/markharrison/data/frequency/.

Part 2

Communism and Defense

Chapter 5

Soviet Industry and the Red Army Under Stalin: A Military-Industrial Complex?

Mark Harrison

Introduction

In recent years it has been hard to pick up a book or article written in any language about Soviet defense without finding reference to the Soviet military–industrial complex.[1] But the term is typically used in two or three quite different ways.

First, there are those who use the term "military–industrial complex" with and without definition. On one side are most Russians and some westerners to whom its meaning is self-evident and does not require any special definition.[2] On the other side are those who take the trouble to define what they mean by it.[3] It appears that the more carefully one defines the military–industrial complex the harder it is to apply to the Soviet Union, to the point where a few writers prefer not to use it at all.[4]

This chapter was first published in *Les Cahiers du Monde russe* 44(2–3), 323–42.
© Editions de l'École des Hautes Études en Sciences Sociales 2003.

[1]This paper was presented to the seminar on "Serving the State: Administrative Practices in the Soviet Union" sponsored by The Centre for Russian, Soviet and Post–Soviet Studies, École des Hautes Études en Sciences Sociales, Paris, with the collaboration of the University of Paris 1–Panthéon–Sorbonne, at the Maison des Sciences de l'Homme, 14–15 December, 2001. I thank the participants for comments and discussion. Thanks also to the Leverhulme Trust, the British Academy, and the University of Warwick for financial support of my research on "Invention, Imitation, and the Birth of Soviet Aerospace", and to Julian Cooper, Jari Eloranta, Paul Gregory, and Peter Law for advice.

[2]For examples, Scott and Scott (1979: 283–301); Aniskov and Khairov (1996); Savitskii (1996); Simonov (1996a; 1996b); and Samuelson (1996; 2000).

[3]For examples, Holloway (1983); Albrecht (1993); Bystrova (1996; 1997; 2000).

[4]For examples, Cooper (1990; 1991); Barber *et al.* (2000); and Harrison (2001).

Second is a linguistic differentiation across which those who avoid definitions can slide unnoticed. "Military–industrial complex" is usually translated into Russian as *voenno–promyshlennyi kompleks* (VPK for short) and conversely. Yet these mean different things. In the English term "military" and "industrial" carry equal weight as adjectival qualifiers of the "complex"; in other words, it refers to a complex made up equally by military and industrial interests. But in the VPK "voenno–" (= military) qualifies "promyshlennyi" (= industrial), not the "complex"; in other words, it refers to a complex made up by the interests of the "military industries", not of the military *and* industry.[5] Which is intended? Perhaps we should be told.

A discussion of such issues may perhaps be too late. The "military–industrial complex" has entered everyday language with the result that this term will be used in many contexts with many meanings that social scientists cannot control. For example "military–industrial complex" is now used as an official designation for some Russian defense industry corporations like *VPK "MAPO"* (Military–Industrial Complex "Moscow Aviation Production Association", formed in 1996 by a merger of *MAPO–MiG* with other defense producers). Social scientists who wish to speak the language of every day must live with such uses. However, there is merit in seeking periodically to check the drift between everyday language and more specialised uses in the light of new knowledge. An opportunity is presented by new research on the Soviet defense industry and its relationship with the Red Army in the 1930s and 1940s. Hence the purpose of this paper is to consider the implications of this research for our understanding of the Soviet army–industry relationship under Stalin.

This paper is organised in six parts. Part 1 reviews past and present–day uses of the concept of a military–industrial complex in relation to the Soviet Union. Part 2 develops a particular aspect of this concept, that of collusion between military and industrial interests in lobbying for resources. Parts 3 and 4 consider evidence concerning Soviet military procurement outcomes and procedures. Part 5 casts this evidence in the framework of a prisoners' dilemma. Part 6 concludes.

[5]For further discussion, see Barber *et al.* (2000: 23–28). The *VPK*, used in the present context as an abbreviation of the military–industrial complex must also be distinguished from the military–industrial commission of the USSR Council of Ministers, also abbreviated to *VPK*.

1. *The Concept of a Military–Industrial Complex*

On 17 January 1961, the retiring US President Dwight D. Eisenhower (1965: 1035–1040) gave a parting warning to the American people.

> Until the latest of our world conflicts, the United States had no armaments industry. American makers of plowshares could, with time and as required, make swords as well. But now we can no longer risk emergency improvisation of national defense; we have been compelled to create a permanent armaments industry of vast proportions. Added to this, three and a half million men and women are directly engaged in the defense establishment. We annually spend on military security more than the net income of all United States corporations.
>
> This conjunction of an immense military establishment and a large arms industry is new in the American experience. The total influence — economic, political, even spiritual — is felt in every city, every State house, every office of the Federal government. We recognize the imperative need for this development. Yet we must not fail to comprehend its grave implications. Our toil, resources and livelihood are all involved; so is the very structure of our society.
>
> In the councils of government, we must guard against the acquisition of unwarranted influence, whether sought or unsought, by the military–industrial complex. The potential for the disastrous rise of misplaced power exists and will persist.

The idea was further developed from seeds already sown by C. Wright Mills (1956: 171–224) in an influential work of political sociology published in 1956. Among the significant themes of *The Power Elite* was that of the new connectedness of America's political, military, and industrial leaders. From these beginnings the idea of the military–industrial complex was developed in various directions, but the common thread was of a coalition, sometimes even a conspiracy, among military and industrial interests and political representatives to lobby for lucrative weapons programmes so as to extract rents from the political process (Rosen, 1973: 3). The argument was not just about pressure, but about collusion: for the "conjunction of an immense military establishment and a large arms industry" to constitute a military–industrial complex, it was necessary that the pressure brought to bear on politicians by military and industrial interests should be coordinated among them rather than competitive between them. Finally, the argument that American policy was dominated by a powerful military–industrial complex was buttressed by a strongly held belief that David Holloway (1982: 293) identified: "that the United States has spent more than is necessary on defense".

A few scholars have investigated the applicability of the concept of the military–industrial complex as a broad military and industrial coalition to the Soviet Union. All have found it difficult to specify, let alone identify empirically, the character of collusion that would have to be found to enable us to confirm that there was indeed a Soviet military–industrial complex in this sense.

Mikhail Agursky and Hannes Adomeit (1978: 6) came to the most sweeping conclusion. They claimed to have found:

> A core of truth in the aphorism that "the USA *has* a military–industrial complex, the USSR *is* a military–industrial complex".

They drew this from a comparison of the United States political structure, open to pressure from a range of interests in society, civilian as well as defense–related, with the closed political structure of the Soviet state from which independent civilian interests capable of resisting military–industrial pressures were excluded. They hesitated to carry the implication to extremes, however, concluding that:

> To consider the whole of the Soviet Union as a military–industrial complex is far too broad to be meaningful.

Vernon Aspaturian (1973: 103) presented two alternative hypotheses or "prototypes" of the Soviet military–industrial complex, one weaker and the other stronger. The stronger sense, he argued, was of:

> An interlocking and interdependent structure of interests among military, industrial, and political figures, that enables or impels them to behave as a distinctive political actor separate from its individual components. A complex of this type [...] would exhibit a high degree of policy unity and act as a single input into the political system.

For the Soviet Union to conform to the stronger hypothesis it would be necessary to identify the influence of the military–industrial complex exerted by its representatives acting in unison. In the weaker hypothesis, its influence would be seen in military, industrial, and political leaders acting separately, but in harmony, to promote shared objectives, in other words through:

> A deliberate and symbiotic sharing of interests on the part of the military establishment, industry, and high–ranking political figures, whose collective influence is sufficient to shape decisions to accord with the interests of these groups at the expense of others in Soviet society.

Aspaturian's own preference was for something in between harmony and unison, "much more than the first prototype and something less than the second".

Peter Almquist (1990: 12–13) concurred with Aspaturian's weaker hypothesis and argued that a model of collusion based on harmony was more realistic that one based on unison. He agreed that shared interest must underlie the idea of a military–industrial complex:

> For a military–industrial complex to exist in a meaningful way, the military and its supporting industries must have, first, complementary interests. By this it is meant that one of the 'partners' generally benefits from the self–interested actions of the other [...]

As distinct from shared *interest*, however, Almquist suggested that shared *purpose* must be capable of independent and separate expression:

> Second, and equally important, both the military and the industry must have a means of influencing the political decision makers. In a military–industrial complex, a "silent partner" is an irrelevant partner [...]

Julian Cooper (1990: 166) stressed the fact that Soviet military and defense–industry leaders led largely separate existences and pursued separate careers that almost never crossed over from one sphere to the other: how then could their interests be enabled to coalesce? This led him to reject the idea of an interconnected military–industrial elite on straightforward evidential grounds.

Most recently Irina Bystrova (1997: 35; 2000: 9), specifically dissociating herself from those who identified the military–industrial complex narrowly with the defense industry and its leaders, has argued for a return to a wider and also stronger conception of the Soviet military–industrial complex as:

> *A powerful corporation that represented the common interests* of social–political groups associated with the provision of the USSR's national security: professional soldiers, party and state officials, representatives of the security agencies, and scientific–technical circles. In our view, the term "power elite" applies even more closely to the Soviet *VPK* than to the American, where the element of [private] property relations and economic dominance somewhat obscured a "pure" portrayal of political power [emphasis added].

As will become clear, Bystrova limits her empirical application of this concept to the post–Stalin period. But then, even Julian Cooper concurs that under Brezhnev the interests of the Soviet defense industry lived through their "golden age".

2. *Collusive Lobbying: A Rationale*

Under what circumstances may either the army or industry have something to gain from investments in lobbying?

Under any constitution, various aspects of military budgeting and procurement make these a suitable target for lobbying activity. The funding principal takes decisions in secret and often does not have to account for them even afterwards in a transparent way. Much military spending has an insurance aspect, but the worst cases against which the principal seeks to insure are not susceptible to actuarial calculation. Where military projects involving new technologies are concerned there is usually intrinsic uncertainty about the timescale and expected value of returns to investment. It may be rationally expected that many projects will fail. There are powerful information biases: agents know the true worth of their projects better than the principal does. All of these impede the principal in the course of both selection and monitoring of spending programmes. Under these circumstances self–interested agents from within both military and industrial organisations can be expected to invest resources in lobbying the principal for funding of their own projects.

This on its own does not mean that more will be spent on defense in the presence of lobbying than without it; it just means that the military budget will be allocated in a way that reflects the effectiveness of lobbying investments. If the resources invested are taken out of the sums nominally allocated to national defense, then it could even mean that true defense spending is less than appears. For such lobbying to increase the total of resources available for national defense in a market economy it appears to be also necessary that the budget constraints on military projects should be relatively soft. Alternatively, in a command system like the Soviet economy where all budget constraints were soft to some degree, it would be necessary for the military budget also to have some degree of priority in the rationing of material supplies, which was in fact the case. Under these circumstances military and defense–industry interests will be able to invest more in lobbying than their civilian counterparts and will anticipate a higher return on lobbying investments. As a result, military procurement will claim more resources than an uncorrupted principal would have allocated.

When rivalrous lobbying can influence the principal, why should agents collude? The reason is that unrestricted rivalry dissipates the rents they extract. In the simplest case, if n risk–neutral agents are competing for a given sum of project funding that offers a net surplus k and each has the same expectation of being awarded the contract, then each should be willing to invest $\frac{k}{n}$ in lobbying to win the contract and the result will be that the entire surplus available is competed away in lobbying costs. However, if two agents secretly collude they can pool their probabilities of winning the contract, reduce their joint costs of lobbying, and so gain.

When collusion pays, why should military and industrial agents collude among themselves rather than with others? For example, the Red Army annually procured very large quantities of food. Farmers and soldiers both stood to gain from increased investments in food production and procurement for the armed forces. Therefore, why not a "military–agricultural complex" in which farmers and soldiers colluded to lobby for increased agricultural investment? One reason might be that soldiers and weapon producers could hide collusion more easily under a common cloak of secrecy than any other coalition one could think of. For another reason, it is commonly held that the success of collusion depends in part on the frequency and repetitiveness of interaction between agents that may collude. From this point of view the interaction between soldiers and industrialists was far more intense and far more likely to give rise to long-term relationships. If the Red Army and Soviet industry could not learn to trust each other, no one could.

Finally, consider the motivation of the funding principal who is the target of lobbying. Why should a principal permit any lobbying? One reason may be that lobbying provides the principal with more information than in its absence, even if the information itself may contain biases. This information is of two kinds: about the relative worth of the spending projects from among which he must choose, and about the relative loyalty of the agents competing for his favour.

A dictator, for example, might intentionally design the budgetary system to distribute rewards to loyal agents. As a result public spending will be larger than on a pure public–good calculation of the costs and benefits. The excess is the signal that loyalty is being rewarded and expected in return; if some extra spending did not result, those receiving the funding would have no reason to offer thanks to the government in exchange since any politician would rationally promise to undertake at least those expenditures that were efficient (Wintrobe, 1998: 31). A dictator like Stalin, already closely linked with military and industrial interests, might rationally choose military spending for one of the channels through which he distributed rewards, and military spending would then also be enlarged.

However, such a dictator should prefer his agents to compete, not collude, in lobbying him. This is for two reasons: first, collusion among agents would restrict the flow of information and probably increase its inherent biases. Second, collusion would enable them to increase both the level of rents extracted, and the proportion extracted not as a reward for loyalty but as a return on lobbying investments. The implication is that, if military and industrial interests succeeded in colluding in such a way as to form a military–industrial complex, then the dictator was failing to rule efficiently in some degree.

3. *Evidence on the Defence Burden*

How can we tell whether there was a Soviet military–industrial complex? I have outlined an argument that collusive lobbying by military and industrial interests should lead to more spending than is necessary on defense. Some have been tempted, therefore, to seek to resolve this question by looking at outcomes. If a particular society shows a particularly high level of military spending, perhaps that is evidence that a military–industrial complex is at work. By this criterion, however, the case of the Soviet Union under Stalin would not be especially clear cut.

It is true that the Soviet Union, although a relatively poor country, allocated somewhat greater resources to defense than other countries at a similar level of development for much of the twentieth century. For most of the interwar period, for example, most other European countries, including the United Kingdom, were spending three percent of their national incomes on defense or less, and the United States much less.[6] In contrast the peacetime share of Soviet military spending rose unremittingly from two per cent in 1928 to six per cent in 1937 and 15 per cent in 1940 (Bergson, 1961: 46).

Even so, the 1937 Soviet figure is only slightly in excess of the five per cent recorded for the Russian Empire in 1913 (Gregory, 1982: 57). Moreover, the argument that the Soviet Union was spending more than was necessary on defense is undermined by two stubborn facts. One is that a

[6]Eloranta (2002: 110), provides figures for the share of military spending in GDP of a number of European and world powers averaged over the years 1920 to 1938. These yield 3.0 per cent for the UK, 2.9 per cent for France, an unweighted average of 2.1 per cent for eight smaller European countries (Belgium, Denmark, Finland, Netherlands, Norway, Portugal, Sweden and Switzerland), and only 1.2 per cent for the United States.

war was coming. In 1937, Germany was already spending 15 per cent of national income on military rearmament (Abelshauser, 1998: 138). Given the relative sizes of the two economies, Soviet real military outlays at this time were probably only half Germany's.[7] The other is that the peacetime burdens were modest by comparison with those imposed by war itself. War drove the defense share far higher: as a share of GDP at pre-war prices the Soviet military burden rose from 17 per cent in 1940 to more than 60 per cent in 1942 and 1943 (Bergson, 1961: 46) before falling back to nine per cent by 1950 (Harrison, 1996: 110).

It is true that after the Korean rearmament the Soviet economy again carried a burden consistently higher than that of its main rival. The Soviet post-war military burden was lower, and shows less variation both across estimates and through time, than is often supposed, but still exceeded figures reported for the United States. Higher estimates of Soviet defense share of GDP vary within a 13 to 16 per cent range from the 1960s through the 1980s, while lower estimates give a range of 10 to 13 per cent (Easterly and Fischer, 1995: 348). United States figures are lower: they show nine per cent in 1960 falling to eight per cent in 1970 and less than six per cent in 1980 (Smith and Smith, 1983: 23). But again it is necessary to take into account that the US economy was never less than twice the size of the Soviet economy in the post-war period.

It has been argued that such ratios understate the scope of the Soviet military–economic effort in comparison with other countries. At the end of the 1970s, for example, Harriet and William Scott (Scott and Scott, 1979: 283) wrote:

> A number of Soviet dissidents argue that over 40 per cent of the Soviet gross national product goes for military purposes. Their rationale may take into account hidden costs for which there are no counterparts in the United States. [...] To [the regular Soviet military program] must be added paramilitary training of the population and premilitary and reserve training. Civil defense programs, hardening of both military and industrial facilities, and dispersion and duplication of industries for defense purposes are additional costs. The costs of highways and railroads constructed primarily for strategic purposes, even though they have some peacetime use, must be considered. Foreign military aid is another factor.

More recent writing in this vein has focused on the special role of *mobpodgotovka* (mobilisation readiness) in the Soviet economy (Simonov,

[7] For size comparisons of GDP here and below, see Maddison (2001).

1996: 115–25; 2000). For example, in his foreword to Lennart Samuelson's recent book on the Soviet military commander M.N. Tukhachevskii, Vitalii Shlykov describes how:[8]

> The Soviet economy very rapidly became an economy of *total preparation for war*. To enable the transition from civilian to military production an elaborate and strictly centralized system of mobilisation readiness was *established at every level of Soviet power and in all the agencies of economic administration down to the factory level.*

In my view too much should not be made of these arguments, certainly where the Stalin period is concerned. For one thing, the Soviet authorities certainly intended to create an all–embracing system of mobilisation preparation, but whether they succeeded in doing so, or merely in creating the appearance of it, has not yet been sufficiently researched. For another, while *mobpodgotovka* was aimed at enhancing the Soviet Union's military potential, the nature of an economic potential is that in the outcome it may yet be used for other purposes. Many Soviet outlays that were defense–related without being counted in the military budget either contributed more to goals other than to defense (for example, the maintenance of large internal security forces the primary task of which was defense of the regime against its internal enemies), or else added to the economy's general potential through the accumulation of fixed assets and the stock of knowledge. In the Soviet Union as in other countries, finding a strategic rationale for something desired for other reasons was always an effective way of pushing it up the budgetary agenda. Finally, there are parallels in other countries from the origins of Germany's *autobahn* network in the consolidation of Hitler's regime of the 1930s to the Defense and Interstate Highways built in the United States after World War II.

The main problem with using high levels of military spending to detect the presence of a military industrial complex, however, is that high Soviet military spending may be explained otherwise than by the collusive lobbying of military and industrial interests. It might be that competitive lobbying was powerful enough to secure additional outlays. It might be that the Soviet dictator designed the budgetary process intentionally to reward loyal agents in the military and industrial sectors. Finally, it might

[8]Samuelson (2000: xii) with emphasis added. Shlykov goes on: "[...] until the publication of this book and Samuelson's other writings these preparations were inadequately evaluated and understood both by Western Sovietologists and by the wartime adversaries of the Soviet Union, especially Germany".

be that high military spending was justified on public–good grounds of Soviet national security simply as a response to external threat perceptions.

Related to the national security argument is the possibility that the Soviet Union spent more on defense simply because it was a dictatorship. Michelle Garfinkel (1994) has argued that incumbent policy–makers in competitive political systems have an incentive to underinvest in national security in the face of a given threat, because the likelihood of being replaced in the short–term by another policy–maker with different goals prevents the incumbent from fully internalising the long–term benefits. Given that threats are not actually given but are determined by interaction between states, the outcome can be a good equilibrium of low military spending in regions where democracies predominate. Garfinkel also shows empirically that democratic states have tended to spend less on defense than non–democratic ones. As a possible explanation of high levels of Soviet military spending this has nothing to do with lobbying or special interest groups.

In summary, if we wish to detect the operation of a military–industrial complex we cannot rely on statistical indicators of the outcomes of military budgeting and procurement. We have to find direct evidence of the process that gave rise to the outcome. This evidence should substantiate the hypothesis of a coalition among military and industrial interests in lobbying for resources.

4. *Evidence on Collusion*

Irina Bystrova (2000: 317) has suggested a useful distinction between lower–level lobbying by agents within ministerial departments, which I will call lobbying for project funding, and the sort of higher–level lobbying that ministers and ministries engaged in where the funding of entire ministerial programmes was at stake.

Recent archive–based studies have thrown light on lobbying for project funding in defense industry. "The activity of creating monopolies is a competitive industry" (Tullock, 1987: 604), and the Soviet aircraft industry was no exception. Mikhail Mukhin's (2000) study of aircraft designers suggests that rivalry, far from being artificially engendered from above, was intrinsic, intense, and amoral. Harrison's accounts of aero engine innovation show that rivalry began with each inventor's struggle to gain a first–mover's advantage over competitors for influence over the funding principal. Their competition was developed through behaviours that aimed to establish and defend a monopoly over intellectual and physical assets of various kinds.

An active market for second-hand assets was expressed through activities such as hostile takeover bids and mergers.[9] While each struggled to create a personal monopoly, there is little evidence of collusive behaviour among agents. Bystrova (2000: 319) considers it typical, for example, that one designer would refuse to take on the former employee of another.

Evidence of intense competition for project funding does not settle the issue of whether or not there was collusion among military and industrial interests at higher levels in the political system in lobbying for ministerial programmes. For example, if military and industrial interests successfully colluded to extract some of the dictator's rent, they might reasonably also agree to use competitive lobbying at lower levels to distribute the surplus among the agents within the military–industrial complex.

In this connection, a story told by Irina Bystrova (2000: 320–1) is suggestive. In the spring of 1946, following the "aviators' affair", the new air force chief K.A. Vershinin reported to Stalin that one reason for the technical backwardness of Soviet aviation was that aircraft designers had too much control over the air force's plan for experimentation, and that they had achieved this control because design work had become too monopolised, with only three bureaux for fighter aircraft and one each for other types. The significance of this case lies in several implications: the designers' competitive struggle to create a monopoly had in part succeeded, perhaps as a result of wartime conditions; previous leaders of the air force and aircraft industry had colluded, but for the sake of a quiet life rather than to distribute rents; such collusion had damaged the interests of the armed forces as a whole; and the new air force chief lobbied the dictator to correct this. Finally, it is significant that the dictator agreed.

Is there direct evidence of collusion between ministries that is more than anecdotal? Consider first the lower levels of daily interaction between the industrial producers and the military consumer of defense products. Here the evidence is almost entirely of tension (Harrison and Simonov, 2000). Producers tried to inflate prices and persuade the consumer to accept and pay for products than fell below contractual standards. Defense ministry regulators in industrial establishments struggled to secure information about costs and verify claims about deliveries and quality standards. Collusion took the form of military regulators accepting bribes from producers to make decisions in favour of industry rather than the armed

[9] Harrison (2003; 2005), reproduced here in Chapters 7 and 8.

forces. The consistent policy of the defense ministry was to legislate against collusion and punish it when it was detected.

If we raise our sights to higher levels of the political system, is there evidence that the commissars for defense and the defense industries, say, colluded to win enhanced funding for ministerial programmes? The archives confirm that the army and heavy industry each separately pressed for additional resources at various times (Samuelson, 1996; 2000; Davies and Harrison, 1997). But did they pursue their interests jointly? Evidence of collusion — for example, that military leaders were prompted or induced to press for increased allocations by industrialists, or conversely — has not been found. In general the record of daily interaction of industrial and defense officials suggests that mutual tensions, frustrations, suspicions, and conflicts between the army and industry were endemic (Harrison, 2001: 103–7).

If anyone attempted to eliminate this conflictual situation in Stalin's time, it was probably Tukhachevskii. Throughout his career, Tukhachevskii repeatedly advocated industrial mobilisation under military management. In the spring of 1927, as chief of the Red Army general staff, he sponsored a project for a military–industrial complex organised not through collusion but by a military takeover of defense industry. His ambition rested partly on a proposal to create a "defense sector" for industrial mobilisation within the state economic planning commission, in which he proposed personally to play the leading role. The Gosplan defense sector was established, but Tukhachevskii was sidelined (Samuelson, 2000: 42–7). At the same time the defense commissariat requested the right to agree appointments to the defense industry, plans and reports of defense producers, and plans for capital investment in the industry. This proposal was rejected (Harrison and Simonov, 2000: 230). Tukhachevskii was resisted in particular by Stalin's defense commissar and loyal agent K.E. Voroshilov. Samuelson has shown that Tukhachevskii's resignation as chief of staff, which followed in May 1928, was most likely a result of the failure of his ambition to control the defense industry (Samuelson, 2000: 55–9). Later, as chief of armament for the Red Army in the early and mid–1930s Tukhachevskii tried to organise a reseach and development empire that was independent of industry under the Red Army administration for military inventions, and to establish a monopoly in such new fields as jet propulsion. Here he was again unsuccessful (see Chapter 7).

After Tukhachevskii's execution no one followed in his footsteps. When we turn to the later Stalin period Bystrova has shown that at key moments the voice of the armed forces was conspicuously absent. For example, when minister for the chemical industry M.G. Pervukhin fought for more resources for the uranium industry after World War II, it was within a framework that excluded the military: the Special Committee appointed by Stalin to take charge of atomic weapons development had no armed forces representatives (Bystrova, 1996: 6). When minister for armament D.F. Ustinov struggled against Moscow's civilian administration for more factory space for jet and rocket armament in the same period Stalin settled the dispute, not pressure from the armed forces (Bystrova, 2000: 247–9).

It appears generally that industry and army had little opportunity to act in concert, and even the influence that each could exert separately was strictly constrained by the political system in which they operated. While the interests of Soviet society were explicitly identified with both military and industrial construction, the concentration of power n the central party organs and the ubiquitous role of the party–state apparatus meant that military and defense–industry interests had little or no freedom of independent action. Civilian leaders from Stalin onwards retained complete authority through pre-war rearmament, World War II, and post-war military confrontations. The political influence of outstanding soldiers was always tenuous from Tukhachevskii, executed by Stalin in 1937, to air force Marshal A.A. Novikov, imprisoned by Stalin in 1946, and Marshal G.K. Zhukov, sacked first by Stalin in 1946, then by Khrushchev in 1957.

If any branch of government developed an organic relationship with the defense industry at this time, it was the security organs under the leadership of the civilian minister for internal affairs and deputy prime minister L.P. Beriia. The latter, like Stalin's postwar commander of ground forces N.A. Bulganin, held the military rank of Marshal, but neither was a professional military man. Boris Starkov (2000: 265) has shown from the archives that Beriia shared Stalin's distrust of the professional soldiers to the point where, in the early 1950s, he even opposed handing over his newly developed nuclear weapons to the armed forces.

What explains the absence of collusion? The structure of individual incentives was evidently such that the private gains to collusion were typically less than the gains from acting in rivalry. In other words the absence of collusion may be explained in terms of a prisoners' dilemma.[10]

[10]Harrison (2001: 91) suggested this without elaborating on it. The prisoner's dilemma is not the only possible game–theoretic interpretation of relations between the Red Army and Soviet industry. For an alternative approach, see Gregory (2002).

5. *A Prisoners' Dilemma*

The concept of the prisoners' dilemma describes a class of games in which the players would gain jointly if they could only cooperate with each other. However, to prefer cooperation they need some means of making a binding commitment beforehand to cooperate and share the mutual gain. Without this mutual trust each is better off cheating on the other.[11]

The dilemma in the Soviet military–industrial complex could be described as follows. Consider two players, Army and Industry, under a Dictator who fixed the demands upon each and allocated the resources with which each must meet these demands. Think of each player as aiming to maximise their own resources relative to the demands placed on them by the Dictator. In other words each player sought to minimise plan tension — the degree of effort or "mobilisation" they must exert to achieve the demands placed on them, but since they could not easily control these demands the best way of achieving this goal was to lobby for additional resources.

A problem for each was that the additional resources must come from somewhere. When the Army, for example, faced the Dictator with a demand for more resources, the Dictator had to decide where to find them, and this must have detracted either from the Dictator's objectives for living standards, or for investment in agriculture or trade, or it must have resulted in harder work and lower pay for Industry. When Industry pressed the Dictator for more resources, the Dictator could similarly choose to find them from agriculture or consumption, or by making the Army wait longer for new weapons. By colluding and lobbying jointly, Army and Industry could make it harder for the Dictator to make the other pay and easier for him to decide to sacrifice consumption and agriculture instead. Thus each might support the other's case for more resources.

Suppose Army and Industry acted on this logic. Before the game each might promise the other to adopt a high plan rather than a low plan. Industry would propose a high plan for the output of military products

[11] Here is the original illustration by Tucker (1950/2001). "Two men, charged with a joint violation of law, are held separately by the police. Each is told that (1) if one confesses and the other does not, the former will be given a reward of one unit and the latter will be fined two units, (2) if both confess, each will be fined one unit. At the same time each has good reason to believe that (3) if neither confesses, both will go clear". Although the total of penalties on the two would be minimised at zero if each remained silent, each player is individually better off confessing regardless of the other's action. If B remains silent, A gains by confessing. If B confesses, by confessing too, A loses but avoids the greater loss from remaining silent.

		Army	
		High	*Low*
Industry	*High*	Army and Industry both gain at the expense of the Dictator (1, 1)	Buyer's market: Army gains at expense of both Dictator and Industry by imposing lower prices and higher quality standards (–1, 2)
	Low	(2, –1) Seller's market: Industry gains at expense of both Dictator and Army by imposing higher prices and lower quality standards	(0, 0) Neither Army nor Industry gains or loses.

Fig. 1. Industry versus Army.

(*plan voennoi produktsii*), and the Army would propose a high plan for the matching procurement of military products (*plan voennykh zakazov*). But in playing the game each could then choose either to honor its commitment to go high, or cheat on the other by going low.

A possible set of payoffs to each player and each strategy is set out in Figure 1. When both went low, each neither gained nor lost but simply received the allocation that the Dictator decided without influence being exerted. When both went high, each gained. This gain was at the expense of the Dictator, or to be precise at the expense of the Dictator's other objectives, say, to develop the civilian sector.[12]

In the other cases one went high and the other low. This shifted the balance of power in the market for military products: whoever cheated was able to shift this balance against the other. When the Army went high and Industry low, it appeared as if the Dictator had favoured the Army by augmenting its budget for weapons and military equipment. But the reality was that the Army faced a seller's market in which the supply of munitions was restricted; the Army found its favourable budget eroded by inflation as producers pushed up prices faster than expected, and the Army could only

[12]Tucker (1950/2001) pointed out that the prisoner's dilemma is a zero–sum game when the state is introduced as a third player. In the Army–Industry game the Dictator does not exercise a choice, but receives a payoff in each case chosen by the other players that is the sum of the players' payoffs with opposite sign.

succeed in procuring weapons in the numbers envisaged by relaxing quality standards. Thus Industry could gain at the Army's expense.

Alternatively Industry went high and the Army low, so it appeared as if the Dictator had favoured Industry by permitting rapid expansion of its production and capacity. But in reality Industry now faced a buyer's market in which the Army could exploit the increased availability of munitions; the Army could pick and choose among producers and products, exert downward pressure on prices, and control quality more effectively by making it more difficult for Industry to realise its plan for deliveries. Thus the Army could gain at Industry's expense.

In short, the joint gain of both players was greatest when both went high: in Figure 1, each gained 1. But Industry gained by cheating whether or not the Army cheated. If the Army went high Industry received 1 if it went high too, but 2 if it cheated and went low. If the Army cheated and went low, Industry received zero if it cheated too, but −1 if it went high. Either way Industry would prefer to cheat. A glance at the figure shows that the position of the Army was symmetrical.

Finally, the textbook analysis of the prisoners' dilemma tells us that the tendency to cheat could be overcome in specific circumstances: if each player could learn to trust the other as a result of repeated interactions. Aware of this, Stalin demanded that his agents be loyal to him, not to each other, and took measures to speed up the circulation of *nomenklatura* posts to prevent the formation of personal ties. Bystrova (2000: 317) refers to this quite correctly as Stalin's policy of "divide and rule". But in the post–Stalin era elite circulation slowed down. For example, research on the post-war membership of the Sovnarkom and Council of Ministers has shown that the leaders of the defense and heavy industries made up the core of continuity from the Stalin era to the post–Stalin succession (Crowfoot and Harrison, 1990: 52). Defense industry leaders were given stable appointments and built long careers (Cooper, 1990: 167–8). Reduced turnover and increased opportunities for military and industrial agents to build personal reputations and form personal ties of trust might well have increased the incentives for them to collude with each other.

This provides some theoretical justification for Bystrova's view that the Soviet military–industrial complex only came into its own after Stalin's death. It may also support her identification of Ustinov personally as the key figure in the post–Stalin military–industrial coalition (Bystrova, 2000: 328). Ustinov was exceptional in being accepted as a leader of both army and industry, and his ministerial responsibilities over more than 40 years ranged from the armament industry and the defense industry as a whole to defense itself (Ivkin, 1999: 564).

If Ustinov is remembered as the father of the Soviet military–industrial complex, more distant ancestors should not be forgotten. In an earlier period of Soviet history Tukhachevskii represented the interests of the army by advocating high plans for defense industry and tried, although with little success, to create institutional means to enforce them. Another forefather of the modern military–industrial complex was from another era and another country: General Erich von Ludendorff, the military author of Germany's "Hindenburg" plan for industrial mobilisation in August 1916 (Feldman, 1966). Ludendorff was not much more successful than Tukhachevskii, but for quite different reasons, and at least he was not executed for it.

Conclusion

To judge from the present–day discourse of Russian history, the Soviet military–industrial complex is an established fact. Speaking more strictly, however, this paper suggests that in the Stalin period the military–industrial complex was not yet fully developed. Despite the development of large and diversified armed forces served by large–scale specialised industrial facilities apparently integrated and coordinated through the party–state apparatus, the incentives for the agents of these interests to compete rather than collude were very strong. The dictator personally disliked it when his agents colluded with each other rather than with him, and took steps to break up and punish collusion when it was identified. Only after Stalin's death were conditions created that may have been more favourable to the emergence of a durable military–industrial coalition.

References

Abelshauser, W (1998). Germany: Guns, Butter, and Economic Miracles. In: Harrison M (ed.) *The Economics of World War II: Six Great Powers in International Comparison.* Cambridge: Cambridge University Press, pp. 122–76.

Agursky, M, and Adomeit, H (1978). *The Soviet Military–Industrial Complex and Its Internal Mechanism.* National Security Series no. 1/78, Queen's University, Centre for International Relations, Kingston, Ontario, Canada.

Albrecht, U (1993). *The Soviet Armaments Industry.* Chur, Switzerland: Harwood Academic Publishers.

Almquist, P (1990). *Red Forge: Soviet Military Industry since 1965*. New York: Columbia University Press.

Aniskov, VT and Khairov, AR (1996). Istoriia voenno–promyshlennogo kompleksa Rossii v regional'nom aspekte: ot nachala pervoi do okonchaniia vtoroi mirovoi voiny. Na primere Verkhnevolzh'ia, Iaroslavl': Iaroslavskii gosudarstvennyi universitet

Aspaturian, Vernon V (1973). The Soviet Military–Industrial Complex: Does It Exist? In: Rosen S (ed.) *Testing the Theory of the Military–Industrial Complex*. Lexington, MA: Lexington Books, pp. 103–33.

Barber, J, Harrison, M, Simonov, NS and Starkov, BS (2000). The Structure and Development of the Defence–Industry Complex. In: Barber J and Harrison M (eds.) *The Soviet Defence–Industry Complex from Stalin to Khrushchev*. London and Basingstoke: Macmillan, pp. 3–29.

Bergson, A (1961). *The Real National Income of Soviet Russia since 1928*. Cambridge, MA: Harvard University Press.

Bystrova, I (1996). *The Formation of the Soviet Military–Industrial Complex*. Stanford University, Centre for International Security and Arms Control.

Bystrova, I (1997). Sovetskii VPK. Teoriia, istoriia, real'nost. *Svobodnaia mysl'*, 6, 30–44.

Bystrova, I (2000). *Voenno–promyshlennyk kompleks SSSR v gody kholodnoi voiny*. (Vtoraia polovina 40–kh — nachalo 60–kh godov), Institut Rossiiskoi istorii Rossiiskoi Akademii nauk, Moscow.

Cooper, JM (1990). The Defense Industry and Civil–Military Relations. In: Colton TJ and Gustagson T (eds.) *Soldiers and the Soviet State: Civil–Military Relations from Brezhnev to Gorbachev*. Princeton, NJ: Princeton University Press, pp. 164–91.

Cooper, JM (1991). *The Soviet Defence Industry: Conversion and Reform*, London: Pinter.

Crowfoot, J and Harrison, M (1990). The USSR Council of Ministers under Late Stalinism, 1945–1954: its Production Branch Composition and the Requirements of National Economy and Policy. *Soviet Studies*, 42(1): 41–60.

Davies, RW and Harrison, M (1997). The Soviet Military–Economic Effort under the Second Five–Year Plan (1933–1937). *Europe–Asia Studies*, 49(3), 369–406.

Easterly, W and Fischer, S (1995). The Soviet Economic Decline. *World Bank Economic Review*, 9(3): 341–71.

Eisenhower, DD (1961). *Public Papers of the Presidents of the United States, Dwight D. Eisenhower*. Volume 8, Washington, DC: US GPO.

Eloranta, J (2002). *The Demand for External Security by Domestic Choices: Military Spending as an Impure Public Good among Eleven European States, 1920–1938*. PhD Dissertation, European University Institute, San Domenico di Fiesole.

Feldman, GD (1966). *Army, Industry and Labor in Germany, 1914–1918*. Princeton, NJ: Princeton University Press.

Garfinkel, M (1994). Domestic Politics and International Conflict. *American Economic Review,* 84(5): 1294–1309.
Gregory, PR (1982). *Russian National Income, 1885–1913.* Cambridge: Cambridge University Press.
Gregory, PR (2002). *Why Was Soviet Defense Planning So Far Off the Mark?* Working Paper, University of Houston, Department of Economics.
Harrison, M (1996). *Accounting for War: Soviet Production, Employment, and the Defence Burden, 1940–1945.* Cambridge: Cambridge University Press.
Harrison, M (2001). Providing for Defense. In: Gregory PR (ed.) *Behind the Façade of Stalin's Command Economy.* Stanford, CA: Hoover Institution, pp. 81–110.
Harrison, M (2003). The Political Economy of a Soviet Military R&D Failure: Steam Power for Aviation, 1932 to 1939. *Journal of Economic History,* 63(1): 178–212.
Harrison, M (2005). A Soviet Quasi-Market for Inventions: Jet Propulsion, 1932 to 1946. *Research in Economic History,* 23: 1–59.
Harrison, M and Simonov, NS (2000). Voenpriemka: prices, costs, and quality assurance in defence industry. In: Barber J and Harrison M (eds.) *The Soviet Defence–Industry Complex from Stalin to Khrushchev.* London and Basingstoke: Macmillan, pp. 223–45.
Holloway, D (1982). Innovation in the Defence. In: Amann R and Cooper J (eds.) *Industrial Innovation in the Soviet Union.* New Haven, CT: Yale University Press, pp. 276–367.
Holloway, D (1983). *The Soviet Union and the Arms Race.* New Haven, CT: Yale University Press.
Ivkin, VI (1999). Gosudarstvennaia vlast' SSSR. Vysshie organy vlasti i upravleniia i ikh rukovoditeli. 1923–1991 gg. Istoriko–biograficheskii spravochnik. Moscow: ROSSPEN.
Maddison, A (2001). *The World Economy: A Millenial Perspective.* Paris: OECD.
Mills, CW (1956). *The Power Elite.* New York: Oxford University Press.
Mukhin, MIu (2000). Sovetskoe rukovodstvo 1920–30–kh godov i aviastroitel'nye inzhenerno–tekhnicheskie kadry: ot liubvi do nenavisti — odin shag. Institut Rossiiskoi istorii Rossiiskoi Akademii nauk, Moscow.
Rosen, S (1973). Testing the Theory of the Military–Industrial Complex. In: Rosen S (ed.) *Testing the Theory of the Military–Industrial Complex.* Lexington, MA: Lexington Books, pp. 1–29.
Samuelson, L (1996). *Soviet Defence Industry Planning: Tukhachevskii and Military–Industrial Mobilisation.* Stockholm: Stockholm School of Economics.
Samuelson, L (2000). *Plans for Stalin's War Machine: Tukhachevskii and Military–Economic Planning, 1925–41.* London and Basingstoke: Macmillan.
Savitskii, IM (1996). Oboronnaia promyshlennost' Novosibirskii oblasti. Opyt poslevoennogo razvitiia (1946–1963 gg.), Novosibirsk: Olsib.
Scott, HF and Scott, WF (1979). *The Armed Forces of the USSR.* Boulder, CO: Westview.

Simonov, NS (1996a). Voenno–promyshlennyi kompleks SSSR v 20–50–e gody. *Svobodnaia mysl'*, 2: 96–114.
Simonov, NS (1996b). Voenno–promyshlennyi kompleks SSSR v 1920–1950–e gody: tempy ekonomicheskogo rosta, struktura, organizatsiia proizvodstva i upravlenie, Moscow: ROSSPEN.
Simonov, NS (2000). Mobpodgotovka: Mobilisation Planning in Interwar Industry. In: Barber J and Harrison M (eds.) *The Soviet Defence–Industry Complex from Stalin to Khrushchev.* London and Basingstoke: Macmillan, pp. 205–22.
Smith, D and Smith R (1983). *The Economics of Militarism.* London: Pluto.
Starkov, BS (2000). The Security Organs and the Defence–Industry Complex. In: Barber J and Harrison M (eds.) *The Soviet Defence–Industry Complex from Stalin to Khrushchev.* London and Basingstoke: Macmillan, pp. 246–68.
Tucker, A (1950/2001). A Two–Person Dilemma. In: Rasmusen E (ed.) *Readings in Games and Information.* Blackwell: Oxford, pp. 7–8.
Tullock, G (1987). Rent Seeking. In: Eatwell J, Milgate M and Newman P (eds.) *The New Palgrave: The World of Economics.* London and New York: Macmillan, pp. 604–9.
Wintrobe, R (1998). *The Political Economy of Dictatorship.* Cambridge: Cambridge University Press.

Chapter 6

Contracting for Quality under a Dictator: The Soviet Defense Market, 1930–1950

Mark Harrison and Andrei Markevich

Introduction

The record of Soviet defense procurement under Stalin presents an apparently standard problem in market organization: contracting for goods when quality matters.[1] What took it out of the ordinary was the institutional setting in which the contractor operated: not a law-governed market economy, but a centralized dictatorship. What sort of market could there be, when the seller was owned by the state, the buyer was a government agency, and both were subject to centralized plans that had the force of law? What kind of contract could there be, when at any moment the dictator could step in, tear up the contract, and shoot the contractor? These turn out to be far from trivial questions.

Contracting for quality is a problem the world over. It was specifically a problem in all branches of the Soviet economy where quality mattered. In the military sector, however, the Soviet defense ministry was willing to incur the significant cost of employing tens of thousands of military procurement officers to work on site with the contractors to enforce contractual standards for the quality of military goods, endowed with sweeping powers to test and

This chapter is previously unpublished. An earlier version was first made available in November 2007 under the title *Quantity Versus Quality in the Soviet Market for Weapons* as Warwick Economics Research Paper no. 822 (University of Warwick, Department of Economics). © Mark Harrison and Andrei Markevich, 2013.

[1]Earlier versions of this paper were presented to seminars at Carlos III University, Madrid, the City University, London, the Universities of Warwick and York, the Hoover Institution summer (2007) workshop on totalitarian regimes, and the Royal Economic Society annual (2008) conference. The authors thank the organizers and participants for advice and comments.

reject deliveries. As a result, the military agents were able to learn about many, if not all, dimensions of the quality of the goods they were accepting.

Recent research has thrown up a puzzle. The military agents were not corrupt, and were genuinely tough on the enforcement of quality standards. But they often colluded with contractors to conceal quantitative violations of the plan for deliveries of weapons, not on a private basis but with their superiors' clear knowledge and approval.

We attribute their behavior to a hold-up problem in the market for weapons. In this paper, we set out how the problem arose and how it worked out, along with the evidence for our reasoning. Our evidence is taken from the records of the Soviet defense ministry, the defense industry, and the Communist Party's "control commission" — Stalin's first-line agency for monitoring the implementation of his commands.

The paper is organized as follows. In Part 1, we discuss the institutional arrangements of the Soviet defense market. Part 2 suggests how the hold-up problem arose and why the advantage lay on the side of the seller. Part 3 presents a model in which there is bargaining power and also collusion to conceal the bargain. Parts 4 and 5 discuss the historical evidence on bargaining power and collusion respectively. The final section concludes.

1. *The Soviet Defense Market*

In all countries, markets for military goods do not work well. This seems to be largely independent of the constitution of the state and, the social and economic system. There is one buyer. Sellers are highly specialized and few in number. Entry is costly. Exit is impeded by the perception that the government will not allow big defense corporations to fail. As a result, defense markets have become notorious for rent seeking and underperformance.

National defense markets are not all the same, however. Jari Eloranta (2009), for example, has researched the historical variation in competitiveness and public accountability of European markets for military goods. The Soviet market had several unique features; both buyer and sellers were state-owned, for example, so that it was to a large extent an artificial or internal market. Despite this, the Soviet economy supported a revealed comparative advantage in military equipment for many decades.

Western observers gave considerable attention to the problem of quality in the Soviet economy. Since the defense sector was shrouded in comprehensive secrecy, their studies began with civilian enterprise.

They found a pervasive privilege for quantity over quality. Sheltered from competition and guaranteed economic survival by state plans, civilian managers faced strong temptations to seek a quiet life for themselves and their employees by fulfilling the plan for least effort (Granick, 1954; Berliner, 1957). The authorities assigned plans in rubles of gross output subject to fixed plan prices and technical specifications. Quality, however, was costly to the producer. Once plans and prices had been written down, the main scope for the factory to economize on effort lay in finding ways to reduce quality that were hard to verify. Kornai (1980) pointed to pervasive shortage and successive waves of mobilisation as conditions that undermined consumer sovereignty and enabled sellers to impose poor quality and narrow assortment on buyers. Davies (1989; 1996) has documented the rapid deterioration of product quality that accompanied the historical transition to the command system.

In the hope of limiting the producers' opportunism, the authorities turned first to industrial self-regulation. Every factory, civilian or military, had its own quality department, following arrangements detailed by Harrison and Simonov (2000) and, Markevich and Harrison (2006). In theory the quality department was responsible for ensuring that the factory's products met contractual standards. Not surprisingly, this was largely ineffective. Managers had only weak incentives to submit to self-regulation. Work in quality assurance brought little pay or recognition; the employees saw themselves as paid mainly to take the rap for management when things went wrong. If they tried to enforce external benchmarks, managers slapped them down.

Western studies of Soviet defense procurement produced a strong contrast with the civilian sector (Albrecht, 1993; Alexander, 1978; Almquist, 1990; Holloway, 1982). Whereas the civilian consumer and industrial user were powerless to refuse substandard goods, the same was not true of the powerful defense ministry, which stationed an army of serving officers on its contractors' facilities — more than 20,000 in number by 1940, as we now know (Harrison and Simonov, 2000). These officers, trained in military engineering and procurement, were empowered to oversee production and assure the quality of deliveries. The management was obliged to support the military agents with necessary accommodation and equipment. Faced with substandard products the military agents could halt acquisition and, if necessary, production itself. Under this regime quality was upheld, although at heavy cost to both seller and buyer, by high rates of mass production and high rejection rates.

Some puzzles remained. One was the extent of corruption among the military purchasing agents. Alexander (1978: 59n) asked, for example, whether the seller would not seek to induce the purchasing agent to betray his superiors and overlook defects, either for reward or just for the sake of goodwill. The Stalin-era archives suggest that the military agents' independence from management was well protected. They were prohibited from accepting rewards or benefits from the side of industry (Harrison and Simonov, 2000). By the standards of the industrial employees alongside whom they worked, they were well paid and not unduly taxed by overtime or other burdens. In fact, corruption allegations were most infrequent; rare reports went straight to the highest level (Markevich and Harrison, 2006).[2]

In short, the defense procurement officials are best seen as loyal agents of their employer.[3] Despite the cost of reserving thousands of serving officers for this work, the defense ministry valued their services highly. In wartime, for example, when the front was crying out for additional commanders, the ministry three times rejected proposals that would have redeployed many military agents to the battlefield.[4]

While clearing up such old puzzles, the archives have presented new ones. One is that the military procurement apparatus was not able to prevent widespread quantitative shortfalls in procurement. There are many possible explanations, but few that stand up to scrutiny. These include planning errors, differences in military and civilian supply chain economics, and a propensity to issue (and accept) a larger volume of defense contracts than could ever be fulfilled.

The scope for planning errors in the defense sector should not be underestimated. Perfect planning was impossible in the presence of uncertainty, asymmetric information, and computational limits. All these factors were certainly at work, and they form the starting point for any account of the general failures of Soviet economic planning. But they do not sufficiently explain the plan failures that were specific to the defense sector. This

[2]What could have been more tempting for any Soviet official in the 1930s than the lure of *la dolce vita*? A review of the work of Soviet military purchasing agents in Italy, dated October 1935, noted that local contractors had offered them bribes to accept substandard goods and that they had rejected the inducements (GARF, 8418/8/175: 95–101).

[3]These were not the only loyal agents in the Soviet economy; for other examples, see Belova and Gregory (2002) and Markevich (2011).

[4]Hoover/RGANI, 6/2/49: 9 (July 7, 1943).

becomes clear when we consider the comparative success of planning in the civilian sector. A striking result from the archives (Gregory, 2003) is that defense procurement plans failed more frequently and by wider margins than civilian plans. This cannot be explained by differences in planning procedures, which were closely similar in military and civilian branches (Markevich, 2008).

Differences in military and civilian supply chains should, if anything, have worked to the advantage of the defense sector. The important factor was the priority system, designed to ensure that, when overall supplies fell short, orders for defense and construction (including defense industry investment) had the first claim on resources (Levine, 1966). Except in wartime, the Soviet defense market was a relatively small part of the economy as a whole. In the mid–1930s, for example, when the behaviours we are studying were already well established, the defense budget was running at one twentieth or one fifteenth of national income, and about one third of the budget went to pay for weapons and equipment (Davies and Harrison, 1997). This element of the budget made disproportionate demands on high quality capital, labor, and materials, but the priority system was designed to protect it. As a result, defense contractors did not have to fight over a fixed pool of resources. For this reason alone, defense targets should have been met more frequently and margins of failure should have been narrower than in the civilian sector.

Lastly, did strategic interaction of some kind lead to systematically more contracts being issued than could ever be actually fulfilled at the rates they were originally planned for? This happened elsewhere, for example in Soviet public investment (Dyker, 1983: 36–39). But in our case, it was not the key thing. Most important, there was no lobbying or drive by the producers to expand the procurement budget beyond the capacity of the economy. Soviet equipment manufacturers did not need defense contracts to survive, did not have to fight to get them, and frequently displayed reluctance to accept them.

When we look more closely we see that defense shortfalls were registered at two stages, corresponding to pre- and post-contract. At the pre-contract stage, the defense ministry was given an annual equipment procurement plan but its agents were typically unable to place enough contracts to meet it. At the end of 1938, for example, the minister of the defense industry asked the government to trim its plan for deliveries in the year ending by 532 million rubles, or more than 4 per cent, appending a list of items that it wished to drop. Among the reasons he gave was that the defense ministry had placed a contract for less than the planned quantity or had failed to place one at all.[5] If defense suppliers were unwilling to accept contracts, it cannot be just that they feared shortages further back in the supply chain, because they knew that a defense contract automatically gave the contractor an entitlement to priority in supply.

Once contracts had been placed, there was a second stage of underperformance. Here, the action of the purchasing agents was often critical; many cases have come to light where they rejected significant fractions of a factory's monthly output on grounds of substandard quality. Again, this cannot be explained by shortages elsewhere in the supply chain; for one thing, if the supply chain was a factor it was the purchasing agent's responsibility to fix it.

When we look closely at post-contract shortfalls, we find a further puzzle. Post-contract shortfalls were often a consequence of the tough action taken by military agents to prevent the acceptance of substandard items. Having taken tough action over quality, however, it was not uncommon for them to collude with industrial suppliers to conceal quantity violations. Why they should have done this is not immediately obvious. It is strongly suggestive of some kind of private renegotiation, but whatever took place was not privately corrupt. On the contrary, we will show that the concealment frequently had the knowledge and approval of superior officials, sometimes on both sides of the market.

Our explanation is rooted in market behaviour, which may seem out of place in a command economy. We do not mean that there was a market relationship between the Red Army and the defense industry as corporate entities. At the highest level there was political bargaining between the defense minister and a few industrial ministers, not a market relationship. The internal market emerged, as Markevich and Harrison (2006) have suggested, at a lower level where military purchasers representing the defense ministry had to match requirements and negotiate terms with individual defense factories.

In the present paper, we suppose that the internal market developed unevenly. By implication, Soviet defense purchasing went on partly within a sphere of command allocation where planners assigned quotas from above, and partly in an internal market that featured search, matching, and voluntary contracting.

---

[5]GARF, 8418/22/463: 46–63 (Mikhail Kaganovich to Molotov, 1938).

Table 1. The Internal Market versus Command Allocation.

Command sphere	Internal market
Established products	New products
Given quality	Variable quality
Random assignment by planner	Search and matching by buyer
Compulsory quotas	Voluntary contracts

The market had less scope where established models were in serial production year after year so that contracts could be planned in advance on the basis of previous experience. But for many lines, including ship, airplane, and vehicle building, defense buyers were continually seeking contractors for new or unique items. Innovation in military machinery was particularly rapid in the mid–1930s and this accelerated the year-on-year turnover of products (Davies and Harrison, 1997). At such times directive planning lost influence because the planner lacked the information to select producers or lay down quality standards and prices. Instead, voluntary contracting took the place of planning. Table 1 summarizes this distinction.

The best evidence of voluntary contracting in the internal market is that it was possible for the seller to refuse a contract. In fact, the defense ministry was forced each year to wage a costly and frustrating military-style "contracts campaign" to place orders for new products with industrial suppliers. Harrison and Simonov (2000: 231) previously identified major obstacles to the contracts campaign as "the difficulty of finding willing suppliers of new defense products, and the desire of industry to secure a relatively homogenous assortment plan that would allow concentration on long runs of main products without a lot of attention to spare parts and auxiliary components, no matter how essential to the customer"; the resistance of potential contractors could go so far as to leave significant orders completely unfilled.

When the placing of contracts ran into difficulties, the government was reluctant to intervene on principle; it insisted that the defense ministry and the defense industry had to sort things out among themselves.[6] When the defense ministry appealed to the government to force state-owned enterprises to accept its orders, the government typically did nothing.[7] Even during accelerated rearmament, industrial firms were able to refuse defense orders with impunity. In 1938, for example, defense industry factory no. 145,

[6]GARF, 8418/4/39: 1 (Rykov to Voroshilov and Kuibyshev, 1930).
[7]GARF, 8418/24/2: 7–9. (Ivanov and Nikolaev to Safonov, 1934).

the sole supplier of two-headed lubricators for artillery, refused a contract on the grounds that its workshop was under reconstruction; meanwhile, it started to sell off unique equipment required for the manufacture of these products.[8]

2. *The Hold-Up Problem*

The most important problem in the Soviet military market was the quality of weapons. Both quantity and quality can be observed. But they differ in the time and effort required to observe and verify them. Quantity is easily observed and verified, and quality less so. Harrison and Markevich (2006) considered the problem that arose when quality was not freely observable before purchase: the defense ministry used its army of purchasing agents to slow down acquisition and learn about quality from delay. In contrast, we will examine the problem that arises when quality, although observable before purchase, cannot be verified: the buyer is aware of substandard quality but cannot prove it.

We find analytical foundations in various literatures, in first place that on contractual incompleteness. The problem of quality arises when quality is variable and contractual quality requirements cannot be written down or, if written down cannot be observed or, if observable cannot be enforced. A considerable literature from Grossman and Hart (1986) and Hart and Moore (1988) has considered the implications. Hart and Moore (2008: 6), for example, "suppose that trade is only partially contractible. Specifically, we distinguish between perfunctory performance and consummate performance, or performance within the letter of the contract and performance within the spirit of the contract. Perfunctory performance can be judicially enforced, while consummate performance cannot."

Contracting for quality is a recognised problem in defense markets. Defense orders for complex machinery fit perfectly the class of contracts that can be fulfilled either consummately or perfunctorily. In practice the economic literature on defense procurement is largely policy-driven and ephemeral.[9]

Hold-up problems in the spirit of Goldberg (1976) and Williamson (1985: 61–63) are a feature of defense markets where, to realize the

[8]RGAE, 7515/1/404: 46–53 (Savchenko to Mikhail Kaganovich, brother of the more famous Lazar, 1938).

[9]But there are exceptions, *e.g.*, Tirole (1986) and Rogerson (1994).

benefits of a potential exchange, buyer and seller sometimes invest in costly relationship-specific assets. In market economies, there is a typical form (Rogerson, 1994). The defense contractor must invest in specialised R&D and production facilities to produce military goods for which there is only one potential buyer. If the relationship breaks down, these assets will be less valuable in their best alternative use. As a result, the contractor becomes vulnerable to hold-up by the government buyer.

The risk presented by the hold-up problem is that, in order not to be held up, agents will avoid investing in the specific assets that make them vulnerable; as a result, society as a whole will lose the gains from trade. The hold-up problem has standard solutions, however, that should bring the incentives of buyer and seller back into approximate alignment (Schmitz, 2001). One is vertical integration, which brings the parties together under a single authority and completely replaces their market relationship by hierarchy. Other solutions retain the market relationship but regulate it by long-term contracts with some combination of joint financing of initial joint costs and contingent rules for distributing the subsequent benefits.

In both world wars, for example, the vast capital expansion of war industries in each major market economy had to overcome business resistance and was largely government-financed for this reason. During World War II, the U.S. Defense Plant Corporation and other federal agencies provided and afterward wrote off capital facilities for war production to the private sector valued by Gordon (1969) at $45 billion (at 1958 prices).

In the Soviet defense market, we see the hold-up problem with a difference. Under capitalism, the defense producer is vulnerable to being help up by the government buyer. Under socialism, it was the other way round. The Soviet government generally made research and production facilities available to every industrial firm free of charge; this weakened the hand of any buyer, including the government itself. The buyer of new high-grade or high-tech products was vulnerable, on the other hand, because of the cost of searching and matching with a producer. The annual "contracts campaign" for new defense orders provides evidence of this.

Suppose the Soviet defense buyer faced a fixed selection cost, which was also the cost of switching from one seller to another. Once formed, the relationship with the particular seller was then worth at least this fixed cost to the buyer and this value is what the buyer stood to lose if the initial relationship broke down. It became part of the quasi-rent that was the profit created by the relationship, but it did not necessarily accrue to the buyer. Who actually received it depended on post-contract renegotiation. The seller could hold up the buyer: by threatening to withdraw from the relationship, the seller faced the buyer with a potential loss at least equal to the switching cost so the buyer would be willing to pay the seller up to this value to avoid the loss. The extent of the transfer depended on the relative bargaining strengths of the two sides; the party with more to lose was more likely to lose it.

We suggest that search costs in the internal market left the Soviet defense buyer vulnerable to hold-up by the seller. Given this, what form did the hold-up take? Under Soviet arrangements, once the contracts campaign was over, the main opportunity for the seller lay in undershooting on quality, knowing that the buyer would have difficulty in taking its business away.

Standard solutions to this problem were unavailable in the context of Stalinist dictatorship. Stalin ruled out vertical integration of the buyer and seller because he did not want to encourage the formation of a powerful military–industrial complex. In fact, when military interests advocated integration with the defense industry, Stalin opposed it. In 1927, army commanders Tukhachevskii, chief of the general staff, and Unshlikht, a member of the Revolutionary Military Council, sought new powers for the Red Army; they wanted to oversee appointments to the defense industry, plans and reports of defense factories, and plans for capital investment in the industry. Stalin rejected these proposals (Harrison and Simonov, 2000: 230; Samuelson, 2000: 42–47). Tukhachevskii's subsequent resignation as chief of staff was most likely because his project to control the defense industry had failed (Samuelson, 2000: 55–59). As for Stalin's motivations, he built his power on divide-and-rule and this included keeping soldiers and industrialists at odds (Harrison, 2003).

An intermediate solution to the hold-up problem is long-term contracting. Under Soviet rules, however, all contracts were rewritten at least once a year (Markevich, 2008); no long term contract was worth more than the paper it was printed on. In any case, Stalin could not credibly promise to uphold long-term agreements between his agents to share gains from trade since he had the power to break any contract and expected to confiscate all gains for the state.

Because of these limitations the buyer could not eliminate the harm threatened by the seller's behaviour. The damage could be limited, however, by an illegal compromise. The compromise required conspiracy between the parties since, if not hidden, it would trigger an intervention by the dictator that both parties wished to avoid.

3. A Model

We illustrate with a simple dynamic game of three moves: the defense buyer's search, the contractor's hold-up, and a readjustment.

There are two players, Defense and the Contractor. Defense aims to maximise the military utility of its available equipment, derived from some expectation of its effectiveness in use on the battlefield. This effectiveness is decided by their quantity and quality. Defense seeks a Contractor to supply units of a particular weapon, a tank or an airplane for example, that can be of high or low quality. The Contractor aims to maximise a surplus over costs. This surplus could be measured equally well in monetary terms or in terms of the reduced effort required to extract Defense's cash; which is more appropriate depends on the form in which rents were shared within the Soviet industrial firm, but does not matter for present purposes.

Off stage is a Dictator, whom we do not model explicitly.[10] The Dictator shares Defense's preferences over quantity and quality; when the players have contracted between themselves, he validates the contract and will punish those contract violations that can be verified.

In our model, only quantity is verifiable. The quality of its execution, therefore, is the difference between consummate and perfunctory performance of the contract. Perfunctory performance can be enforced, while consummate performance cannot. In our model consummate performance is more costly to the producer, and perfunctory performance is less valuable to the buyer.

Our supply side has two sectors. There is a sphere of command allocation in which a high-priority buyer such as the defense ministry can routinely procure goods of low quality; that is, on being notified of military requirements, the planner can assign the quota to a producer that, randomly selected, provides low quality and low cost. Items of high quality are available only from an internal market that the buyer must search so as to match with a contractor by mutual agreement.

Before the game, the Dictator endows Defense with a fixed sum of rubles for the procurement of weapons. We normalise this budget to $1 + \bar{S}$ where $\bar{S}$ is the fixed cost of searching for a high-quality contractor; $\bar{S}$ is payable if and only if Defense chooses to search.

Figure 1 illustrates the timing of the game. At various points Defense must take its budget, or what is left of it, either to the command sphere (to

[10] Markevich (2007) suggests how this might be done.

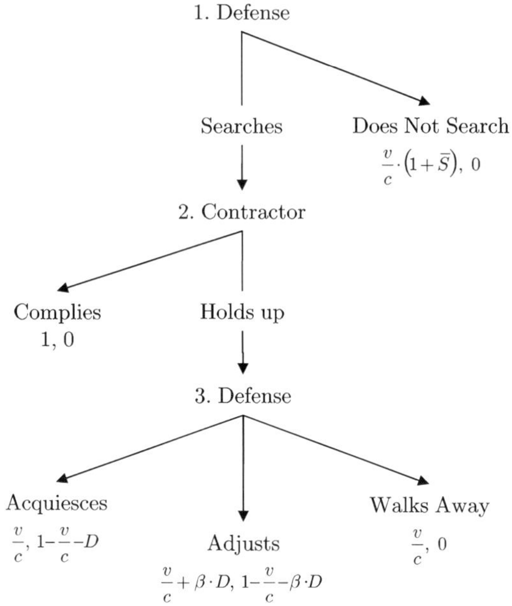

Note: Payoffs are (Defense, Contractor).

Symbols:

β	Bargaining power of Defense
c	Contractor's relative cost per unit of low quality: $1 > c > 0$
D	Deadweight loss from hold-up
$\bar{S}$	Defense's fixed cost of search
v	Defense's relative utility per unit of low quality: $1 > c > v > 0$

Fig. 1. Timing and Payoffs.

the right) or to the market (down and to the left). At the first move, Defense can seek to allocate this budget to weapons of High and Low quality, and Low has the advantage over High that it is available without search. If Defense chooses at the outset to procure goods from the command sphere without searching, its entire budget of $1+\bar{S}$ can be spent on Low. Under the prevailing law of contracts and prices, the low-quality items are acquired at cost, so the producer receives a zero surplus.

Alternatively, Defense can enter the market for High by paying the search cost $\bar{S}$ as an entry fee. Down this path, Defense and the Contractor agree terms for a contract, but $\bar{S}$ is now a sunk cost so Defense can commit only 1 to the purchase of High.

Down the path to the internal market, the Contractor must deliver units of High to Defense in contractually agreed quantities and qualities, for example, 500 tanks of a specific model, subject to agreed technical and performance standards, at prices fixed on the basis of the Contractor's production costs. In that case, the Contractor again covers its costs and receives zero.

For the Contractor, *consummate fulfilment is no better than loss of the contract.* Perfunctory fulfilment is the only way the Contractor can extract a surplus of revenue over costs, and this must be at the expense of Defense. The Contractor's surplus could be gained by delivering the agreed number of tanks produced with lower standards of care or quality of materials, for example. As a result the quality standards agreed in the contract would be violated, and the tanks would be of less military value than Defense anticipated.

In the second move along this path, the Contractor chooses between consummate and perfunctory execution of the contract, and perfunctory execution is its dominant strategy.

Given perfunctory fulfilment, the third move belongs to Defense, which at first sight must either acquiesce (to the left), or walk away (to the right), leave the market and return to the command sphere where it can spend its remaining budget on Low. This option provides Defense with its reservation utility, which turns out to be v/c. Defense will acquiesce provided its loss from perfunctory fulfilment is kept within a limit set by this reservation utility.

The Dictator will also acquiesce provided that the total number of units of any quality remains the contracted number. It is true that he shares Defense's preference for consummate fulfilment, and in principle Defense could appeal to the Dictator for justice, but the latter could observe only the quantity, not the quality of fulfilment. Or, in other words, we assume that the probability of proving perfunctory fulfilment in court was zero.

We will show that the Contractor's gain is less than the harm done to Defense; there is a deadweight loss D. Based on an intuition of Harrison and Markevich (2008), our argument is that the hold-up involves the substitution of Low for High, and the relative cost of Low must exceed its relative utility ($c > v$) since otherwise Low would always be preferred. This creates an opportunity for Defense to lessen the harm done and offer a benefit to both parties in further renegotiation. A final round of bargaining eliminates the deadweight loss by restoring quality at the expense of quantity. The two parties share the gain in proportion to their mutual bargaining power, β for Defense, so $1-\beta$ for the Contractor. Because bargaining is at the expense of quantity, however, there is no longer even perfunctory fulfilment. The fact that the contract is no longer fulfilled in quantity risks the intervention of the Dictator. Only Defense can give permission for this to go forward, and must collude with the Contractor to hide it.

We now analyse the players' possible payoffs. At the outset, the Dictator gives Defense a budget of $1 + \bar{S}$ rubles to procure a stock of a particular weapon. Defense has preferences over weapons that can be of high or low quality based on their military applications, so its (military) utility function is:

$$V = H + v \cdot L \tag{1}$$

Here H and L are the (non-negative) quantities of High and Low respectively. For calibration each unit of H gives Defense one unit of utility, and v is the relative utility of a unit of L. Defense maximises V, subject to procurement costs C remaining within its budget, but the amount of its budget that will be available to spend on procurement depends on whether or not Defense engages in the costly search for a high quality supplier. Thus:

$$\text{No search: } C \leq 1 + \bar{S} \tag{2}$$

$$\text{Search: } C \leq 1 \tag{2a}$$

Meanwhile, procurements costs are also the Contractor's production costs:

$$C = H + c \cdot L \tag{3}$$

Again for calibration, each unit of H costs the Contractor one ruble and c is the relative ruble cost of a unit of L. We assume that the unit value and cost of High exceed the unit value and cost of Low respectively and, without loss of generality, that the relative unit cost of Low to the seller exceeds its relative value to the buyer:

$$1 > c > v > 0 \tag{4}$$

How much utility is available to Defense depends, first, on its choice to search or not. If Defense declines to search and sets out to procure only Low, given the cash available, and combining equations (1) and (3) with inequality (2), it follows that its utility and procurement will be:

$$V_{no\text{-}search} \leq \frac{v}{c} \cdot (1+\bar{S}) \quad \text{and} \quad L_{no\text{-}search} \leq \frac{1}{c} \cdot (1+\bar{S}) \tag{5}$$

If Defense searches the market and finds a Contractor for High, in contrast, the utility available to it from a contract is found by combining equations (1) and (3) with inequality (2a):

$$V_{search} \leq 1-(c-v) \cdot L \tag{5a}$$

From (4), and using stars to denote the terms of Defense's optimal contract, this expression is maximized when $L = 0$, so:

$$V^* = 1; \quad H^* = 1; \quad L^* = 0 \tag{6}$$

We find the motivation for the subsequent hold up in the Contractor's problem: since its contract revenue cannot exceed 1, it can create a surplus only by reducing total costs (3) below the contracted level. The Contractor will covertly substitute units of Low for the same number of units of High, one for one, up to a limit set by two constraints. One is Defense's participation constraint, and the other is the Dictator's intervention constraint. We assume that, while the Dictator shares Defense's preference for high quality, he does not observe the substitution; he sees only the number of items delivered. He has no basis to intervene as long as this number matches the quantity H^* stipulated in the contract. The Contractor is safe while it delivers $H + L \geq H^*$. Recall that $H^* = 1$, so we will write the Dictator's *intervention constraint* as:

$$H \geq 1 - L \tag{7}$$

As for keeping up the contractual relationship, the Contractor must maintain Defense at or above its *reservation utility*. This is the utility that would be available if Defense now stopped trying to procure High, walked away from the contract, returned to the command sphere, and asked the planner to assign a supplier of Low. Combining equations (1) and (3) with inequality (2a) and setting $H = 0$ yields the utility from spending 1 in this way as v/c and this is Defense's reservation utility, which in turn leads to Defense's *participation constraint*:

$$H \geq \frac{v}{c} - v \cdot L \tag{8}$$

Combining equations (7) and (8), and using primes to signify the values obtaining in the hold-up, Defense will remain within the relationship *and* the Dictator will not intervene as long as:

$$L' \geq \frac{1}{c} \cdot \frac{c-v}{1-v} \quad \text{and} \quad H' \geq \frac{v}{c} \cdot \frac{1-c}{1-v} \tag{9}$$

In the hold-up the Contractor reduces costs by shading on quality, substituting units of Low for High, one for one, to the maximum extent permitted by inequality (9). At this point, Defense must know it is being cheated but will not walk away; the Dictator may well observe customary bickering among his agents but will not intervene.

Payoffs to each side now vary from those anticipated in the contract. The relative magnitudes are driven by inequality (4): the comparative cost of Low exceeds the comparative utility of High, and so $1 > c > v > 0$. Military utility was previously 1. The Contractor's perfunctory fulfilment drives it down to the reservation level v/c. The Contractor's utility was previously zero. Now it increases because the substitution of Low for High reduces total costs that, under consummate fulfilment, were also previously equal to 1. The reduction in total costs, which is the Contractor's gain, may be computed from equations (3) and (9) as $1 - \frac{v}{c} \cdot \frac{1-c}{1-v} - \frac{c-v}{1-v}$.

The chief proposition of the model is that in the hold-up, total costs fall by less than the reduction in military utility, so there is a *deadweight loss*. The loss is found by subtracting the Contractor's gain from Defense's loss, that is:

$$D = 1 - \frac{v}{c} - \left(1 - \frac{v}{c} \cdot \frac{1-c}{1-v} - \frac{c-v}{1-v}\right) \tag{10}$$

which simplifies to $\frac{(c-v)^2}{c \cdot (1-v)}$. From $1 > c > v > 0$ we can check that $1 > \frac{(c-v)^2}{c \cdot (1-v)} > 0$ and so D is a positive fraction.

Anticipating such losses, what could Defense do? Deprived of the remedies available in a law-governed market economy, Defense's last resort was to strike a bargain with the Contractor to restore quality at the expense of quantity, eliminate the deadweight loss, and share the gain from doing so. The gain was shared according to the players' relative bargaining power. Defense was still held up by the Contractor, but less inefficiently than otherwise. In the process the Dictator's intervention constraint was violated and so the readjustment was sustainable only if the players colluded to conceal the violation.

First, we define limits on the bargain. At one limit, defense could keep the entire gain by holding the Contractor's costs constant. Total costs at hold-up, producing some low quality items, were $\frac{v}{c} \cdot \frac{1-c}{1-v} - \frac{c-v}{1-v}$ (equation 10). We use double-primes to denote the values resulting from the readjustment. After eliminating Low altogether from the package, the Contractor could increase the delivery of High to $\frac{v}{c} + D$ without any increase in cost. At the other limit, after eliminating Low, the Contractor could increase the delivery of High to v/c without any loss of utility to Defense. The actual outcome can be thought of as a weighted average of the two bounds arising from the exogenous bargaining power β of Defense, so:

$$H'' = \frac{v}{c} + \beta \cdot D; \quad H^* > H'' > H' \tag{11}$$

Whatever bargain is struck, however, the plan is being violated in quantity since $H^* > H''$, and the final readjustment moves both players into a region where the Dictator would intervene on being informed. The Contractor faces a clear risk: after the bargain, Defense can denounce it to the Dictator for breaking the contract in quantity, and the Dictator can confiscate the Contractor's surplus or worse. To be willing to strike the bargain, the Contractor must bind Defense not to denounce it afterward, and Defense must be willing to be bound. There has to be a mechanism for collusion: Defense must join a conspiracy that hides not just the hold-up but also the bargain that restores quality at the expense of quantity. Otherwise, both sides will have to accept the deadweight loss and remain at the hold-up.

Weighing up all possible outcomes of the game, we learn about Defense's best initial choice. Having chosen to search for high quality, the best outcome for Defense is consummate fulfilment of the contract, which pays 1. But only perfunctory fulfilment pays the Contractor a surplus, so that is what the Contractor will always deliver. Backward induction then tells Defense whether or not to search in the first place. Given its own preference for high quality and that the Contractor will never deliver it in the spirit of the contract, Defense should search anyway if it can still claw enough back from the illicit bargain that follows the search. The unconditional expected payoff from search, contract, hold-up, and renegotiation, $\frac{v}{c} + \beta \cdot D$, must exceed the payoff $\frac{v}{c} \cdot (1 + \bar{S})$ from going straight to the command sphere without searching first; simplifying,

$$\text{Search if } \beta \cdot c \cdot D > v \cdot \bar{S} \tag{12}$$

In words, Defense should play Search in the market for High only when its relative bargaining power, the relative cost of Low, and the potential deadweight loss (which stands for the scope for bargaining after the hold-up), are large enough to offset the small relative utility of Low and the high cost of searching.

4. *Bargaining Power*

What decided bargaining power at the stage of renegotiation? The defense ministry's main bargaining instrument in the market for weapons was its purchasing officers, their inside knowledge of the contractor, and their capacity to threaten a quantitative shortfall in the contract.

The archives testify with notable consistency (Harrison, 2003; Markevich and Harrison, 2006) that industry's officials and representatives saw the military agents in an adversarial light. This hostility arose because the military agents made frequent use of the main instrument at their disposal for enforcing quality, their right to reject substandard deliveries. When they rejected deliveries they endangered the ability of the defense contractors to show compliance with supply plans and contracts, and correspondingly threatened the careers and personal security of the industrial managers.

Military agents were certainly willing and able to exercise these powers; many cases are reported when high proportions of monthly deliveries were returned or scrapped, sometimes up to 100 per cent. In January and February 1934, for example, the Tula gun factory produced 3,000 carbines and 106 ShKAS machine guns, but only 800 rifles were accepted for the defense ministry and no machine guns at all. The 3,000 carbines "were presented for acceptance 23,000 times, almost eight times per carbine on average."[11] In March 1938, the military agents rejected the entire monthly output of defense industry factory no. 205 "in view of the totally unsatisfactory installation of electric plugs in all articles supplied."[12] This degree of screening was much tougher than that arising from industrial self-regulation. Among the aircraft that the quality department of factory no. 126 passed in 1940, for example, the military agent found up to 80 defects.[13] In the first nine months of 1940, of 6.6 million shell cases produced at munitions factory no. 184 the quality department scrapped

[11] Hoover/RGANI, 6/1/22: 34 (1934); original emphasis omitted.
[12] RGAE, 7515/1/404: 158 (Savchenko to M. Kaganovich, 1938).
[13] Hoover/RGANI, 6/2/27: 108 (1940).

less than 3 per cent; after that, the military agent scrapped a further 10.5 per cent.[14]

The power of the military agent to reject on grounds of quality was nonetheless more limited than might appear at first sight. One reason was the inability to verify observed quality to others. Military agents rarely looked to higher authority to impose punishments for low quality, and when they did they were typically unsuccessful. In 1933, for example, a military agent tried to use the party committee of aircraft factory no. 24 to bring to account those responsible for "malicious toleration of defective parts", but without success.[15] We have found only one case that, of naval armament factory no. 347, where a military agent took the managers to court on criminal charges of supplying substandard goods; the court cast doubt on the accusations and the file was returned for further enquiries. A review by the ruling party's control commission found that the judicial route was inappropriate and substituted dismissal for the criminal charges.[16]

Another limitation on the authority of the agent was that the contractor had avenues of appeal, and not infrequently succeeded in shifting the blame for delivery problems onto the military agents. In the case of the rejected output of the Tula gun factory in 1934 control commissioners concluded that "discord between management and representatives of military acceptance on the score of product quality" lay behind persistent plan breakdowns.[17] In 1944, the control commissioner for the Khabarovsk region reported that "vexatious litigation", with managers on one side and the quality department and military agents on the other, had taken hold of aircraft factory no. 126 on the issue of parts and components that did not conform to the blueprints. "These disputes . . . sometimes drag on for weeks . . . while business stands still."[18] If agents demanded inflexible adherence to standards, they laid themselves open to criticism for excessive zeal or caution. A control commission report of 1940, for example, condemned the quality department and military agent at aircraft factory no. 126 for "a tendency to over-insurance."[19] Surveying the work of military agents in 1943 the control commission demanded that "the military agent should in most cases rule on

[14] Hoover/RGANI, 6/2/34: 158–159 (1940).
[15] Hoover/RGANI, 6/1/91: 10 (1934).
[16] Hoover/RGANI, 6/6/1616: 128 (1941). For more about the party control commission, see Markevich (2001).
[17] Hoover/RGANI, 6/1/22: 36 (1934).
[18] Hoover/RGANI, 6/2/27: 108–109 (1940).
[19] Hoover/RGANI, 6/2/27: 109 (1940).

the acceptability of one or another deviation [from standards] so as not to delay products for the front."[20] Thus, while military agents may have tried not to accept unserviceable goods, there was pressure on them to tolerate some level of defects.

Finally, the army needed weapons. In the years of urgent prewar rearmament, equipment supplied to military units often turned out to be unfit for service although the military agents had previously passed them as acceptable. Since high-ranking officials of the defense ministry were sometimes complicit in this, it must be supposed that for the armed forces, too, quantity was sometimes more important than quality. Chief of the air force purchasing administration Efimov, for example, was accused of colluding with malpractices: "not only did [he] not take measures to restore order but [he] even suppressed criticism of the defects, describing the communists who raised the criticisms as 'cry-babies' and threatening them with dismissal."[21] Further confirmation is found in evidence that standards were allowed to slip further in wartime (Markevich and Harrison, 2006).

In short, when military agents did not rigorously enforce defense ministry guidelines on substandard equipment, the main reason was that, as loyal agents of the armed forces, they could not reject everything that industry supplied.

5. *Collusion*

To summarise, the defense ministry tried to reach across the market for weapons by deploying thousands of military engineers to the factories of the defense industry. These agents had a dual role. Their first duty was to prevent the defense ministry from being held up and to enforce its contracts. They monitored the process of contract fulfilment with special regard to quality, and aimed to reject items for purchase when their quality fell below some threshold level. The work of the military agents made the quality of military goods to a large extent observable at the time of purchase. When contractors tried to cut the supply of high-quality items, the defense ministry sought to prevent their replacement by low-quality items, and this opened contractors up to penalties for defaulting on quantity.

In practice, however, the military agents and their superiors tended not to make trouble for the contractors over quantitative shortfalls. This

[20] Hoover/RGANI, 6/2/49: 9 (1943).
[21] Hoover/RGANI, 6/2/17: 47 (1939).

suggests that the contractor's cooperation was available at a price: the defense ministry had to accept shortfalls on quantity and help conceal them from the dictator's prying eyes.

In setting out the hold-up problem in the defense market, we made two predictions. First we suggested that, when held up by the seller, the buyer would find it more important to uphold quality than quantity. Second, we suggested that buyer and seller would be jointly interested in collusion to conceal the resulting shortfall on quantity. Consistent with these expectations, we find that military agents typically took a harder line over quality than quantity. They were ready to offer some leeway to contractors over quantitative fulfilment as the price for maintaining quality. The outcome was a bargain which fell short of the initial contract but, by restoring quality at the expense of quantity, was more efficient than the contractor's initial post-contract offer. The defense ministry was still held up, but less inefficiently than would have happened otherwise, and in return allowed its agents to help conceal the contractor's otherwise verifiable shortcomings.

Procurement delays were often concealed. The control commission archive contains many cases of reports falsified by both civilian and defense enterprises. Usually, output was overstated within the accounting period by including *pripiski*, goods that did not exist yet but would be produced in the next period. In short, *pripiski* allowed the enterprise to claim fulfilment of the plan and the associated reward by "borrowing" future deliveries.

This practice involved criminal deception. A single enterprise could not undertake it successfully in isolation, therefore; ministerial superiors had to know about it and the customer had to go along with it in silence. The wider the circle involved, the greater were the risks of disclosure. Despite such risks, however, in the seller's market for civilian goods the power of suppliers was often enough to win the cooperation of both superiors and purchasers (Berliner, 1957).

A report of 1946, for example, claimed that a tank factory director "is systematically engaging in the *pripiska* of goods that have not finished production" and that his superior, although aware of this, "has not only not prevented but has even rewarded it."[22] Similarly, the control commission found that in 1944 officials of the armament ministry told a factory director "to report inflated information to the ministry."[23] In September 1944, the

[22]Hoover/RGANI, 6/2/98: 81, 85 (1946).
[23]Hoover/RGANI, 6/2/67: 11 (1944).

control commission acknowledged that *pripiski* were widespread: in 1943 and 1944 an armament factory had "continually reported falsely inflated information about the fulfilment of the factory's program, typically using from five to 20 days of the following month to complete production"; an aircraft factory had reported "incorrectly inflated information about plan fulfilment" in 1943 and for the months of January, February, and March 1944; the managers of a tank factory "have also been deceiving the government and ministries by reporting false information on the fulfilment of the production program."[24] There were even *pripiski* in a vehicle repair factory of the defense ministry itself; the ministry's vehicles administration, while "aware of all the factory's shortfalls and lack of management, took no measures to overcome them."[25]

Widespread *pripiski* indicate a systematic tendency for industry to ignore delivery deadlines: goods were regularly delivered to the defense ministry a month or more late. The military agents had to be aware of this; it could never have happened without their knowledge. Yet, they virtually never took independent action to enforce deadlines. Of all the cases of *pripiski* that the control commission uncovered, only two were reported by military agents. In September 1941, a military engineer reported an unacceptable delay in an order for gas protection equipment placed with the ministry of general engineering.[26] The control commission's intervention secured a new deadline for the order, but no penalty for the delay. In 1943, a military agent and his senior technician reported on "deception and irregularities" at an electrical factory; this led to a special audit commission which confirmed the various violations.[27] We have found no other cases.

Auditors from the control commission themselves uncovered other *pripiski*. When they did so, they found that the military agents had colluded in the deception. In 1942, for example, the control commissioner for Sverdlovsk district found evidence of large-scale *pripiski* for September, October, and November at the Uralmash factory not just "with the ministry's knowledge" and "on the instruction" of the minister and deputy minister, but also with the collusion of the military agent.[28] This collusion could extend into the upper reaches of the defense ministry itself. In

[24]Hoover/RGANI, 6/6/1583: 10–13 (1944).
[25]Hoover/RGANI, 6/6/1583: 31 (1948).
[26]Hoover/RGANI, 6/6/47: 18 (1941).
[27]Hoover/RGANI, 6/2/55: 1–2 (1943).
[28]RGAE, 8752/4/108: 151-151ob (1942).

1944, for example, the military agent joined the director of an armament factory in signing a cable reporting 101.5% fulfilment of the April program when both knew this to be false since it included items delivered only in May. Significantly, higher officials representing *both seller and buyer* had approved the *pripiska* by 30 April.[29] This was clearly not an isolated case. The same officials justified their action on the basis of precedent; the defense ministry representative noted that he had approved similar arrangements in other cases "to avoid a breakdown of the plan and provision for the needs of the troops."[30]

In short, deadlines for the supply of armament seem to have caused little anxiety to military agents; even their superiors were ready to approve delays. The military agents did have to *look* as if they supported firm deadlines. This led them to collude with enterprise managers in falsifying reports of plan fulfilment. In return, they extracted cooperation over quality.

Contracting for quality in civilian branches of the Soviet economy suffered from similar difficulties but appear to have lacked the same opportunity to substitute quality for quantity through post-contract renegotiation. Civilian users of industrial equipment, for example, had little institutional capacity for search or power to bargain. Civilian supply ministries had no analogue to the statutory powers of the defense ministry's purchasing agents; while they could and did employ private agents, these agents had no right to reject deliveries, and consequently had little or nothing to offer for the purposes of post-contract renegotiation. Any attempt to renegotiate was likely to give the civilian buyer fewer goods of the same low quality as before. In short, civilian users and consumers typically had to take what the system gave them, regardless of quality. Finally, the dictator was less troubled by the problem of contracting for quality in civilian branches since the price was paid by his citizens in their living standards; this affected his security indirectly, rather than directly in the case of defense.

Conclusions

We have looked at the problem of contracting for quality in the Soviet defense market. We have shown that in this market it took the typical form of a hold-up: the seller held up the buyer by defaulting on quality.

---

[29] Hoover/RGANI, 6/2/63: 159–160 (1944).
[30] Hoover/RGANI, 6/2/63: 21 (1944).

The historical setting of a harsh, personal dictatorship ruled out the conventional solutions recommended by economic theory: vertical integration was not in Stalin's political interest, and long-term contracting was ruled out by the discretionary logic of allocation under a dictator. Imperfectly informed, however, Stalin was unwilling to intervene directly to enforce quality standards. Instead, he allowed the defense ministry to manage its problem by other means, deploying agents through industry to observe quality, reject substandard goods, and threaten industrial contractors with an easily verifiable shortfall on quantity.

The defense ministry willingly employed thousands of purchasing agents and paid them well for their loyalty, even in the midst of a total war. These agents, however loyal, still had to reach a compromise with the industrial contractors. Although they were often the injured party, the military agents turned out to be better off bargaining behind Stalin's back than asking him for redress. In return for greater cooperation on quality, they would typically agree to overlook the contractor's quantity violations — and, by concealing them, to keep Stalin out of the picture.

Although the defense market lay at the heart of the Soviet command economy, and was relentlessly scrutinized by overlapping auditing and monitoring agencies that reported directly to the dictator, it turns out to reveal some of the limits on Stalin's power. Although he could and did direct the lives and deaths of millions, he did not have absolute power over the implementation of a military contract.

This story also throws light on the evolution of institutions. According to North (1990), institutions are rules of the game, designed and reformed through politics; being imposed from above on economic agents, they constrain future development irrespective of their efficiency. In contrast Greif (2006) has analysed institutions as equilibrium outcomes of games. In this view they evolve through trial and error, rather than arising from acts of deliberate creation. The institutions in our case appear to reconcile both approaches. The dictator creatively designed the apparatus of procurement agents to overcome the difficulty of contracting for quality in the Soviet defense market; where analogous problems existed in civilian branches, attempts to solve them in the same way did not work or were not allowed to work. The mechanism of renegotiation that led to the solution in the defense market, however, was far from that initially projected. There can exist, in the words of Hayek (1973), a "made" order of society, and at the same time the equilibrium solution can evolve in directions other than the one the designers intended.

References

Archival Collections

GARF. State Archive of the Russian Federation (Gosudarstvennyi Arkhiv Rossiiskoi Federatsii), Moscow.

Hoover/RGANI. Hoover Institution (Stanford, California), documents from the Russian State Archive of Recent History (Rossiiskii Gosudarstvennyi Arkhiv Noveishei Istorii), Moscow.

RGAE. Russian State Economic Archive (Rossiiskii Gosudarstvennyi Arkhiv Ekonomiki), Moscow.

Published Works

Albrecht, U (1993). *The Soviet Armaments Industry.* Chur (Switzerland): Harwood Academic Publishers.

Alexander, AJ (1978). *Decision Making in Soviet Weapons Procurement.* Adelphi Paper no. 147/148. London: International Institute for Strategic Studies.

Almquist, P (1990). Red Forge: Soviet Military Industry Since 1965. New York: Columbia University Press.

Belova, E and Gregory PR (2002). Dictator, Loyal and Opportunistic Agents: the Soviet Archives on Creating the Soviet Economic System. *Public Choice,* 113(3–4): 265–286.

Berliner, JS (1957). *Factory and Manager in the USSR.* Cambridge, MA: Harvard University Press.

Davies, RW (1989). *The Industrialisation of Soviet Russia: The Soviet Economy in Turmoil, 1929–1930.* Volume 3. Basingstoke, England: Macmillan.

Davies, RW (1996). *The Industrialisation of Soviet Russia: Crisis and Progress in the Soviet Economy, 1931–1933.* Volume 4. Basingstoke, England: Macmillan.

Davies, RW and Harrison, M (1997). The Soviet Military-Economic Effort under the Second Five-Year Plan (1933–1937). *Europe-Asia Studies*, 49(3): 369–406.

Dyker, DA (1983). *The Process of Investment in the Soviet Union.* Cambridge: Cambridge University Press.

Eloranta, J (2008). Rent Seeking and Collusion in the Military Allocation Decisions of Finland, Sweden, and the UK, 1920–1938. *Economic History Review*, 62(1): 23–44.

Goldberg, VP (1976). Regulation and Administered Contracts. *Bell Journal of Economics*, 7(2): 426–52.

Gordon, RJ (1969). $45 Billion of U.S. Private Investment Has Been Mislaid. *American Economic Review*, 59(3): 221–38.

Granick, D (1954). *Management of the Industrial Firm in the USSR.* New York: Columbia University Press.

Gregory, PR (2003). Soviet Defence Puzzles: Archives, Strategy, and Underfulfilment. *Europe-Asia Studies*, 55(6): 923–38.

Greif, Avner (2006). *Institutions and the Path to the Modern Economy: Lessons from Medieval Trade.* Cambridge University Press.

Grossman, SJ and Hart OD (1986). The Costs and Benefits of Ownership: A Theory of Vertical and Lateral Integration. *Journal of Political Economy,* 94(4): 691–719.

Harrison, M (2003). Soviet Industry and the Red Army Under Stalin: A Military-Industrial Complex? *Les Cahiers du Monde russe,* 44(2–3): 323–42.

Harrison, M and Markevich, A (2008). The Market for Weapons. In: Harrison M (ed.) *Guns and Rubles: The Defense Industry in the Stalinist State.* New Haven, CT: Hoover Institution and Yale University Press, pp. 156–179.

Harrison, M and Simonov NS (2000). Voenpriemka: Prices, Costs, and Quality Assurance in Interwar Defence Industry. In: Barber J and Harrison M (eds.) *The Soviet Defence-Industry Complex From Stalin to Khrushchev.* Basingstoke, England: MacMillan, pp. 223–45.

Hart, OD and Moore, J (1988). Incomplete Contracts and Renegotiation. *Econometrica,* 56(4): 755–785.

Hart, OD and Moore J (2008). Contracts as Reference Points. *Quarterly Journal of Economics*, 123(1): 1–48.

Hayek, FA (1973). *Law, Legislation, and Liberty.* Volume 1. Chicago: University of Chicago Press.

Holloway, D (1982). Innovation in the Defence Sector. In: Amann R and Cooper J (eds.) *Industrial Innovation in the Soviet Union.* New Haven, CT: Yale University Press, pp. 276–367.

Kornai, J (1980). *The Economics of Shortage.* 2 vols. Amsterdam: North-Holland.

Levine, HS (1966). Pressure and Planning in the Soviet Economy. In: Rosovsky H (ed.) *Industrialization in Two Systems: Essays in Honor of Alexander Gerschenkron.* York: Wiley, pp. 266–286.

Markevich, A (2007). *The Dictator's Dilemma: to Punish or to Assist? Plan Failures and Interventions under Stalin.* The Warwick Economic Research Papers no. 816. University of Warwick, Department of Economics.

Markevich, A (2011). How Much Control is Enough? Monitoring and Enforcement Under Stalin. *Europe-Asia Studies*, 63(8): 1449–1468.

Markevich, A (2008). Planning the Supply of Weapons: the 1930s. In: Harrison M (ed.) *Guns and Rubles: The Defense Industry in the Stalinist State.* New Haven, CT: Hoover Institution and Yale University Press, pp. 78–117.

Markevich, A and Harrison, M (2006). Quality, Experience, and Monopoly: the Soviet Market for Weapons Under Stalin. *Economic History Review,* 59(1): 113–42.

North, DC (1990). *Institutions, Institutional Change, and Economic Performance.* Cambridge: Cambridge University Press.

Rogerson, WP (1994). Economic Incentives and the Defense Procurement Process. *Journal of Economic Perspectives*, 8(4): 65–90.

Samuelson, L (2000). *Plans for Stalin's War Machine: Tukhachevskii and Military-Economic Planning, 1925–41.* Basingstoke, England: Macmillan.

Schmitz, PW (2001). The Hold-Up Problem and Incomplete Contracts: A Survey of Recent Topics in Contract Theory. *Bulletin of Economic Research*, 53(1): 1–17.

Tirole, J (1986). Contract and Renegotiation. *Journal of Political Economy*, 94(2): 235–259.

Williamson, OE (1985). *The Economic Institutions of Capitalism*. New York: The Free Press.

Chapter 7

A Soviet Quasi-Market for Inventions: Jet Propulsion, 1932–1946

Mark Harrison

1. Introduction

This paper is about an artificial market within a centralized command economy.[1] It is generally thought that markets allocate resources more efficiently than hierarchies when information is dispersed and there is no easy way of making bureaucrats pay for mistakes, in other words, in most circumstances (Hayek, 1945). But when political and economic power is already concentrated in the hands of an authoritarian ruler there is also no easy way for society to promise him compensation in return for surrendering power (Açemoglu and Robinson, 2000). In this case excessive hierarchy is likely to persist and society will continue to lose from it. Beyond a point, however, the losses may detract from the income not just of society but also of the ruler. In that case, why should the ruler not gain by selective delegation to others of those decisions that he is least equipped to make efficiently?

This chapter first appeared in *Research in Economic History*, 23: 1–59 (edited by Alex J. Field), © Elsevier Ltd 2005.

[1]I thank the Leverhulme Trust, the British Academy, and the University of Warwick Research and Innovations Fund for financial support of my research on "Invention, Imitation, and the Birth of Soviet Aerospace"; the University of Warwick for study leave; the staff of the State Archive of the Russian Federation (GARF), the Russian State Economics Archive (RGAE), and the Russian State Military Archive (RGVA) for access to documents; and Mike Berry, Alex Boyd, Leonid Borodkin, Keith Dexter, Jari Eloranta, Avner Greif, Malcolm Hill, Valery Lazarev, Vera Mikhaleva, Andrei Miniuk, Liudmila Selivanova, Asif Siddiqi, Paul Stoneman, Elena Tiurina, and Ivan Rodionov and the journal referees for valuable comments, advice, exchange of ideas, and other helpful assistance and collaboration.

From the early 1930s, Soviet officials began to discuss whether it was possible to nest the informational and incentive advantages of markets within the hierarchical structures of the command economy behind closed doors (Davies, 1996: 201–228). From the 1960s, they pursued this quest openly under the banner of socialist economic "reforms." These reforms aimed to devolve use rights over selected assets to selected agents who would be converted from hierarchical subordinates into financially independent stakeholders, and from rent-sharers to profit-makers. Given a clearer interest in the results of their own efforts, they would be motivated to achieve the leaders' objectives at lower cost as a result.

The reforms were ultimately a failure, however. The literature understands this as the result of a dynamic commitment weakness. From one point of view it is about taxation. Litwack (1991) describes the problem as the inability of Soviet planners to commit to a long-term scheme to tax the results of initiative; rather, the tax on tomorrow's productivity is determined only after observing that of today. As a result, today's effort goes into negotiating tomorrow's taxes, not improving productivity. From another point of view, selective delegation is not sustainable in the long run. Williamson (1996: 17) considers the problem of "credible selective intervention" where intervention is the opposite of delegation: a ruler who is in a position to intervene selectively, i.e. at some particular point, cannot commit not to intervene at any point. If a hierarchical principal were able to commit to a policy of hands-off when decentralization yields the first-best outcome, and to intervene only when the market outcome was second best, then socialist planning could have the best of everything. But it cannot.

Up to this point the problem is framed in terms of control over streams of expected future taxes on and rewards to effort. Control over future income streams is also an aspect of property rights. For subordinate agents to become autonomous stakeholders, and to change from rent-sharing to profit-making, they must seize their stake and acquire irrevocable disposal rights over it. As long as their control over assets remained at the discretion of the ruler, they would rationally fear future expropriation. By the same token, the ruler, as long as he stood above the law, could not bind himself not to take back his assets and the profits earned on them in the future (Williamson, 1996: 113). He had specific production objectives for the sake of which he remained willing to override his own previous decision to delegate. The use rights that he delegated could only be temporary in principle, even if continually extended in practice. Consequently agents who were granted these temporary rights continued to seek to share the rents generated in the command system rather than make profits from production. Moreover, they could exploit these rights to extract a rent from the dictator up to the level of the cost saving that he aimed to achieve by decentralizing.

At first sight, this seems to imply that selective delegation can never work. Yet, sometimes it does. One area where it seems to have worked was in the invention phase of Soviet military research and development. The invention phase was marked by technological uncertainty and information bias: while nobody yet knew the answers, the specialists knew the problems better than the officials who funded them. Under these conditions inventive activity could not be regimented from above. Instead, rival inventors competed for funding and rewards (Holloway, 1982a: 317–319). Decentralization seems to have worked in the sense that Soviet designers did keep pace with the global technological frontier as it expanded and did occasionally expand it themselves. The results included high-quality tanks and aircraft, atomic weapons and intercontinental missiles, satellites in space, and a remotely guided vehicle on the surface of the moon. In short, the Soviet command system could apparently solve some problems through selective delegation even though permanent disposal rights over assets were not devolved and future taxes remained undetermined. An empirical question then arises: what mechanism made selective decentralization work in this specific context?

The present paper is a study of decentralization in the Soviet command system. I consider the process that led to a new aerospace technology based on jet propulsion. The period is the Stalin era, before decentralizing "reforms" became fashionable. I begin at the dawn of the process in 1932 and I conclude in 1946 when the invention phase had come to an end. Throughout this period there existed an artificial "quasi-market" for inventions. The problem of high-speed, high-altitude aviation was placed in this quasi-market at the beginning of the 1930s when uncertainties and information biases were at their greatest. The problem was removed from the quasi-market and returned to direct hierarchical regulation between 1944 and 1946 when the concept had been implemented and the uncertainties removed.

I explore this topic on the basis of the recently declassified records of the defense industry held by the Russian State Economics Archive (RGAE), supplemented by those of the Red Army held by the Russian State Military Archive (RGVA), and a few records from the State Archive of the Russian Federation (GARF), all in Moscow. These are supplemented by available memoirs and a wide secondary literature that narrates the stories of lives and deeds. The official documents are marked, however, by qualities that the available narratives tend to lack: they are written without hindsight, and they are not selected to tell a story with an uplifting moral or to show someone in a good or bad light. Of course, they are selective in other ways. For present purposes the archives make it possible to do something that is new: to describe the Soviet market for inventions as an economic institution, and analyse the conditions for it to succeed.

A note on Soviet terminology seems inescapable. Soviet ministries were called "people's commissariats" until they were renamed ministries in 1946. I call them ministries throughout. I generally refer to the USSR Council of People's Commissars (Ministers) as "the government," and its chairman as the prime minister. Where necessary, I use Russian official acronyms as follows:

KB, OKB: Design Bureau, Experimental Design Bureau
NII: Scientific Research Institute
NKVD (MVD): People's Commissariat (Ministry) of Internal Affairs

It may also help to know of the major reorganizations of industrial ministries in the late 1930s: in December 1936, the ministry of heavy industry was split among several specialized branches of which one was the defense industry; two years later, in January 1939, the ministry of the defense industry was itself split into new ministries of the aircraft, ammunition, armament, and shipbuilding industries.

The paper is organized as follows: Much of it is description. Section 2 sets out the technological problem that confronted the Soviet economy. Section 3 evaluates the scale and scope of the effort invested in solving it. Section 4 discusses the nature of quasi-markets and describes the structure of the quasi-market for inventions within the framework of a command system in terms of principals and agents. Section 5 illustrates the role of the agents in forming the market, entering it, and securing initial funding. Section 6 looks at the interplay between principals and agents when projects came up for refinancing. How did principals intervene in the market? To what extent did agents' competition limit or embody rent seeking? Section 7 shows that the quasi-market for inventions gave rise to an informal secondary market in long-lived research assets. As Section 8 briefly recounts, once the technological uncertainties had been removed, the authorities closed the quasi-market for inventions down. Section 9 returns to analysis of the problem posed in the introduction, that of credible selective delegation: how was successful invention motivated in this temporary market, and how could agents enforce their payoffs for success? A final section concludes.

2. *The Problem of Jet Propulsion*

In the interwar period, the airscrew propeller driven by a reciprocating piston engine reached its limits in terms of speed and altitude of aircraft performance (Grigor'ev, 1994: 189). This prompted intensive efforts in several countries to develop new types of aero engine based on a continuous thermal cycle. Some efforts succeeded and others failed. The success story was that of jet propulsion. The parallel story of the aviation steam turbine described by Harrison (2003a) is also of interest but ended in failure.

Jet propulsion or reaction is the principle underlying both rocket motors and jet engines: action and reaction are equal and opposite. A rocket is a jet that does not need to breathe the air. Small solid-fuelled rockets had been used for hundreds of years in many countries for display and signals. A survey prepared in the USSR ministry for ammunition in 1939 reminds us that rockets were used for the first time in European warfare in the Napoleonic wars as an incendiary siege weapon and later as ammunition against troops; by the middle of the nineteenth century most armies carried substantial stocks of rocket artillery.[2] As a result of improvements in conventional artillery rockets were little used in World War I, but interest in them revived in the interwar period in connection with new fuels and artillery uses. Accompanying developments in aviation played a role because rockets could be used for both air-to-surface and surface-to-air artillery.

In the 1930s, rockets also began to be used in a new way as aviation boosters. Since rocket fuel contained its own combustion ingredients, rockets were capable of performing at limitless altitudes but for limited duration. To create a primary aviation power plant therefore required a rocket motor of unprecedented size and complexity, using more powerful liquid fuels with pressurized or pumped fuel delivery; in turn these required substantial advances in material and fuel sciences and control systems. In the interwar years it was unclear whether a rocket aircraft would ever be practicable.

The air-breathing jet engine was a more recent concept. The simplest version was a hollow tube or ramjet; when the tube was moving at high

---

[2]RGAE, fond 7516, opis' 1, delo 324, folios 6–11 (hereafter 7516/1/324, 6–11) (no date but 1939).

speed air would enter it at one end and be compressed, mixed with fuel, and burnt; the exhaust gases left the other end in a jet stream that drove the tube forward. The ramjet was no good as a primary power plant because something else had to accelerate it to high speed before ignition. Thus a primary power plant had to be able to suck in and compress air when stationary. In Frank Whittle's patent of 1930 this need was met by a supercharger or compressor attached to a turbine driven by the exhaust gases; hence turbojet.

In the 1930s, theorizing ran far ahead of practice. In theory jets were capable of higher speeds at higher altitudes than airscrew engines, though not of space flight, and for much longer duration than rockets. The theory of the gas turbine was also not a problem; it arose naturally from existing applications of steam turbines, principally in electricity generation and marine engineering (Voronkov, 1984: 115). The other major concepts that would power military and commercial aircraft for the next half century were also worked out at this time. For example, the thermodynamic efficiency of the turbojet was already understood to be poor at low speeds and altitudes, so designers were already thinking about a turboprop in which the gas turbine would drive both the compressor and reduction gears linked to an airscrew for slower long-range aircraft. Between the turbojet and turboprop lay the turbofan or turbojet engine with a bypass chamber, universally applied in modern jet airliners; a bypass engine was patented by the Soviet turbojet designer Arkhip Liul'ka in April 1941 (Liul'ka and Kuvshinnikov, 1981: 91).

To put a gas turbine into a jet engine in practice, however, raised requirements on material and fuel sciences and control systems that were far above the level of the time. In the interwar years, no one knew for sure that a turbojet could be made to work. If anything, it looked further from realisation than the idea of a rocket aircraft.

Expectations, positive and negative, were very important. In each country faith in the future of the turbojet was reinforced by the belief that rivals in other countries were making equal or greater efforts. These beliefs helped to make the turbojet a reality and in this sense (MacKenzie, 1996: 57) it was a self-fulfilling prophecy. But there were many failures en route; these were double-edged. In the long run the failures were positive: they promoted learning and were a necessary cost of ultimate success. In the short run, however, the setbacks were just that; they set the process back and fuelled scepticism and conservatism as even committed believers understood. As an investigation into faulty parts for a Soviet prototype gas turbine reported in 1938: "to start the engine in the form produced by the factory — if it were possible — could only end in an accident and destroy the idea of building a gas turbine at its very inception."[3] In Britain, Frank Whittle (1953: 78) feared that each mishap would destroy his sponsors' faith in himself and his engine. Scepticism was an obstacle everywhere. According to Liul'ka disbelief persisted in the Soviet Union until the end of 1943 (Liul'ka and Kuvshinnikov, 1981: 89). In the United States in June 1940, the National Academy of Sciences announced that the turbojet was technically infeasible; a passage from its report reproduced by Golley (1987: 114) bears Whittle's handwritten comment: "Good thing I was too stupid to know this." Actually, a Heinkel turbojet aircraft had already flown in secret in Germany the year before the American report. Even in Germany, as the inventor Hans von Ohain and the BMW engineer Peter Kappus recalled independently (Ermenc, 1990: 8, 89), the scepticism evaporated only after the principle had been demonstrated in flight. But it may be that inventors also puffed up their reputations by stressing the resistance they had to overcome.

A natural response to the great gap between theory and practice was to compromise. To get around the fundamental difficulty of the turbocompressor, clever and inventive people all over Europe were exploring a variety of intermediate steps and hybrid solutions; the Soviet Union was perhaps unusual in the number of alternatives that were pursued simultaneously (Egorov, 1994: 424–436; Serov, 1997; Gordon and Dexter, 1999).

The simplest stopgap was to strap auxiliary ramjet boosters onto an otherwise conventional aircraft, but there was little practical gain because the boosters did not add much speed and caused large aerodynamic losses. A more advanced compromise was to use an auxiliary piston engine to supercharge a jet engine. This hybrid engine resulted in a primary power plant that was less demanding than the turbojet because the elements were all established technologies; Whittle (1953: 39) filed a patent for such an engine in the early 1930s. There was also interest in refining the existing steam turbine technology to drive a conventional airscrew more cheaply and reliably (Harrison, 2003a): if oil-fired steam turbines had replaced the reciprocating engine at sea, why not in the air as well?

Finally, there was still rocketry: if the turbojet turned out to be impractical, then rocket motors must be developed far beyond their existing limits.

[3] RGAE, 8328/1/995, 111 (December 9, 1938): emphasis added. The prototype was an Uvarov gas turbine (see Table 1).

Thinking along this line remained influential even after the turbojet had become a proven concept. For example, concluding a long report to Stalin's deputy Malenkov in July 1944, minister of the aircraft industry Shakhurin compared the respective potentials of jet and rocket aviation as follows:

> Aircraft with motors of the liquid-fuelled [rocket] type have a bigger future in terms of achieving greater efficiency [a larger KPD, coefficient of useful action] and a sharp increase in the tactical flight characteristics of jet-propelled devices. The power of such a motor does not depend on altitude and it can operate (even better) in airless space. However, extremely high fuel expenditures at the present stage of their development (cf. the Me[sserschmitt]-163) sharply limit the great potential of this type of aircraft. This is why existing jet-propelled aircraft have big engines of the air[-jet] type.[4]

From a mission standpoint, what were the appropriate responses to the extent of technological uncertainty? The best chances of progress would result from an open-ended commitment to advance on many fronts at the same time. Many problems demanded simultaneous technological solutions. Many applications would not be detected without free-ranging exploration of new technologies. The state needed to fund many projects, accepting a high probability of failure in any one of them, in order to ensure that at least some successful projects would be included. As Joel Mokyr (1990: 176–177) has taught us, many failures could be expected as part of the cost of success.

As far as the scale of effort is concerned, no government would accept a completely open financial commitment, but there were different degrees of open-handedness across countries. For each country the feasibility of a given commitment varied with economic size, development level, and mobilization capacity. The interwar British, German, and Soviet economies were of about the same size in gross domestic product (GDP). Britain and Germany were more developed in science and industry than the USSR; the Soviet economy could claim an advantage, however, in its superior mobilization capacity (Harrison, 1998).

3. *Scale and Scope*

This section gives a brief account of the scale and scope of jet propulsion research for aviation in the Soviet Union from 1932 to 1946. Table 1 lists 18 major projects over the period; the research in rocketry, limited to aviation-related activities, excludes numerous artillery projects some of

[4] RGAE, 8044/1/1182, 147 (July 26, 1944).

Table 1. Major Soviet R&D Jet Propulsion Projects for Aviation, 1932–1946.

	1932	1933	1934	1935	1936	1937	1938	1939	1940	1941(1)	1942(2)	1942	1943	1944	1945	1946
Rocket motors																
Smaller Glushko rocket motors and aviation boosters	GDL	≫	RNII	≫	≫	NII-3	≫					KB Z-16	≫	≫	≫	≫
Larger Glushko rocket motor leading to RP-218 aircraft						RNII	NII-3									
KB-7 rocket motor								KB-7								
Dushkin, Isaev rocket motors leading to RP-318 and BI fighter								NII-3, OSK Z-1	≫	NII-3	≫	GIRT, OKB-293	≫	NII-1		
NII-3 rocket booster											NII-3	GIRT				
Dushkin rocket motors															NII-1	≫
Jet engines																
VTI gas turbine VTI	VTI															
MAI gas turbine		MAI														

(*Continued*)

Table 1. (*Continued*)

	1932	1933	1934	1935	1936	1937	1938	1939	1940	1941(1)	1942(2)	1942	1943	1944	1945	1946
Uvarov gas turbines leading to turboprop engine					VTI	≫	≫	≫		TsIAM	≫		TsIAM	≫	≫	≫
KB-7 ramjet							KB-7									
Liul'ka turbojet								TsKTI, Z-18	SKB-1	≫	≫	OKB-293	OKB-293, TsIAM	NII-1	≫	≫
Pobedonostsev, Merkulov ramjet	GIRD	≫	RNII	≫	≫	NII-3	≫	NII-3, OSK Z-1, Z-18	NII-3, OSK Z-1	≫	≫	GIRT	Z-84			
NII-3 hybrid jet								NII-3	≫	≫	≫	GIRT				
Bas-Dubov, Zaslavskii ramjet										Z-28	≫					
Abramovich hybrid jet													TsAGI			
Fadeev, Kholshchevnikov hybrid jet														TsIAM	≫	≫
Tolstov hybrid jet														TsIAM	≫	≫

Source: The appendix, supplemented by Egorov (1994: 424–436). The documentation supporting the appendix comprises plans, reports, and memoranda of the ministries of defense, internal affairs, heavy industry, the defense industry, ammunition, and the aircraft industry.

which were highly successful. It is compiled mainly from items reported in the plans, reports, and memoranda of the ministries of defense, internal affairs, and the heavy, defense ammunition, and aircraft industries; these are listed more completely in the appendix.

The "major project" as a unit of measure is convenient but simplified or fuzzy in some dimensions. First, research strategies varied; for example Glushko, Merkulov, and Uvarov were often to be found engaged in multiple designs for power plants of varying size and capacity based on similar principles at the same time, whereas Liul'ka committed everything to a single design that evolved through time. To make sense of the data, I have counted the work of each as a single major project, and this has imposed some aggregation on the "minor" ones. (I have made an exception for Glushko whose collaboration with Sergei Korolev on a rocket aircraft in 1936 and 1937 was clearly different from his other projects in scale and significance.)

Second, the official documents do not capture the informal or peripheral involvement of designers before they moved to the centre of the field. The documents must also fail to reflect the disruption or continuity in designers' work as they moved from one organization to another. It is not clear, for example, what Uvarov did in 1940 or whether, when he left VTI in 1939 and reappeared at TsIAM in 1941, he brought with him his assistants and equipment or only his personal intellectual capital.

The great advantage of Table 1 is that the data underlying it were created without foresight. In contrast, the memoirs and biographies of the designers such as Liul'ka and Uvarov, who were eventually successful, naturally pay scant attention to the projects of their lesser rivals. This makes the table substantially more complete than previous narrative accounts.

Table 1 shows six major projects in aviation rocketry and 12 in jet aviation over the period, but this does not give a clear impression of relative importance since some projects were long-lived, while others appeared and disappeared within a year. Figure 1 measures the overall investment in cumulative "major project-years." It shows that until 1937 research in rocketry and jet engines advanced more or less in step; after that, the balance shifted away from rocketry and by 1946 jet engine development had accomplished almost two-thirds of the 76 project years accumulated in total.

Also of interest is the turnover of projects, especially the termination of those regarded as unsuccessful., Figure 2 illustrates this for the aviation

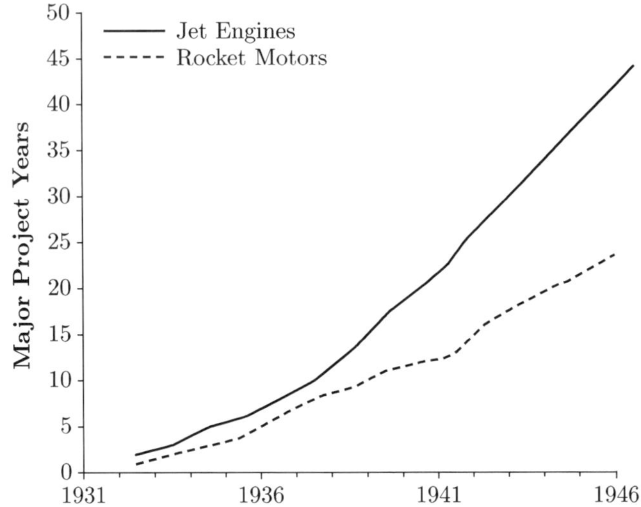

Fig. 1. Soviet R&D in Jet Propulsion, 1932–1946: Cumulative Investment in Major Project Years.

Source: Table 1.

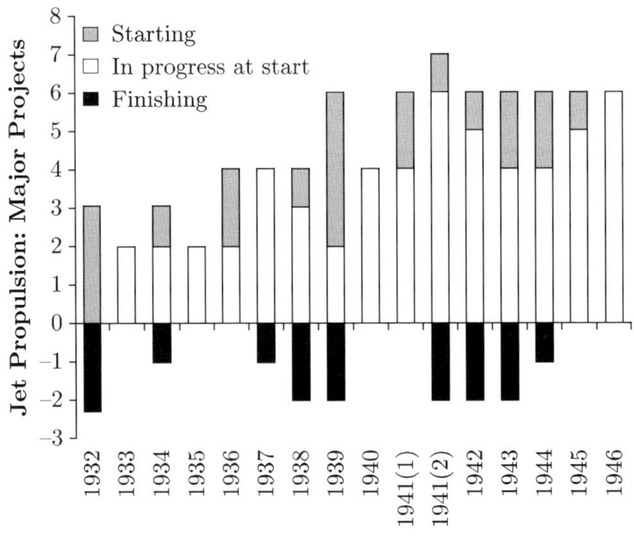

Fig. 2. Soviet R&D in Jet Propulsion, 1932–1946: Major Projects.

Source: Table 1.

field as a whole. It shows clearly the build-up of research and design activity in the second half of the 1930s. It shows that project closures were concentrated in 1937–1939, and in the first two and a half years of the war. Closer inspection of Table 1 reveals that there was a complete break in rocketry development at the end of 1938, when all projects then current were terminated. The figure concludes with the year 1946 when, with activity now at a high level, no projects were closed down and no new ones were started.

Table 1 showed that some 15 distinct research establishments were involved in developing jet propulsion for aviation; these are detailed in Table 2. The true total of establishments listed in the table is less than those that appear in the table because there were many reorganizations and changes of name, usually driven by a desire to break with the errors of the past. Behind these are a strife-torn story of the fall and rise of the idea of jet-propelled aviation. The central organization in this story was the Jet Propulsion Research Institute (RNII) formed in 1934. Its turbulent history has been described recently by Harrison (2000: 127–130) and Siddiqi (2000: 1–14). The low point fell between 1937 and the second half of 1941. In 1937, discredited by association with "enemies of the people" charged with wasting state funds on useless dreams of rocket-powered flight, RNII was converted into NII-3, an adjunct of the ammunition industry, and ordered to concentrate on designing rocket shells and mortars. Interest in rocket aviation was revived in August 1941 and in jet engines in July 1942. NII-3 was converted at first to a State Institute for Jet Propulsion Technology (GIRT) reporting directly to the central government. It was fully rehabilitated in the spring of 1944 as the Research Institute for Jet-Propelled Aviation (NIIRA) and then Research Institute No. 1 (NII-1) of the aviation industry. Its many transformations are shown for reference in Figure 3.

The organizations involved were of modest size. Table 3 shows that RNII-NII-3 had 400 or so staff in the mid-1930s of whom around one quarter could be classed as specialist "engineering and technical employees." By the end of the decade, this number had risen to more than 800. In the same period, the RNII budget trebled, although not all of this was real growth; price and wage inflation was especially rapid between 1937 and 1940 (Bergson, 1961: 367–368, 422). In comparison with other establishments of a similar profile, RNII-NII-3 was medium-sized; the other main rocketry establishment, KB-7 was much smaller. This is shown in Table 4, which

Table 2. Soviet Jet Propulsion R&D Organizations and their Fund Holders, 1932–1946.

Acronym	R&D Organization	Fundholder	Period involved	Location	Notes
GDL[a]	Gas Dynamic Laboratory	Red Army Administration for Military Inventions	**A** To 1934	Leningrad	Merged with GIRD into RNII in 1934
GIRD[a]	Jet Propulsion Study Group	Society for Cooperation in Air and Chemical Defense; transferred to Red Army Administration for Military Inventions in 1933	**A** To 1934	Moscow	Merged with GDL into RNII in 1934
GIRT[a]	State Institute for Jet Propulsion Technology	USSR Council of People's Commissars	**D** 1942–1944	Moscow	Formerly NII-3
KB Z-16	Factory no. 16 Design Bureau	NKVD Fourth Special Department	**D** From 1942	Kazan'	
KB-7[a]	Design Bureau no. 7	Red Army Administration for Military Inventions, then Artillery Administration; transferred to Defense (later Ammunition) Industry in 1938	**A** 1938 and 1939	Moscow	Spin-off from RNII in 1935

(Continued)

Table 2. *(Continued)*

Acronym	R&D Organization	Fundholder	Period involved	Location	Notes
MAI	Moscow Aviation Institute	Heavy (later Defense, later Aircraft) Industry[b]	**I** 1934	Moscow	
NII-1[a]	Research Institute no. 1	Aircraft Industry	**I** From 1944	Moscow	Formerly NIIRA
NII-3[a]	Research Institute no. 3	Defense (later Ammunition) Industry[b]	**I** 1937–1942	Moscow	Formerly RNII
NIIRA[a]	Research Institute for Jet-Propelled Aviation	Aircraft Industry	**I** 1944	Moscow	Formerly NII-3
OKB-293[a]	Experimental Design Bureau no. 29	Aircraft Industry	**I** 1942–1944	Khimki, Moscow district; evacuated temporarily to Bilimbai, Sverdlovsk district, October 1941–January 1942	Merged into NIIRA-Nii-1 in 1944
OSK Z-1[a]	Factory no. 1 Department of Special-Purpose Designs	Aircraft Industry	**I** 1939–1942	Moscow	

(Continued)

Table 2. *(Continued)*

Acronym	R&D Organization	Fundholder	Period involved	Location	Notes
RNII[a]	Jet Propulsion Research Institute	Heavy (later Defense, later Aircraft) Industry[b]	**I** 1934–1937	Moscow	Formed by GDL and GIRD in 1934
SKB-1	Special Purpose Design Bureau no. 1	Defense (later Aircraft) Industry[b]	**I** 1940–1941	Leningrad	
TsAGI	Central Aero-Hydrodynamic Institute	Aircraft Industry	**I** 1943	Stakhanovo (later Zhukovskii), Moscow district, evacuated temporarily to Kazan' and Novosibirsk, 1941–1942	
TsIAM	Central Institute for Aeroengine Building	Aircraft Industry	**I** 1941 and from 1943	Moscow	
TsKTI	Central Boiler and Turbine Institute	Electricity Generation Industry	**I** 1939	Leningrad, with a subsidiary in Podol'sk, Moscow district	

(Continued)

Table 2. *(Continued)*

Acronym	R&D Organization	Fundholder	Period Involved	Location	Notes
VTI	Dzerzhinskii All-Union Thermal-Technical Institute	Electricity Generation Industry	**I** 1932 and 1936–1939	Moscow, evacuated temporarily to Kererovo, 1941–1943	
Z-18	Factory no. 18	Defense (later Aircraft) Industry[b]	**I** 1939	Voronezh, evacuated to Kuibyshev in October 1941	
Z-28	Factory no. 28	Defense (later Aircraft) Industry[b]	**I** 1941	Moscow, evacuated to Sverdlovsk in October 1941	
Z-84	Factory no. 84	Defense (later Aircraft) Industry[b]	**I** 1943	Khimki, Moscow district; evacuated to Tashkent in October 1941	

Note: **A** (Army); **D** (Dictator); **I** (Industry).
Source: Table 1 and Figure 1.
[a]See Figure 1.
[b]On December 8, 1936 the ministry of heavy industry was divided into a number of specialized branches of which one was the defense industry; on January 11, 1939 the ministry of the defense industry was divided into new ministries of the aircraft, ammunition, armament, and shipbuilding industries.

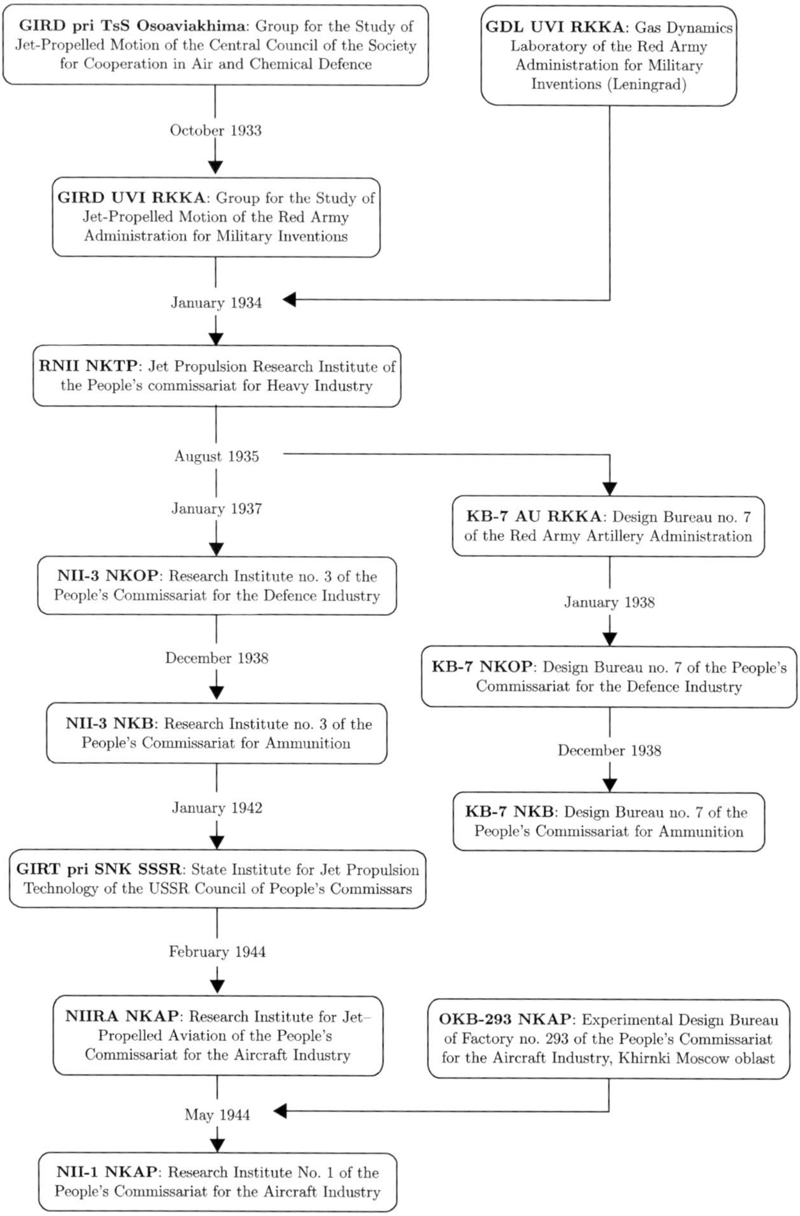

Fig. 3. The Evolutionary Path of RNII, 1932–1944.

Sources: For details see Siddiqi (2000, pp. 1–18). In addition: GIRD, originally sponsored by Osoaviakhim, was taken over by the Red Army administration for military inventions in October 1933 before being merged with GDL, renamed RNII, and transferred to the ministry of heavy industry (RGVA 4/14/1171, 33: memorandum dated 23 January 1934). Siddiqi states that NII-3 was handed over to the ministry of the ammunition industry in November 1937, but this ministry was only created on the dissolution of the ministry of defense industry on 11 January 1939. Various documents indicate that the ammunition industry also acquired KB-7 from the Red Army's artillery administration at the beginning of 1938. A memorandum from Deputy Defense Minister Fed'ko to Prime Minister Molotov dated 15 February 1938 refers to "the former KB no. 7 of the AU RKKA [Red Army artillery administration], transferred to NKOP [people's commissariat of the defense industry]"(RGVA, 4/14/1925, 22), and KB-7 is listed among the establishments of the thirteenth chief administration of the ministry of the ammunition industry in its report of work for the year 1938 (RGAE, 8162/1/89, 101). KB-7 was apparently dissolved in 1939. GIRT is described as "pri SNK SSSR [attached to the USSR Council of People's Commissars]" in its deed of transfer to the ministry of the aircraft industry, not dated but in 1944 (RGAE, 8044/1/1182, 11-16).

Table 3. RNII-NII-3, 1935–1941: Personnel by Employment Status and Gross Value of Output.

	1935	1936	1937 plan	1938	1939	1940	1941 plan
Engineering & technical employees	—	102	118	—	—	—	215
Manual employees	—	196	295	—	—	—	385
Nonmanual employees	76	84	88	—	—	—	—
Of which:							
Accounting and clerical	—	—	—	—	—	—	125
Production and planning	—	—	—	—	—	—	79
Junior service personnel	37	64	63	—	—	—	32
Total	403	446	476	514	799	—	836
Gross value of output (thousand rubles)	—	3,377	4,482	6,111	11,434	11,233	—

Source: Employment: 1935 from RGAE, 8162/1/16, 16, and 1936–1937 from ibid., 4 (no date but about February 1937); 1938–1939 from RGAE, 8162/1/240, 32 (January 13, 1940); 1941 from RGAE, 8162/1/449, 144 (November 18, 1941). Gross value of output: for 1936 and 1937 plan see RGAE, 8162/1/16, 2–3 (February 28, 1937), for 1938–1939 RGAE, 8162/1/240, 32 (January 13, 1940), and for 1940 RGAE, 8162/1/449, 3 (January 14, 1941). The figure for 1940 that had been planned and approved by KO was slightly higher at 11,725 thousand rubles. However, towards the end of that year an investigation disclosed that the responsible fourth chief administration of the ministry of the aircraft industry had illegally planned a much higher figure of 13,162 thousand rubles; see RGAE, 7516/1/692, 3 (November 21, 1940).

Table 4. Research Institutes and Design Bureaux of the Thirteenth Chief Administration of the Ministry of Defense Industry, 1938.

	Budget, thou. rubles	Planned research topics	Scientific workers
NII–24	12,764	178	60
—Leningrad filial	11,052	81	55
KB–47	8,006	94	55
NII–3	5,667	39	44
KB–7	1,200	9	13
KB–31	700	6	19
Total	39,389	407	246

Source: RGAE, 8162/1/299, 9 (no date but 1938). The thirteenth chief administration of the ministry of defense industry was the future ministry of the ammunition industry.

compares them with other research outfits of the ammunition industry in 1938.

In considering these figures, it is important to bear in mind that most of the work done in RNII-NII-3 and KB-7 was concerned with rocket artillery, not aviation. Aviation projects accounted for two-fifths of the value of research and experimentation planned by RNII in 1937.[5] By 1940 their nominal value had fallen absolutely, while their share in the much larger NII-3 budget was now only six per cent.[6]

These research expenditures were a very small fraction of overall defense procurement. In the late 1930s, they ran at less than a million rubles a year. Equipment orders for the army and navy in 1937 were 5.7 billion rubles (Davies and Harrison, 1997), and 14.5 billion rubles for the army alone in 1940 (Harrison, 1996: 281).

The Soviet outlays appear trivial in comparison with the resources that Germany devoted to the development of V-weapons. A post-war American estimate puts the total development costs of the jet-powered V-1 cruise missile to Germany at approximately $200 million in wartime US prices, and those of the V-2 rocket at 10 times that amount, about the same as the $2 billion cost of the Manhattan Project (Ordway and Sharpe, 1979: 242, 253). Given that one pre-war ruble was worth at most 35 US wartime cents (Harrison, 1996: 275), the resources invested in Soviet aviation jet propulsion in the 1930s and 1940s can hardly have totalled more than $10 million.

[5]RGAE, 8159/1/6, 74 (December 1936).
[6]RGAE, 8162/1/300, 65–66, 80–81 (November 17, 1940).

4. *Dictator, Army, and Industry in a Quasi-Market*

In Section 4, I discuss the differences between "real" and quasi-markets and provide a simplified description of the structure of the market for inventions. We think of real markets as formed by buyers and sellers who enter the market independently, motivated by their own self-interest. In real markets, prices are set by interpersonal negotiation or impersonal bidding, or are preset by one side in the presence of market power. At least one equilibrium is possible and the interaction of supply and demand generally leads to it. Contract disputes are resolved by custom or law. In the outcome, the market steers resources in the general direction of their most profitable use.

Quasi-markets, in contrast, are created by the state to allow its own agents to engage in decentralized transactions with each other. The agents enter the quasi-markets because they are told to. They are not supposed to behave in an independently self-interested way but to follow contingent rules. If they find themselves in dispute, the principal determines whether or not to intervene and which side to uphold. Prices and incentives in quasi-markets are formed by the principal's decision; the process is not equilibrating and adjustment typically involves "false" or cross-trading, described by Morishima (1984: 15). The principal usually tries to calibrate incentives in advance so when his agents allocate resources in detail the results will conform to broad limits already set out in centralized plans.

The idea of quasi-markets is widely applied to describe decentralized allocation within large private and public-sector organizations, for example, the British welfare state since Margaret Thatcher (Le Grand, 1991; Le Grand and Bartlett, 1993; Bartlett *et al.*, 1998). Its conceptual origins, however, lie with Ludwig von Mises (Mises, 1949/1998: 701–706), who employed the term to describe the market-like rules for decentralized allocation that contemporary socialists such as Oskar Lange wanted to embed within a planned economy. Mises regarded the idea as inherently unworkable; he argued that, since the agents would have no property of their own to lose, quasi-markets would be dominated by the "audacity, carelessness, and unreasonable optimism" of "the least scrupulous visionaries or scoundrels". We shall find that, while he had an element of prophetic truth on his side, the reality of the quasi-market for inventions was much more complex and interesting than this would suggest. In particular, the quasi-market began to blur into a real market at two points: the sellers played a significant role in influencing the principal to create it; and enforcement relied significantly on a reputation mechanism that was independent of the principal.

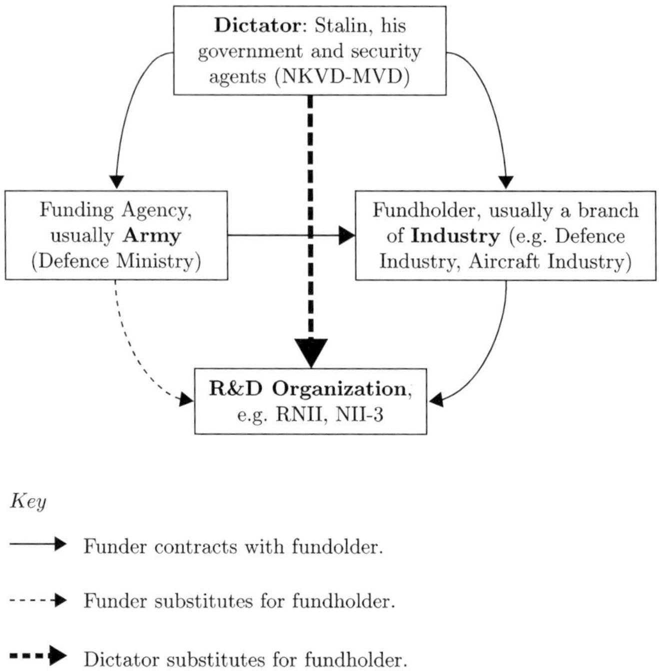

Fig. 4. Players in the Quasi-Market for Inventions.

The structure of the market was roughly as follows: There were four main categories of players and Figure 4 illustrates the quadrilateral relationships among them. Stalin, the dictator, was personally represented by central government. On Stalin's behalf, high-level government agencies established the framework within which the other players operated by issuing strategic directives from time to time. Formally, these were usually the government subcommittee's responsibility for defense matters or, in wartime, the war cabinet. In reality, regardless of formal authority, Stalin decided many of these things personally in consultation with a varying circle of members of the party politburo, usually after receiving representations from other stakeholders (Khlevniuk *et al.*, 1995; Khlevniuk, 1996). Examples of the major directives affecting R&D for jet propulsion during the period under review are shown in Table 5.

The actors at the next level down were the defense agencies to whom Stalin delegated the power to fund research, and the specialized fund holders or suppliers of R&D services. The defense ministry was the prospective purchaser of jet propulsion technology and acted as the funding agency.

Table 5. Major Directives of the Soviet Government Concerning Jet Propulsion, 1932–1944.

Date	Issuing authority	Decision
July 1932	Government Defense Commission	Expand research on jet propulsion
September 1933	Council of Labor and Defense	Establish RNII
July 1940	Government Defense Committee	Develop the Liul'ka turbojet
August 1941	State Defense Committee (the war cabinet)	Develop the BI (Berezniak-Isaev) rocket fighter
November 1942	State Defense Committee	Produce the 302 (Kostikov) rocket fighter
February 1944	State Defense Committee	Reorganize NII–1 as a research institute of the aviation industry
May 1944	State Defense Committee	Develop a number of rocket and jet aircraft and engines

Source: Danilov (1981: 71).

The funder commissioned R&D services from the fund holder who was paid to supply them. The funder's contract could implement a government decree using centralized funds earmarked from the USSR state budget. Alternatively, as a budgetary institution the defense ministry could enter into de-centralized contracts with industrial institutes and design bureaus for R&D services on its own initiative.

The fund holder was the legal owner of the R&D organization. The main fund holders supplying jet propulsion research are listed in Table 2. If the list appears complex at first sight, the detail can be simplified by grasping the three main types of fund holder, indexed in the table as A for Army, D for Dictator, or I for Industry. The "normal" arrangement of the quasi-market for inventions was that Industry, i.e. the ministry of heavy industry and its successors responsible for the various branches of the defense industry, in particular the aircraft and ammunition industries, was the fund holder. But other industrial interests could also take on side-line responsibilities, for example, the electricity generation industry. In varying circumstances the Army and the Dictator intervened. As the principal funder, the Red Army made various attempts to bypass the quasi-market for inventions and substitute itself for industry as the fund holder by establishing its own in-house jet propulsion research and development facilities, for example, KB-7. In Figure 3 this is shown by the fine dashed arrow. Second, Stalin's NKVD sometimes acted directly for him by substituting itself for the fund holder, seizing R&D personnel and assets and managing them on a prison basis in a sharaga or sharashka (Albrecht, 1993: 133–135; Starkov, 2000: 255–260); this is indicated by the heavy dashed arrow. The fund holder of factory no. 16 in Kazan', for example, was the ministry for the aircraft industry, but its aero engine design bureau was a sharaga staffed by prisoners and run by the NKVD fourth special department.

Normally, however, funding departments and fund holders formulated independent operational plans that were then coordinated through a contracting process. The most important planning horizon was annual. The Red Army had an annual plan for the development of military inventions most of which it contracted out to other organizations through the quasi-market for inventions. Industrial ministries also had their own R&D plans, for example, the annual plan for aero engine research and experimentation to be carried out by the institutes and bureaus of the aviation industry, part of which was made up by contracts accepted from the Red Army. But some R&D was also financed by the industry.

The point is illustrated by the sources of funding of a design organization such as NII-3. Table 6 shows that NII-3 planned to undertake 11,725 thousand rubles of expenditure on research in 1940. This sum was to come from three sources, one part from the state budget, another from decentralized contracts with outside funding departments, mainly the Red Army, and the rest from the NII-3 fund holder, the responsible chief administration in the ministry of the ammunition industry.

These arrangements imposed the following structure on competition among designers. There were many designers and many design organizations. Since R&D projects had to fall under one or other design organization, the design organization was the main vehicle for this competition. Designers competed for funding from a limited number of sources. In principle the defense industry, later the aviation industry, was the monopoly fund holder but in practice its monopoly was limited and threatened by other parties: industries with side-line interests in potential diversification,

Table 6. Planned Funding of NII-3, 1940.

Source of funding	Number of projects	Thousand rubles
State budget	18	5,790
Contracts	15	3,440
Chief administration	12	2,495
Total	45	11,725

Source: RGAE, 8162/1/449, 3 (January 14, 1941).

military men interested in the scope for vertical integration, and the dictator who could revoke the delegation of his powers to industry at any time and impose direct control under the NKVD.

5. *Action and Reaction*

The quasi-market for inventions was not atomistic or impersonal. Rather, it was driven by the designers. David Holloway (1982a: 288) has suggested that the balance between supply side or "discovery-push" factors and "demand-pull" factors determined how major innovations were diffused through the Soviet defense sector. The stereotype of a command economy might lead one to expect that the active side of the market was the demand side in the sense that the funding principal issued compulsory contracts on the basis of a high-level plan, and the fund holders then complied with the contracts. The difficulty with this stereotype is that, given the technological uncertainty surrounding the future of aviation in the 1930s, the principal did not know what contracts to offer. The specialist agents on the supply side knew the answers better than the principal, and the command system had to adapt to this reality. Thus the active side of the market was the supply side and decision-making on the demand side was mainly reactive.

In Section 5, I describe briefly the supply side activity, which took the form of proposals for funding from three groups of actors: established designers, backyard inventors, and foreign specialists. I note separately the uses that these proposals made of information from the foreign press and foreign commercial information.

The reactive character of decision-making on the demand side is shown by the character of funding decisions: reports and resolutions that consolidated or cancelled existing rival projects greatly outweighed the number of decisions that authorized new ones. Designers' control of the initiative resulted in a tendency for projects to proliferate that the funding principals found difficult to control. We will look more closely at this in Section 5.5. But the scope of the designers' initiative went beyond this; it was not just that they held the initiative within the market, they also helped to create it. In this sense, the distinction between real and quasi-markets was not completely watertight.

5.1. *Established Designers*

How did projects win a place in the plan? There were a variety of routes, but in each case the initiative lay with the designer. This was not a process whereby all-seeing and all-knowing planners identified needs from above, sought out designers, and put them together with resources to meet the needs identified. Rather, proposals came first from below. As minister of the aircraft industry Shakhurin explained to deputy prime minister Voznesenskii in February 1941, "Work on the creation of jet propulsion engines at home in the USSR [...] began on the initiative of a few engineers taking the form of inventors' proposals."[7] This was the case in other defense projects too. For example, the first movers were the designers in both the failed attempt to build an aviation power plant around a steam turbine (Harrison, 2003a) and the successful project to build a Soviet atomic bomb (Holloway, 1994: 72–95).

Liul'ka, father of the Soviet turbojet, began his work in 1936 on the back of a steam turbine project at the Khar'kov Aviation Institute (Egorov, 1994; Sultanov, 1998); this experience soon convinced him that steam power had no future in aviation. The following account is based on Berne and Perov (1998: 78–81). Liul'ka wanted to develop work on a turbojet but his path was strewn with obstacles. His superiors in the Khar'kov Institute refused to back his work directly and referred him to Moscow. There he met Uvarov, who helped him get some funding from the aviation industry. But money opened few doors without adequate technical support; back in Khar'kov his project languished. In 1938, Liul'ka contrived "with great difficulty" to meet the minister for the defense industry, Mikhail Kaganovich, who was impressed enough to convene a night-time meeting with his deputies; the outcome was to offer Liul'ka the facilities of SKB-1 at the Leningrad Kirov factory. Back in Khar'kov, Liul'ka began to build a new engine with an axial compressor in the expectation of Air Force sponsorship. When this failed to materialize he wrote to prime minister Viacheslav Molotov in March 1939. At the end of this year he finally secured proper funding from the aviation industry and a proper base at the Central Boiler and Turbine Institute (TsKTI) of the electrical industry (see Table 1). During 1940 and 1941, Liul'ka designed an engine and experimented with its turbine, compressor, and combustion chamber. However, within a few days of the outbreak of the war this work was mothballed in favour of an alternative plan to build a liquid-fuelled rocket fighter, the BI, pursued to a catastrophic conclusion in early 1943.

[7]RGAE, 8044/1/460, 59 (February 5, 1941).

Designers worked to secure ministerial approval and the funding that followed. If refused at one level, they appealed to the next. If necessary they began work without waiting for authorization; they illegally diverted resources of their own design organizations that had been allocated to other uses and then used the preliminary results to support subsequent attempts to gain official backing. The principle is illustrated by an anecdote (Perakh, 1998: Section 9.1): in the 1930s a group of young scientists who wanted to embark on research in atomic physics approached A.F. Ioffe, director of the famous Fiztekh, the Institute of Physics and Technology in Leningrad. Ioffe saw the potential of their proposal but was under pressure from above to give more resources to applied research instead. He realized it might be difficult to justify the proposal to the party authorities, and resolved to go ahead by means of a ruse. He gave laboratory space to the atomic physics project on an unofficial basis, and posted a sign on the door: "Stockroom." As he expected, at the next inspection the party officials walked straight past without curiosity. The project was safe until the time came for Stalin to recognize the importance of atomic physics.

Such behaviour caused projects to proliferate in an uncontrolled way. The consequences were outlined by NII-3 director A.G. Kostikov at an internal meeting in May 1942 when the strain on resources was at its most intense:[8]

> As an example of how we are forced to diffuse the attention of our cadres I will take the first research department. There are 26 [research] topics for 10 engineers. Some of these topics are incidental to our institute and do not match its profile or specialization. These topics arose because there were people to put them forward and instead of passing them on to those organizations for whom such topics were more appropriate we engaged in them ourselves. [...] It's characteristic of such topics that working on them involves unnecessary investigations since [we have] no corresponding experience. Often what is done is done many times, and all because we took on what was not our business, because we have neither experience nor cadres to work on items that don't match the profile of our institute.

Successful proposals required investments in lobbying. Such investments could bring the designer not only success with an individual proposal but also a privileged long-term relationship with the government officials responsible for funding. To win support for their projects and adoption of their designs, designers had to be "heterogeneous engineers" capable of

[8]RGAE, 8162/1/574, 101 (May 7, 1942).

reshaping organizational as well as technological constraints (MacKenzie, 1996: 13). To create a demand for new designs they had to build coalitions with soldiers or industrialists to overcome interests vested in markets for products that already existed (Holloway, 1982a: 292). In particular, they had to overcome the preference of industry for the undisturbed mass production of weapons in long serial runs, which was often at odds with radical product innovation and the risks and requirements of continual upheaval in production (Berliner, 1976: 534–538; Albrecht, 1993: 195–197, 207–208). This could make it difficult to establish where the initiative lay and blurred the whole distinction between discovery-push and demand-pull. It also provides an exception to the rule that Soviet producers did not need to hire marketing agents, only supply facilitators or tolkachi.

The aircraft designer Sukhoi, for example, is said to have won success in having his designs adopted only after he took on a partner, E.A. Ivanov, who had the political and bureaucratic skills to push his product through the military and party-state apparatus (Ozerov, 1973: 53). Almquist (1990: 70–73) has described the political "connectedness" of successful post-war Soviet designers. The German turbojet pioneer Hans von Ohain prospered on account of the alliance he forged with the aircraft manufacturer Otto Heinkel; according to Kappus (Ermenc, 1990: 91), the support of Heinkel was critical to Ohain's success with the turbojet in Germany and the lack of similar support explains why Whittle took twice as long in the UK. Finally, Siddiqi (2000: 7) has described the rocket pioneers' alliance with Marshal Mikhail Tukhachevskii.

Tukhachevskii, Red Army chief of armament from 1931 to 1936, was the most important patron of jet propulsion in the Soviet Union between the wars.[9] At the outset he took the Leningrad Gas Dynamics Laboratory (GDL), founded in 1929, under the aegis of the Red Army's administration for military inventions to develop solid-fuelled rocket ammunition. In Moscow in 1931, a voluntary society of rocket scientists, the Jet Propulsion

[9]In November 1929 the post of Chief of Armament of the Red Army was created to help carry through its equipment modernization. The first Chief of Armament was Army Commander Uborevich, followed in 1931 by Army Commander, later Marshal Tukhachevskii. Among the departments reporting to the Chief of Armament was an administration for military inventions. In 1936, the post of Chief of Armament was abolished, its place taken by a chief administration for supply of weapons and equipment, and under the latter a department for inventions (see Holloway, 1982a: 321). On Tukhachevskii and Red Army rearmament generally, see Samuelson (1996; 2000) and Stoecker (1998).

Study Group (GIRD), began to promote the cause of space exploration based on liquid-fuelled rocketry.[10] The group was led by Sergei Korolev, the future chief designer of ballistic missiles and space launch vehicles, and was sponsored by the civil defense organization Osoaviakhim. In September 1933, Tukhachevskii sponsored a merger of GDL and GIRD in a new establishment subordinated directly to him, the RNII.[11] He seems to have hoped to monopolize the development of jet propulsion as both funder and fund holder. He was frustrated by a decree of the Council for Labor and Defense which almost immediately transferred the new establishment to Sergo Ordzhonkidze's ministry of heavy industry.[12] This followed the precedent of arrangements for TsAGI and TsIAM, recently established to focus research in aircraft and aero engine design, respectively.

Although no longer the fund holder, Tukhachevskii personally, and the Red Army as funder, retained close involvement with RNII, contracting for much of its R&D output. RNII was an unhappy marriage, and divisions soon emerged between the weapons specialists of GDL and the space enthusiasts of GIRD. The new director I.T. Kleimenov, formerly head of GDL, curtailed work on liquid-fuelled rockets on the grounds of its low expected military utility, side-lining Korolev and the other GIRDers. The result was a huge row that embroiled RNII with the local party organization and pitched Korolev and Tukhachevskii against Kleimenov and Ordzhonikidze.[13] In 1935, Tukhachevskii exploited these divisions to recruit some former GIRDers led by Korneev, a dissident engineer whom Kleimenov had sacked, to set up KB-7 as a Red Army design bureau for liquid-fuelled rocketry (Siddiqi, 2000: 8). Thus, Tukhachevskii succeeded in becoming a fund holder in jet propulsion by other means, although without a monopoly.

For several reasons unrelated to this theme, Tukhachevskii was arrested in May 1937 and was executed as a traitor with many other officers (Stoecker, 1998; Samuelson, 2000). If there is a connection, it is that Stalin distrusted Tukhachevskii in part for his monopolizing ambitions which were strongly suggestive of a rent-seeking military-industrial lobby (Harrison,

[10]GARF, 8418/6/243, 35–37 (May 14, 1933).

[11]RGVA, 4/14/1171, 33 (January 23, 1934); also Siddiqi (2000a: 4–7).

[12]RGVA, 34272/1/146, 134 (October 31, 1933).

[13]RGVA, 34272/1/177, 5–10 (May 27, 1934), 17–19 (May 29, 1934), 20–21 (June 1934), 1–2 (July 26, 1934), and 33 (September 13, 1934); also Siddiqi (2000a: 7–9; 2000b).

2003b). After this, the cause of aviation jet propulsion lacked a high-level sponsor until Malenkov began to take an interest in 1943 and was briefed on the issues by aircraft industry minister Shakhurin.[14]

5.2. *Citizen Initiatives*

Many ordinary citizens with and without technical qualifications wrote to the Red Army with unsolicited ideas and suggestions for work on high-speed, high-altitude aviation, a few of which were taken up.

The files of the Red Army administration for military inventions show, for example, that in April 1932 and again in March 1936, E.A. Blau submitted proposals for different aero engines based on jet propulsion. These proposals were reviewed and rejected. One referee judged that "despite the fact that the author is an engineer [his designs] are distinguished by their naivety and demonstrate a complete absence of elementary information concerning jet propulsion".[15] Another file collects 51 proposals for jet, rocket, or turbine engines, and airframes that were submitted in 1937, some professionally executed, some handwritten and childishly illustrated. One author proposed to lighten his super-heavy airframe by filling the wings with hydrogen, neglecting the fact that a cubic metre of hydrogen generates only one kilogram of buoyancy; another proposed a winged cruise missile but omitted to allow for a guidance system or automatic stabilization. Depending on its merits the response to each proposal was either a request for further information or a curt rejection.[16]

A special case is presented by GIRD, a voluntary society until it was merged with GDL to form RNII in 1933. The GIRDers were civilian engineers drawn together by a common interest; an official report describes them as "enthusiasts for the cause of rocketry who had no material base and no staff".[17] At first they won backing from Osoaviakhim, a state-sponsored voluntary association for civil defense. Their great achievement was the successful launch of the Soviet Union's first liquid-fuelled rocket in August 1933.[18] Then Tukhachevskii tried to take them over for the Red Army. This was the planned-economy equivalent of a friendly corporate takeover of a private high-technology start-up.

---

[14]RGAE, 8044/1/984, 264–275 (October 22, 1943).
[15]RGVA, 29/56/349 (1932–1936).
[16]RGVA, 29/56/354 (1937).
[17]RGVA, 34272/1/146, 145 (November 16, 1933).
[18]GARF, 8418/6/243, 42 (August 22, 1933).

5.3. *Foreign Specialists*

In the late 1920s and early 1930s, sympathetic foreigners with a technical interest wrote to the Red Army drawing its attention to the military significance of work on rocketry going on abroad and offering to promote such work in the Soviet Union. In 1932, for example, Rolf Engel, a German specialist in rocketry and communist party member or sympathizer, was referred to Tukhachevskii. According to biographical notes he had worked in German astronomy and as a member of the Verein für Raumschiffart (Association for Space Travel) at its test firing range outside Berlin. Engel volunteered a report on developments in rocketry in Germany and abroad, emphasizing the breadth and depth of German developments. He also proposed to bring a group of specialists to the Soviet Union to collaborate with Soviet rocketeers.[19]

5.4. *The Foreign Press*

Established designers monitored the foreign press and worked up the information they found in order to demonstrate foreign progress, promote the cause of aerospace experimentation, and support bids for funding. If foreign press information was lacking, however, they still argued for increased funding on the grounds that foreign powers were evidently forging ahead of the Soviet Union in secret.

The Red Army chief of armament's files for 1931 testify to the pressure from below to take note of progress abroad and emulate it at home. This pressure was clearly related to funding decisions. For example, in May 1931, GDL director Petropavlovskii reported to the Red Army on work on rocketry abroad, mainly in Germany and the United States.[20] A number of German research groups and firms, including Junkers and Opel, were described as competing for patents and funding under the umbrella of a voluntary society for space travel including armed services representatives. The American scene was said to be characterized by a similar mix of commercial and military motivations. Petropavlovskii noted that rocketry could be applied to aviation as well as artillery, with the possibility of an aircraft with a primary rocket power plant capable of speeds of 1,000 or more kilometers per hour. In a similar survey submitted at the same time the GDL rocket engineer Glushko

[19] RGVA, 34272/1/146, 28–39 (no date but 1932).
[20] RGVA, 34272/1/105, 91–94ob (May 20, 1931).

emphasized that there was intense activity in western countries and that the basic difficulties in building rocket aircraft were close to solution; he concluded that "in the West both industrial and, particularly, military circles are keenly interested in the question of creating rocket shells and apparatuses".[21]

In the mid-1930s, the absence of foreign press information was used to promote bids to fund foreign commercial trips. This was the reason that RNII director Kleimenov, for example, used three times in 1936 when asking permission to send his engineers abroad generally, on a tour of Germany, France, Britain, and America, and to the Paris air show.[22] The foreign information available, although limited, was analysed exhaustively to support funding claims. In 1939 an article on developments in rocketry that had appeared 3 years earlier in the Italian journal *Revista Maritima* finally reached NII-3 director Slonimer.[23] He cited the article to demonstrate intensive German work on rocket munitions and asked for more money, supplies, and engineering personnel. He requested that the "appropriate organizations", presumably diplomats and spies, should seek out more information abroad, and applied to send a delegation from the institute to an armaments exhibition in New York.[24] The next year he made a similar request to send two specialists to Germany to find out more about work on rocketry there, and again he cited the *Revista Maritima* article in support.[25]

A survey of the historical applications of rocketry written in the ministry of the ammunition industry in 1939 noted that a veil of secrecy had descended over most military aspects of rocketry abroad; the little that was being published pointed to intense international rivalry in rocket technology. The examples cited were from a French work translated into Russian in the defense ministry.[26] A translation of an American article, also from 1939, listed the potential uses of rockets as ranging from field artillery

---

[21] RGVA, 34272/1/105, 118–120 (May 1931).

[22] RGAE, 8159/1/149, 220 (July 26, 1936), 219 (September 29, 1936) and 218 (October 13, 1936).

[23] The reference was to Revista Maritima, 1936, 6, 421–439; see RGAE, 7516/1/ 324, 12–42, for the translation. The article was mainly about rocket artillery; it raised possible applications to aviation on the last page.

[24] RGAE, 7516/1/324, 1–4 (April 9, 1939).

[25] RGAE, 8162/1/305, 30 (April 16, 1940).

[26] RGAE, 7516/1/324, 10 (no date but 1939).

to intercontinental bombardment and space exploration, and emphasized their ease of construction and use.[27]

The outbreak of war did not cut off press information. Following the maiden flight of the Whittle jet-powered Gloster E.28/39 in April 1941, Flight magazine published a series of articles about jet propulsion in London. These articles were collected in a booklet and republished by the magazine editor (Smith, no date). It seems to be in this version that they reached the Soviet Union. Their impact was significant. The booklet was circulated among designers; according to Gordon and Dexter (1999: 150), the description of the Italian hybrid jet of 1940, the Caproni-Campini N.1, encouraged staff at TsAGI in a similar design (see Table 1). In July 1944 minister of the aircraft industry Shakhurin copied many original Flight drawings to illustrate both existing and futuristic jet and turbine projects in a long briefing report for Malenkov.[28]

5.5. *Foreign Commercial Information*

Trade links gave some information to designers, but its value in lobbying for funding of jet propulsion projects was limited. In 1935, the aircraft designer Tupolev visited the United States for a second time, his first visit having taken place in 1929/30. He toured a number of aircraft factories. He saw nothing of American progress in military rocketry and his report was silent on the whole issue of jet propulsion.[29] This does not reflect secrecy; there was nothing to report because the Americans had nothing to show or hide. Surveys of the German aircraft industry were also carried out in the framework of the August 1939 nonaggression pact and these were reported to the ministry of the aircraft industry in September 1940. Here secrecy did play a limiting role; the Soviet delegations did not catch the least glimpse of the immense German activity in relation to new jet and rocket aircraft and artillery.[30]

---

[27] RGAE, 7516/1/323, 1–18 (no date but 1939), translates "What Can We Expect of Rockets?" by Major James Randolph of the US Army artillery reserve, published in Army Ordnance 19(112), January–February 1939, 225 ff.

[28] RGAE, 8044/1/1182, 123–147 (July 28, 1944).

[29] RGVA, 29/38/96, 1–479 (June 10, 1936).

[30] RGAE, 8044/1/359, 1–187 (September 27, 1940); 8044/1/358, 1–9 (September 29, 1940).

There is no indication in the files that Soviet spies gained any information about progress in jet engines or rocketry in other countries. If they did, it did not reach the aeroengine designers.

6. *Refinancing*

When projects are long term, projects in progress require periodic refinancing. Alternatively, they must be discontinued. In this section, we look at refinancing decisions affecting projects in progress to learn more about the incentives facing designers and funding principals and the calculations they made. When refinancing decisions were disputed, there was also a mechanism for conflict resolution: designers could and did appeal adverse decisions to higher authority, ultimately right up to Stalin.

6.1. *Project Evaluation and Soft Budget Constraints*

There were good reasons for governments to ration the funding of military aviation R&D. Most important was the fact that funding opportunities attracted both bad and good projects, which those allocating research funding could not tell apart beforehand. This made it efficient for principals to ration the available funding across projects and through time. The various research establishments reported regularly to higher authority on each project in progress.[31] Through time, funders could compare the progress of alternative projects in the hope of identifying the bad projects that should not be refinanced. In this way a form of rivalry similar to yardstick competition (Shleifer, 1985) could be exploited to increase principals' information.

By monitoring the progress of long-lived projects frequently, the funder could always obtain more information about the quality of projects than was available initially. Looking forward, by providing funding in instalments and tying refinancing decisions to intermediate progress reports, the funder aimed to use the additional information to restrict financing to good projects. However, the funder could not always act on the information obtained. This was because of a weakness in the funder's commitment to act

[31] For example RGAE, 8159/1/137, 2–28 (no date but 1937), 8162/1/240, 9–63 (January 9, 1940), and 8162/1/449, 2–61 (January 14, 1941) for the annual reports of RNII-NII-3 in 1936, 1939, and 1940, respectively. In 1967, the annual reports for 1937 and 1938 were transferred from RGAE to the archive of the USSR Academy of Sciences where they can no longer be traced. I thank Leonid Borodkin for looking.

on this information after the event. Faced with poor intermediate results it could still be efficient to go on paying for a project that, in hindsight, the funder would prefer not to have initiated in the first place (Dewatripont and Maskin, 1995). This was because of sunk costs: since part of the project had already been paid for, the likely return was now increased relative to the fraction of costs not yet sunk. One result was that despite the funder's intentions budget constraints became soft ex post. Another was adverse selection: it gave R&D agents an incentive to understate needs and overstate expected returns so as to obtain the first instalment of funding. Once the first instalment was paid and had become a sunk cost, the payment of the next instalment became more likely. This meant that projects could continue to be refinanced even when they were known to be bad.

Soviet funding arrangements thus offered a degree of protection for self-serving interests. There was clearly rent seeking; was it intentionally tolerated at any level? Were bad projects deliberately fostered, for example, to share rents and promote loyalty? Some allegations of this nature concern the rocket designer and NII-3 director Kostikov. The background is important: Kostikov remains a controversial figure. His accusers resent the fact that he took the public credit for developing the famous Katiusha rocket mortar from its true inventor Langemak who was executed (Medvedev, 1978: 36–37). They argue that Kostikov was not an accidental beneficiary of the purge at RNII but a willing instrument of Ezhov and Stalin, a renegade GIRDer who turned against his former comrades. They hold him at least partly responsible for the repression of Korolev and others (Siddiqi, 2003). According to Golovanov (1994: 512), Korolev carried a lifelong grudge against him for this reason. But Raushenbakh (1998: 66) considers the charge of complicity in the purge to be unfounded.

Serov (1997: 4) has suggested that Kostikov was unduly favoured by Stalin in the wartime allocation of project funding. In November 1942, Stalin authorized the development of Kostikov's unproven design for the 302 rocket fighter, at a time when the development of new weapons in other fields was being ruthlessly suppressed in favour of mass production of existing ones. Serov notes the "practically unlimited financial possibilities" at Kostikov's disposal: 25 million rubles for NII-3-GIRT in 1943, compared with a similar sum for the Iakovlev and Mikoian aircraft OKBs put together. It is true that subsequently, Kostikov was punished for the 302's failure: in the spring of 1944 he was sacked, then arrested. On the other hand his punishment was mild: he was released after a year in prison, and retained his military rank and medals. Golovanov (1994: 511) claims that "Stalin needed

Kostikov, since [the latter] was one of the bearers of the Stalinist world order." Whatever that means, however, what we know of Stalin suggests that by this time he did not regard anyone as indispensable.

6.1.1. Adverse Decisions

In the quasi-market for inventions, projects developed out of initiatives at lower levels. The role of funding principals and planning decisions was reactive and tended to validate these initiatives. Consequently the refinancing of projects in progress was normal and we do not usually see explicit decisions to that effect except in those rare cases when verbatim minutes of discussion meetings were preserved (Harrison, 2003a).

A decision not to refinance a project in progress is illustrated from a file of the Red Army department of inventions.[32] In October 1937, engineer R.G. Sergeev of the design department of aircraft factory no. 22 at Fili submitted a proposal to design a 500–1,000 kg thrust rocket motor for an auxiliary flight booster, aircraft launcher, or rocket fighter. He based his proposal on a suggestion by the German specialist Eugen Sãnger that had been published in an unnamed Swiss journal in 1936. He signed an agreement with the department of inventions on August 15, 1938 for the sum of 5,000 rubles. He failed to complete the work promised, so his expenses were paid off in the sum of 1,000 rubles only on September 26, 1940. We learn from this that small sums were easily written off.

Bigger decisions were more complicated. This is illustrated by the turbulent history of RNII. Frustration with the results of military R&D boiled over in the purges of 1937–1938 (Harrison, 2000: 128–130; Siddiqi, 2000: 10–11; for previous accounts see Medvedev, 1978: 34–37, 42–43; Holloway, 1982b: 387–388). In May 1937, Tukhachevskii was arrested. The purge of RNII began in October with the arrest of director Kleimenov, Glushko, and others including the rocket mortar designer G.E. Langemak. In June 1938, work on the Korolev–Glushko rocket glider was suspended, the reasons given being the need to concentrate resources for rearmament on projects of more immediate military utility. A few days later Korelev was arrested, accused of being a Trotskyist saboteur, and sentenced to 10 years' forced labor. Impatience with the lack of results of Korolev's work on rocket

[32] RGVA, 29/56/361 (1937–1940).

aviation was clearly a factor. The testing of liquid-fuelled rocket aircraft was suspended while the rocket artillery programme was stepped up.

Why did these conflicts flare with such intensity? The conflict between artillerymen and space enthusiasts at RNII had simmered through the mid-1930s before the purge of 1937–1938 swept the GIRDers away, taking several of their opponents with them. Siddiqi (2003) suggests that the technological uncertainties were simply too large to be settled scientifically on the basis of the limited funding provided by principals. This heightened the risks of R&D activity, and high stakes plus limited resources fed back into bitter infighting.

The end of KB-7 was decided by a combination of factors. The arrest of its sponsor Tukhachevskii created an immediate threat. At first, KB-7 director Korneev staved off repression by joining in the destruction of the leading figures of RNII; he sent slanderous allegations to Stalin about Kleimenov.[33] In January 1938, with Tukhachevskii gone, KB-7 was taken away from the Red Army and handed over to the ministry of the defense industry where, like RNII (now NII-3), it was attached to the thirteenth chief administration for ammunition. But Tukhachevskii had devised KB-7 for the far-off development of liquid-fuelled rocketry, not the quick results now sought for immediate armament. For 3 years, KB-7 had produced nothing to show for its outlays. The annual report of work of the thirteenth chief administration listed the projects completed under each institute or bureau and the weapons officially adopted by the Red Army for armament as a result. Under KB-7 the report for 1938 says only: "for armament in 1938 nothing supplied, in view of the long-term [perspektivnyi] character of work".[34]

In the atmosphere of the time KB-7 became an easy target. In early 1939, the Red Army resolved to close down a wide range of projects in ammu-nition R&D, not just those concerned with aviation jet propulsion, but including the work it was funding in KB-7. According to reports made to ammunition minister Sergeev, and forwarded by him to deputy defense minister Kulik, navy minister Frinovskii, and Prime Minister Molotov, the aggregate plan for ammunition research and experimentation for 1939 had

[33]RGVA, 4/14/1628, 123–128 (June 15, 1937). Kleimenov had a history of personal conflict with Tukhachevskii's subordinates, which might have helped him; see for examples RGVA, 34272/1/177, 1–2 (July 26, 1934) and 33 (September 13, 1934). But he was isolated by Ordzhonikidze's suicide.

[34]RGAE, 8162/1/89, 125 (no date but 1939).

been agreed among these ministries with the general staff the previous year.[35] In the course of disaggregating the plan and agreeing to individual contracts with R&D establishments, however, the Red Army had unilaterally reneged on commitments worth 40 million rubles (out of 52.5 million) and the Navy on 7.5 million (out of 25 million). Even after immediate cutbacks, 25 million rubles worth of research and experimentation remained without a sponsor, including two institutes that were entirely without funding. One was KB-7. Among the projects without funding at NII-3 were the Korolev–Glushko rocket glider and a ramjet project.

The effectiveness of these cancellations is not clear-cut. We do not know that either Voroshilov or Molotov gave Sergeev's protests a hearing. By the end of the year, KB-7 had been closed down; according to Siddiqi (2003) the staff, starved of funding, turned on each other and eventually on Korneev too, who was arrested and imprisoned. On the other hand the rocket glider and ramjet projects at NII-3 were evidently reinstated, and the Korolev–Glushko RP-318 made its maiden test flight in 1940 although the designers were absent and others got the credit. To judge from the annual returns shown in Table 3, NII-3 continued to expand during 1939 at a high rate, perhaps by absorbing the staff of KB-7. But the funding of NII-3 was squeezed in real terms in 1940 and its expansion was brought to a sudden halt.

In short, in 1939, the funder lost patience with the designers' lack of results and tried suddenly to enforce a harsh constraint on the budget for research and experimentation. In the short term, this attempt was only partly successful because the constraint was softened again by the fund holder's lobbying to reverse cutbacks, or by drawing on the fund holder's budget, or by some combination of the two. The efforts to squeeze R&D outlays may have continued in 1940.

Aircraft design yields a few cases of design organizations that were closed because of lack of results (Albrecht, 1993: 214–215). More frequently, the chief designers were imprisoned along with their teams, for example, Bartini, Grigorovich, Miasyshchev, Petliakov, Polikarpov, Sukhoi, and Tupolev. Kalinin was executed (Albrecht, 1993: 133–136). These and episodes such as the purge of RNII gave credibility to subsequent threats of extreme penalties for failure. Those charged with designing the Soviet Union's first atomic bomb, for example, all expected to be arrested if

[35] RGAE, 8162/1/299, 36–54 (March–April 1939). Commissar Sergeev was unconnected with engineer Sergeev mentioned above.

it failed to detonate (Holloway, 1994: 215; Simonov, 2000: 154). The credible penalization of individuals for R&D failures helped to compensate for the weakness of the principal's commitment to penalize organizations financially for lack of results.

6.1.2. Appeals against Adverse Decisions

Some designers appealed to higher authority against threats to project financing. In a related context, Markevich and Harrison (2006) note that "written appeals to higher authority were a general feature of life in a society with underdeveloped legal enforcement, and citizens in all walks of life used them to seek truth and justice". Disputes within RNII were a plentiful source of petitions. In May 1934, for example, both Korneev and Korolev complained to party and military authorities over RNII director Kleimenov's suppression of liquid-fuelled rocket projects.[36] After the RNII purge, Korolev appealed from prison to the prosecutor (Raushenbakh, 1998: 61–64), he also wrote to both Beriia and Stalin personally (Golovanov, 1994: 286–289) to protest his innocence and ask to be allowed to return to work.

For a petition to carry weight at higher levels the appellant had to invest something in the outcome; thus it was normal to support an appeal by listing the writer's progress in the cause at hand. In effect the appellant offered her specialist reputation as a hostage (Williamson, 1996: 120–144) to support the transaction sought. But most inventors do not usually have much reputation in their lifetimes. Most people are lucky to have even one great idea, let alone carry it out; as a result, the past is a poor guide to an inventor's future performance and this is true both before and after she has actually invented something. At a given point in time most inventors either have nothing to show for their efforts or they are failing to live up to the promise of past achievements. Thus an inventor's reputation is hard to establish and harder to maintain.

Some other kinds of reputation that could be brought into the equation were also fragile, for example, a reputation for loyalty to superiors. In the 1930s, all were familiar with the figure of the careerist who accumulated this reputation strategically, so as to spend it later. Winning a reputation for loyalty to vertical superiors was also a good way of making enemies out of horizontal rivals. Those with more energy than talent, who risked their

[36]RGVA, 34272/1/177, 5–10 (Korneev to the Okt0iabrskii party raikom, May 27, 1934), and 17–19 (Korolev to Tukhachevskii, May 29, 1934).

credit with higher levels without the talent to back it up, invited destruction as the fate of Korneev in 1939 suggests. According to Serov (1997: 4), Liul'ka was known as a "complainer" (zhalobshchik). Having already failed in an appeal to minister Shakhurin of the aircraft industry, he wrote to Stalin in May 1942 asking to be allowed to go back to work on the turbojet. This appeal is a rarity: unlike most, it had the desired result.[37] Serov (1997: 3) suggests that in Liul'ka's case it may have suited rivals to go along with his petition in order to raise the profile of the issue in their own interest. Korolev's similar appeals from prison met with the more usual response: they were ignored. It is true that Korolev was subsequently transferred from the Kolyma labor camps to the NKVD sharaga at factory no. 16 in Kazan' but this, according to Raushenbakh (1998: 66), was entirely the result of Tupolev's desire to recruit him.

7. *The Secondary Asset Market*

When projects are long term their need for refinancing has the necessary effect of creating a secondary asset market. This section gives a brief account of the market for R&D projects and the kinds of behaviour that can be observed.

According to the principle of the command economy the ownership of each project by a ministerial fund holder could only be transferred by a centralized decree. In reality there were substantial incentives for officials to mount takeover or merger bids for projects of other fundholding authorities. The attraction of a project in progress lay in the sunk costs that had already been incurred at the expense of others to whom the new fund holder did not have to pay compensation. Even if the transfer was of people rather than physical assets, the project personnel brought with them accumulated tacit knowledge, which formed significant intangible capital. The costs of takeover were political rather than financial. First, a bid required the payment of direct lobbying costs. Second, it required the expenditure of reputation; a successful bidder made promises for which he might later be held to account. Third, it weakened the centralized enforcement of ownership rights over assets on which all fund holders ultimately relied.

[37] RGAE, 8044/1/817, 19–25 (Liul0ka to Stalin, May 18, 1942). Stalin took it up with Malenkov, who took it up with Shakhurin (Berne and Perov, 1998: 86). Liul'ka's group was absorbed into OKB-293 in July 1942 (RGAE, 8044/1/817, 18 (August 10, 1942).

However, circumstances could easily arise in which it was more dangerous to abstain from the secondary market than to enter it.

These can be readily translated into the costs and benefits to a proprietary dictator. In Olson's (1993) metaphor, Stalin resembles a bandit chief who settles on a territory and monopolizes it so as to maximize the rents from it. But in secondary markets that formed under his regime, lesser bandits roved. Up to a point this could benefit the economy; it reallocated resources to those who would put them to better use. But the standard of valuation of "better use" was private not social. Since the losing side did not recover the social value of their assets the fear of expropriation weakened the dynamic incentive to invest rather than seek rents. Finally, even if static allocation improved, the dictator's control over the production and sharing of rents was likely to be weakened.

In the secondary market for R&D assets the transactions that we observe were of two kinds. First, there were horizontal mergers and takeovers, sometimes hostile. Second, from time to time the NKVD swooped down from above to confiscate projects using its powers of arrest and confinement.

Any organization could take part in the secondary market, and small organizations could be as aggressive as large ones. There were few advantages to being small in the command system other than the chance of being overlooked. As an example of the latter, to work in a small outfit like KB-7 in 1937 meant two more years of life expectancy compared with working in the larger and more prominent RNII. Usually, however, it was better to be large. Large units could realize significant economies of scope; they were less reliant on outsiders for essential goods and services. There were also managerial economies of scale. Ministerial officials tended to promote larger units on the grounds of alleged economies of scale in research and this was reflected in the frequent calls to concentrate effort and eliminate duplication or "parallelism." Whether these economies really existed is another matter. Rationalization and concentration were regarded as progressive almost beyond debate and their advantages were seldom questioned, especially when comparisons were made with the scale of R&D establishments in aero engineering abroad.[38] But as long as ministers believed in them, bigger organizations had the advantage. As a result,

[38]RGAE, 8044/1/460, 49–51 (December 31, 1940): an explanatory memorandum by People's Commissar for the Aircraft Industry, A.I. Shakhurin, on the 1941 plan for aero engineering research and experimentation.

larger units were continually on the lookout for favourable rationalization opportunities, while smaller units also had to grow, if necessary at the expense of others, or risk being swept up by rivals at any time.

The logic of the takeover bid was a call to write off the past. Consider a failing project, one that had incurred significant sunk costs without giving results on schedule. Was the project intrinsically bad, underfunded, or poorly led? If the lack of results could be reasonably attributed to lack of resources or organization, then it was efficient not only to write off the sunk costs but also to refinance the project under new management. This logic may have been stronger when the scope of activity and the number of projects were increasing because growth was likely to mean an increasing number of potentially weak projects. Thus proposals for takeovers and mergers were particularly evident in the years 1937 and 1938 during and after the purge of RNII.

The purge sparked a bid to rationalize research on jet engines. Staff of the Academy of Sciences Institute of Theoretical Geophysics wrote to Prime Minister Molotov at the end of December 1937; Molotov's secretary forwarded it to both defense minister Klim Voroshilov and deputy minister for the defense industry Kaganovich for comment.[39] The authors highlighted the unmet needs of Soviet aviation in contrast to the resources being devoted to jet aero engine development by the "capitalist countries," the designs being pursued by Breguet and Junkers in France and Germany, and the veil of military secrecy, which was hiding real progress abroad. They ascribed resistance to jet designs in the Soviet Union to a coalition of "enemies of the people" including designers such as Langemak of RNII and soldiers such as Efimov, chief of the Red Army artillery administration. As for the established jet engineers such as Merkulov, they charged them with "creating 'conditions' of work bordering on mockery" (the phrase "bordering on" could have been significant: not actually mockery, just bordering on it). They called for pure and applied research encompassing ramjets, pulse-jets, and hybrid engines to be scaled up and personnel and projects concentrated in KB-7, which the Red Army should hand over to the defense industry.

Although the Institute of Theoretical Geophysics had no clear interest in the fortunes of KB-7 its intervention was probably not altruistic. The chances are that someone had put them up to it. The bid failed, however.

[39] RGVA, 4/14/1925, 16–18 and RGAE, 7515/1/378, 304–306 (both December 31, 1937).

Kaganovich called on the new NII-3 director Kostikov for comment. The latter presented a strongly argued case for his own institute to be the new centre for jet engine R&D, based on a short scientific review of jet concepts and experimental results. He concluded that it was essential to draw into this line of work people "closely involved with aviation technology" as opposed to those "incidentally showing an interest"(this was a slighting reference to KB-7); Kaganovich in turn supported the NII-3 position.[40] So did the Army: the new air force chief Loktionov wrote to Voroshilov supporting the writers of the Institute of Theoretical Geophysics on the principle of giving more priority to jet engines but rejecting the case for KB-7 on grounds that the latter lacked the necessary research and production equipment. He recommended NII-3 as the new centre for jet engine development, and deputy defense minister Fed'ko relayed these arguments to Molotov adding a proposal that NII-3 absorb relevant personnel of KB-7.[41] This was the eventual outcome, although KB-7 survived until the end of 1939.

As has been seen, the years 1938 and 1939 saw high mortality among aviation jet propulsion projects; in 1939/40 there were several new start-ups. Therefore it is no surprise that, in his memorandum of February 1941 to deputy prime minister Voznesenskii, minister of the aircraft industry, Shakhurin, listed the various ongoing projects and proposed "to concentrate all the work in progress in [NII-3] of the ministry for the ammunition industry [...] and transfer the institute to the ministry of the aircraft industry", enclosing a draft decree to that effect.[42] This particular bid failed for the moment, or was overtaken by events; after the outbreak of war, NII-3 was first subordinated directly to the central government, and handed over to the aircraft industry only in 1944.

Finally, when the R&D agents were seen to have gone too far their own way, Stalin used his security agency to force them back into conformity with the priorities of state. As is well known the NKVD arrested a number of aircraft and aero engine designers in the purges of 1936–1938 and used them to formulate proposals for implementing new designs. A number of aircraft designers from TsAGI, TsIAM, and the Tupolev design bureau, including Tupolev himself, were held at factory no. 156 (Ozerov, 1973), then

[40] RGAE, 7515/1/378, 298–303 (no date).

[41] RGVA, 4/14/1925, 21–21ob (February 4, 1938) and 4/14/1925, 22–22ob (February 15, 1938).

[42] RGAE, 8044/1/460, 60–57 (February 5, 1941).

reorganized as TsKB-29. Of the RNII personnel arrested in 1937, some were shot and the rest sent to labor camps; some survivors were subsequently recalled from the Kolyma and put to work in the aero engine sharaga at factory no. 16 in Kazan' (Golovanov, 1994: 318–328).

To summarize, was the secondary asset market "real" or just a quasi-market? The dictator did not intentionally create or authorize it; the buyers and sellers were independently self-interested; and the dictator seems to have made little attempt to align their interests with his own. But there was no equilibrating process, and enforcement rested exclusively with the dictator. For these reasons it would be a mistake to think of it as a real market, but still the lines are blurred; although only a quasi-market, the secondary asset market had some strongly "realistic" features.

8. *Closing the Quasi-Market for Inventions*

This section describes the end of the Soviet quasi-market for inventions in jet propulsion. The market was closed down between 1944 and 1946. By 1944 Germany had revealed the breakthroughs that had been in preparation for so many years, launching jet fighters and bombers and a rocket fighter into aerial combat and firing jet powered cruise missiles and ballistic rockets at London. The fog of technological uncertainty was gradually blown away, and this gave the authorities the information they needed to recapture centralized control over aerospace innovation.

The new phase began on February 18, 1944 when Stalin's war cabinet, the GKO, resolved to give new attention to jet propulsion technology. There was a certain amount of the usual reorganization and re-designation, so GIRT, under direct control of the central government since January 1942, was renamed the Research Institute for Jet-Propelled Aviation (NIIRA) and then Research Institute no. 1 (NII-1) of the aviation industry. A further decree dated May 22, 1944 commissioned a number of aircraft designers to fit rocket and jet boosters to existing propeller-driven aircraft; Lavochkin was to build a turbojet aircraft; Glushko, Dushkin, and Isaev were to work on new rocket motors, and Liul'ka and Uvarov were to work on jet engines; by 1945 Liul'ka was roughly where Whittle had been in 1939, with an engine, the S-18, ready for ground testing but not for flight (Egorov, 1994: 413–424, 431–436).

From this point there began a marked change in the organization of aviation R&D. Regimentation and drilling increasingly took the place of initiative and networking. This process was aided by a political shock to the designers, fund holders, and funders: the "aviators' affair" of early 1946 in which leaders of the Air Force and aircraft industry were arrested and imprisoned. At the end of the war Stalin concluded that the Soviet Union was lagging in aircraft and aero engines because relationships between the Air Force and the leading aircraft design bureaus had become too cosy, enabling favoured designers to establish monopolies and relax the pace of development (Bystrova, 2000: 320–321). The purge also had wider political dimensions that do not concern us (Pikhoia, 1998: 45–47).

A meeting in the ministry of the aircraft industry in March 1946 whipped the design sector of the industry into line. The new minister Khrunichev promoted the solution as "raising the initiative of the chief designers" and "lifting the design bureaus and research institutes towards more initiative-led work [initsiativnaia rabota]."[43] What this meant in practice was a demand for fewer failures, less seeking of rents, and more accountability:[44]

> [...] we should not take the path of adventurism, we don't have to spend money on any project [just] in order to convince ourselves that the expenditure was pointless [...] Until now we convinced ourselves of the lack of profit in one or another project or new design [only] after the development work [...] In future we must change the system and test a person's ability over a fixed period of time. Here I have to say that a thousand objections will be raised, they'll say "They're hindering us, they're not giving us the chance to work." In the context of a rational state approach we must sweep aside all such discussions so that this business isn't taken over by demagogy. We have to find the grain that will give the state the necessary yield.

The aircraft designer Il'iushin, first in discussion, developed the point:[45]

> Our resources are limited. We must review all the aircraft and engine types, throw all our resources into them, and move the matter forward. We shouldn't throw a single extra kopeck where it isn't needed. For this, our aviation industry has to stop looking at design bureaus as "free" organizations in the spirit of: "if something works out that's good, and if it doesn't no one has to account for it." The ministry of the aviation industry is accountable for the work of the design bureau. That's clear to all.

The outcome was that during 1945 the German jet engines and aircraft were inventorized and allocated for further investigation and development to research institutes and design bureaus grouped under a

[43]RGAE, 8044/1/1342, 17 and 21 (March 1, 1946).
[44]RGAE, 8044/1/1342, 7 (March 1, 1946).
[45]RGAE, 8044/1/1342, 31 (March 1, 1946).

new chief administration for jet-propelled aviation within the ministry of the aircraft industry.[46] As the scope of research and design work widened, a relentless process of regular target-setting and monitoring set in. Government decrees set targets. The industry's progress towards each target set by the decree of April 22, 1944 was rigorously monitored.[47] By 1947 the industry plan for experimental work on aero engines for 1947 listed 48 separate projects of which 38 were for jet or rocket engines; against every project was noted the government decree that authorized it, the technical parameters set for it, and the deadlines for completion and handover for external assessment.[48] There was still rivalry, but it is hard to imagine anything more different from the designer-led, uncoordinated rivalry of the 1930s.

It took nearly two years to close the quasi-market completely; the length of time required may seem surprising. The reasons are to be found abroad: just at the time the authorities were trying to close down the market at home, two completely new markets for inventions sprang up outside the country. To add to the confusion, each market was specialized in a different, competing version of the turbojet.

One market was in the Soviet zone of occupied Germany and this market continued to grow actively for more than a year after the end of the war. On offer in this market were BMW and Junkers turbojets with axial-flow compressors on the same scheme as Liul'ka. The German designs were low powered and unreliable, however. The German market was not closed until October 1946, when Stalin authorized the wholesale deportation of nearly 3,000 rocket and aviation specialists from eastern Germany. Once on Soviet territory they came under the direct control of the MVD, which accommodated them for several years and managed their work in a number of specialized sharashki (Sobolev, 1996: 58–118; Harrison, 2000).

The other market was a real international one. The original Heinkel and Whittle jets had used a radial compressor, which gave superior reliability although with a large front profile and limited scope for scaling upwards. Further down the same road, Rolls Royce was now manufacturing reliable engines that were efficient for their size. In 1946 the British government

[46]RGAE, 8044/1/1318, 21, 22 (1945). This file contains the materials for a report by the Commission for Study and Assimilation of German Jet Propulsion Technology headed by minister for the aircraft industry Shakhurin.

[47]RGAE, 8044/1/1496, 317–323 (December 1945) and 274–284 (January 5, 1946).

[48]RGAE, 8044/1/1637, 230–235 (July 15, 1947).

agreed to the sale of 25 Rolls Royce Nene engines to the Soviet Union, followed by 30 Derwent V engines (Shavrov, 1988; Egorov, 1994).

At the end of the process, the Soviet Union had a jet aircraft industry with engines that came from three genetic branches: German, Soviet, and British. In 1946, the BMW and Junkers axial-flow engines were copied and installed in the Soviet Union's first jet aircraft, the experimental Iak-15 and twin-engine MiG-9, but were not developed further. In 1947 a Soviet jet engine, Liul'ka's axial-flow TR-1, was at last developed for the twin-engined Su-11 fighter. Also in 1947 the Rolls Royce radial-flow Derwents were rapidly assimilated and developed for the Iak-19 fighter, while in 1948 the Nenes were installed in the Soviet Union's first serial production jet fighter, the MiG-15.

9. *Payoffs, Reputation, and Enforcement*

To summarize: for a decade and a half the field of jet propulsion was marked by technological uncertainty and information bias; specialists knew more than the officials who funded them. Under these conditions centralization was out of the question. Instead, a quasi-market for inventions evolved with a secondary quasi-market for research assets alongside it. Both markets had significant "realistic" features, and rivalry in both was regulated with difficulty. When the uncertainty was resolved, however, the market was closed down.

In Section 9, I address the issue that remains: how, in the period of uncertainty, could this rivalry have produced any successes at all? The rewards that funding principals offered to inventors were small, distributed as much for time serving and loyalty as for achievement, and could be taken away at any time. In such a setting we would expect agents to continue to invest their efforts in seeking to share the available rents rather than in making profits from inventing something new and useful.

As a first pass at this problem, it would seem that successful designers were strongly motivated by two things that were very valuable and could not be confiscated. One was intellectual curiosity and a desire to solve the intrinsically interesting problem at hand, strengthened in some cases no doubt by the desire to escape from a crazy world into a bubble of scientific rationality. The other was the reputational gain attached to priority in invention. A significant aspect of this reputation is that it was not the gift of the Soviet state, although we shall find circumstances in which the state could steer its allocation; rather, it was awarded by the "Republic of

Science." I borrow this phrase from Dasgupta and David (1994) but the context suggests a departure from their framework. Dasgupta and David contrast motivations and incentives in Science with those in Technology: in their terminology scientists do research so as to gain priority in discovery, and technologists engage in innovation, which is replicative of scientific discovery, so as to make profits. In the present case it is obvious that technologists could achieve priority not by discovery but by invention. Where invention is concerned, therefore, Technology is like Science.

I do not mean that the position and salary that the state provided were completely unimportant. Specialists at the global frontier of military technology had material aspirations for themselves and their families just like other people, and research and design work could help them towards such goals. The cash value of the potential rewards was substantial. To give a rough idea, the average monthly pay of specialist ("engineering and technical") workers at NII-3 in the first quarter of 1941 was 818 rubles, roughly two and a half times the average industrial wage of 1940.[49] More detail is available for the 250 "management and administrative" staff of Liul'ka's aircraft factory no. 165 in August 1946; this category included everyone from the chief designer (6000 rubles a month) to the floor sweepers in the labs (200 rubles). The median monthly wage was 875 rubles, compared with 626 rubles for the average industrial wage in December of that year (Filtzer, 2002: 235). It was enough to staff the factory fairly fully; there were only eight vacant posts, of which seven offered less than 400 rubles.[50]

Nominal pay was just the start. Work in military R&D also gave access to plentiful bonus payments and awards. Director Slonimer of NII-3, for example, is quoted as having received 19,250 rubles on top of his salary in 1939 and the first half of 1940.[51] The evidence is fragmentary and we have no clear picture of how such sums were fixed or allocated. On the face of it officials made recommendations to mark significant achievements, transitions, and anniversaries.[52] To put it another way, it seems possible

[49]RGAE, 8162/1/449, 87 (April 10, 1941).

[50]RGAE, 8044/1/3079, 82–91 (August 27, 1946).

[51]RGAE, 7516/1/692, 3 (November 21, 1940).

[52]Transitions: To mark the transformation of GIRD into RNII, and in light of their achievements including the first Soviet liquid-fuelled rocket, chief of the Red Army administration for military inventions Terent'ev asked Tukhachevskii to set aside 2,500 rubles to be distributed among the GIRDers as bonuses (RGVA, 34272/1/146, 145: November 16, 1933). Achievements: to mark the successful exploitation of its rocket

that any excuse would have done. This is certainly the impression given by pre-war investigations, which threw up many alleged cases of unjustified side payments and awards. A finance ministry audit of defense industry research establishments in 1938 found that TsIAM, the lead organization for aero-engine development, was running no less than 19 separate incentive schemes on which it had spent 1.2 million rubles in 1937 along with another 200,000 rubles on rest cures and sickness benefits.[53] An audit of NII-3 2 years later not only threw up Slonimer as a case of unjustified side payments but also alleged that Slonimer had used incentive schemes to pay off his colleagues and bosses.[54] Whether we should fully trust such accusations is another matter; a lot of unfounded allegations were flying around at that time.

Also of significance equal to nominal pay in a shortage economy was the privileged consumer provisioning available to those whose jobs gave them the right to a Moscow residence permit. Mukhin (2004) has shown that in the 1930s Moscow-based aviation specialists could be persuaded to relocate to the provinces only with great difficulty, and then only when the ministry guaranteed their future right of return. This applied whether "the provinces" were hundreds of kilometres away or only just outside the city limits.

In short there is no doubt that working in aviation R&D offered good pay and a reasonable life style for the time. But that is not the point. The point is: was this the mechanism that motivated inventive effort rather than time serving?

This seems unlikely for three reasons. First, Frey (1997) has analysed labor markets where agents are strongly motivated by intrinsic incentives such as satisfaction or reputation. He concludes that, where intrinsic incentives are strong, they are weakened by external attempts to control motivation by substituting cash. In short, when morale matters, manipulation is demoralizing. Thus, supposing the Soviet state had been capable of

shells in combat against Japan and Finland, NII-3 director Slonimer asked ammunition minister Sergeev to decorate his most outstanding staff, not named (RGAE, 8162/1/306, 186–187: July 22, 1940). This request then became evidence in the charges subsequently levelled against him that I describe in the text. Anniversaries: in relation to the tenth anniversary of the prison design bureau OKB172, armament minister Ustinov and NKVD chief Kruglov wrote to Stalin to request that he award commemorative decorations to the former "enemies of the people" working in it (GARF, 9401/2/170, 213–228: July 13, 1947).

[53] RGAE, 7515/1/379, 134–137 (April 19, 1938).

[54] RGAE, 7516/1/692, 1–7 (November 21, 1940).

targetting monetary incentives accurately on inventive effort, the result was likely to have been counterproductive. Second, as a matter of historical fact the state does not seem to have had this capability. Rather, the incomes and rights of residence of aviation designers went with position rather than success. Third, it is clear that the state could and did confiscate income, rights of residence, and position at any time.

There were always safer ways of earning a living than by designing military equipment. If that was what one had to do, then it was safer to be paid less, not more. In 1950, for example, Stalin suddenly accused his favourite aircraft designer Iakovlev (2000: 395) of diverting state funds into excessive salary and bonus payments: "Do you know what they say about you behind your back? They tell me you're a thief". What saved Iakovlev was the support of his boss, minister of the aircraft industry Khrunichev, who proved to Stalin that Iakovlev's design team and production workers were fewer in number, lower paid, and less well equipped than those of the other designers.

In short, while position and pay should not be neglected, they can hardly be regarded as credible payoffs for an inventor's lifetime effort. Something else that was very valuable had to be available to overcome the fear of expropriation. This is why I look for another mechanism to explain their motivation, and I find it in the one thing that could be credibly offered, since it was in the gift of the Soviet Union's rudimentary civil society rather than of the Soviet state: reputation for priority.

To demonstrate the high value and decisive importance of reputation for priority in practice, I offer four kinds of evidence. First, when it turned out that a given problem had already been solved abroad we find that Soviet designers were reluctant to put effort into replicating foreign experience or to accept foreign advice. When Stalin ordered the deportation of thousands of German specialists in atomic science and aviation, to the Soviet Union in 1946 it appeared that his country was gaining priceless human capital, but in fact it was difficult and at times impossible to persuade Soviet designers to collaborate effectively even when the Germans were willing (Harrison, 2000). In the nuclear industry, Soviet designers were reluctant to copy from American designs without introducing their own trademark innovations, not all of which were successful (Lebina, 2000). Notably, Beriia tricked the designers of the first Soviet atomic bomb into replicating the work done at Los Alamos by making them believe that they had the priority; he had their leader Kurchatov guide their work on the basis of intelligence reports in such a way that they thought they were breaking all the ground themselves and did not know that they were merely following in Oppenheimer's footsteps (Holloway, 1994). The low status of replication generally (Wible, 1998: 23–42) then helps to explain why a country so rich in invention could be so poor at innovation (Berliner, 1976): the same mechanism could not promote both.

Second, we find that Soviet designers feared plagiarism before the event. Harrison (2003a) provides two cases: in one, a steam turbine designer defied a direct order from a ministerial superior to share progress with rivals and so accelerate progress; those involved were clear that the desire to protect his personal priority was the motive. In another case, the gas turbine designer, Uvarov put up a barrage of excuses to shut interested observers out of his work; he argued that it was too secret to share with naval designers who wanted to find out what he was up to. A cynic might wonder whether these specialists just wanted to hide their own lack of progress, but Uvarov at least was a serious pioneer who had real claims to protect.

Third, it is evident that nothing caused more lasting personal bitterness in the "Republic of Science" or gave rise to a deeper desire for personal vindication than the exceptional cases in which the temporal state intervened in the process of attributing priority and caused a reputation for priority to be suppressed or enabled it to be stolen. A claim that was suppressed was to the Soviet Union's first and the world's second rocket aircraft, which the future chief missile designer Korolev developed in the late 1930s; he was arrested and imprisoned in 1937, while it was in development, leaving it to others to fly it for the first time in 1940. A claim that was stolen was to the Katiusha, the famous rocket mortar of World War II; Langemak developed it but the new NII-3 director Kostikov took the public credit for it after Langemak was arrested and shot in 1937 (Siddiqi, 2000: 25).

Fourth, the high value of reputation for priority is confirmed by the fact that, when it ceased to be available as an incentive for the aviation designers, the government had to replace it with some other inducement and this took for the form of side payments that were extravagantly large by comparison with anything previously offered. By 1944, it was clear that the Germans and British had solved the principal problems of jet reaction. Thus, Soviet designers could no longer win a worldwide reputation for priority. Korolev and Liul'ka could be the fathers of Soviet rocketry and the Soviet turbojet but no more than that. As far as Stalin was concerned the priority was now to copy the west; replication had to come before invention. Government decrees of April and June 1946 and May 1947 now set the designers in an organized competition with each other, setting design targets and offering enormous cash prizes to the design teams and chief designers that met them. First prize in the aircraft competition included 700 thousand rubles for the chief designer together with an Order of Lenin, a Stalin Prize, and a luxury ZIS-110 private car, and many more hundreds of thousands of rubles, apartments, cars, and medals to be shared among his deputies and design staff.[55] Something similar was also on offer to the aero engine designers; in the spring of 1948 Liul'ka was given a Stalin Prize (third class) and was personally awarded 600,000 rubles, or 100 times his monthly pay in 1946, with a further 800,000 rubles for his design team; he got another Stalin Prize the next year, upgraded to first class.[56]

In short, the designers themselves regarded a reputation for priority in their field as extremely valuable, were ready to take considerable risks to establish and protect it, and regarded attempts to infringe on it as one of the most heinous crimes, equivalent to a deadly physical assault. To obliterate a person's achievement was as bad as to destroy them physically. When money took the place of reputation, enormous sums were required.

Finally, markets for inventions had some common features in all countries. Fears of plagiarism and confiscation were ever present in Whittle's calculations. Because he was a serving officer, the British government as funding principal held "Free Crown User" rights over his patents (Whittle, 1953: 47) and could and eventually did do with them as it wished. Whittle had hoped (1953: 102–103) to profit from the development and manufacturing. He also feared theft by Rover, an interim collaborator (1953: 115–116, 205–206), but in the end the government handed the business to Rolls Royce. For his ultimate reward Whittle had to make do with £100,000 from Parliament, a knighthood from George VI, and an FRS from the Republic of Science. As for Germany, according to Ohain (Ermenc, 1990: 30, 40–41), Heinkel saw no protection in patents and defended himself by keeping the German air ministry in ignorance until the first test flight; Ohain also regarded patents as worthless. In the end the big German contracts for wartime mass production of jet engines went to BMW and Junkers, not Heinkel; this was less an act of confiscation than a punishment for pursuing a radial-flow concept when German aviation officials had

[55]RGAE, 8044/1/1795, 94 (March 26, 1948).

[56]Stalin Prizes for 1947: RGAE, 8044/1/1962, 94 (March 31, 1948). Prize money, RGAE, 8044/1/1795, 79 (no date but April 1948). Stalin Prizes for 1948: RGAE, 8044/1/1965, 7 (no date but March 1949).

decided, correctly, that the future lay with the axial compressor (Kappus in Ermenc, 1990: 72–75).

Conclusions

To conclude: Jet propulsion R&D was carried out in the context of a vertically organized command system. In the interwar period technological uncertainty was so great that it could be processed only in a market-like context. There was a quasi-market for inventions with horizontal rivalry, competitive rent-seeking, and attempts to bar entry and create monopolies. A secondary quasi-market in research assets also sprang up that involved takeover and merger activity.

Designers created the quasi-market for inventions and held the initiative in it. There were more initiatives than the authorities were willing to fund. Inventiveness was not in short supply. The authorities' main problem was to control it, not promote it.

In the Soviet Union, jet propulsion R&D was an artisan industry. The resources available to fund research were extremely limited and funding was rationed. Budget constraints on individual projects in progress tended to become soft, however. Once a project had been selected for funding it had a good chance of its funding being continued until aggregate limits on the funding principals' resources and patience were breached.

It was difficult or impossible for the authorities to know whether they were getting value for money. Designers who succeeded in getting initial funding and subsequent refinancing were "heterogeneous engineers." They invested resources in lobbying and political reputation to ensure that their projects were selected for funding and, once selected, to protect them against termination from above or takeover by rivals in the name of rationalization.

When faced with adverse funding or career decisions or takeover threats designers retained the option of appealing to higher instances in the vertical hierarchy. The success of such appeals rested in part on technological reputation, but an inventor's reputation was difficult to establish and appeals were rarely successful.

A reputation for priority was one reward for successful invention and it was the only reward that the state could not easily steal or confiscate after the event. This is because reputation for priority was bestowed by the "Republic of Science" (or Technology), not by the state. Some evidence of the value of a reputation for priority is that, once it could no longer be won, the state had to replace it with individual incentives valued at hundreds of monthly pay cheques.

The state's exploitation of the value of reputation for priority explains why the market for inventions worked, despite the fact that it was only a quasi-market, not a proper market. Selective decentralization was effective under these conditions partly because the incentive mechanism on which it relied was assured and enforced by an external party: the Republic of Science. By definition, however, the quasi-market for inventions was the only one that could be made to work this way; its particular reputation mechanism would be powerless in any activity that involved replication, that is, in most areas of economic life.

Appendix

Tables 1 and 2 are supported by an appendix, available in spreadsheet form at http://warwick.ac.uk/markharrison/data. The data in the appendix are extracted from plans, reports, and memoranda of the ministries of defense, internal affairs, heavy industry, the defense industry, ammunition, and the aircraft industry.

References

Official Papers

GARF (Gosudarstvennyi Arkhiv Rossiiskoi Federatsii): State Archive of the Russian Federation, Moscow.

RGAE (Rossiiskii Gosudarstvennyi Arkhiv Ekonomiki): Russian State Economics Archive, Moscow.

RGVA (Rossiiskii Gosudarstvennyi Voennyi Arkhiv): Russian State Military Archive, Moscow.

Publications

Açemoglu, D and Robinson, J (2000). Political Losers as Barriers to Economic Development. *American Economic Review*, 90(2): 126–130.

Albrecht, U (1993). *The Soviet Armaments Industry.* Chur, Switzerland: Harwood Academic Publishers.

Almquist, P (1990). *Red Forge: Soviet Military Industry Since 1965.* New York: Columbia University Press.

Bartlett, W, Roberts, JA and Le Grand, J (eds.) (1998). *A Revolution in Social Policy: Quasi-Market Reforms in the 1990s.* Bristol, England: Policy Press.

Bergson, A (1961). *The Real National Income of Soviet Russia Since 1928.* Cambridge, MA: Harvard University Press.

Berliner, JS (1976). The Innovation Decision in Soviet Industry. Cambridge, MA: MIT Press.

Berne, LP and Perov, VI (1998). Istoriia sozdaniia pervogo otechestvennogo turboreaktivnogo dvigatelia (K 90-letiiu so dnia rozhdeniia A.M. Liul'ki). In: *Iz istoriia aviatsii i kosmonavtiki,* Volume 72. Moscow: Institut istorii estestvoznanii i tekhniki RAN, pp. 77–94.

Bystrova, I (2000). *Voenno-promyshlennyk kompleks SSSR v gody kholodnoi voiny. (Vtoraia polovina 40-kh — nachalo 60-kh godov).* Moscow: Institut Rossiiskoi istorii Rossiiskoi Akademii nauk.

Danilov, B (1981). Iz istorii sozdaniia reaktivnoi aviatsii. *Voenno-istoricheskii zhurnal,* 3: 70–75.

Dasgupta, P and David, P (1994). Towards a new economics of science. Research Policy, 23(5): 487–521.

Davies, RW (1996). *The Industrialisation of Soviet Russia, Volume 4: Crisis and Progress in the Soviet Economy, 1931–1933.* Basingstoke, England: Macmillan.

Davies, RW and Harrison, M (1997). The Soviet Military-Economic Effort Under The Second Five-Year Plan 1933–1937. *Europe-Asia Studies,* 49(3): 369–406.

Dewatripont, M and Maskin, E (1995). Credit and Efficiency in Centralized and Decentralized Economies. *Review of Economic Studies,* 62(4), 541–555.

Egorov, IA (1994). Zarozhdenie reaktivnoi aviatsii v SSSR. In: Biushgens GS (ed.) *Samoletostroenie v SSSR (1917–1945),* Volume 2. Moscow: TsGAI, pp. 394–436.

Ermenc, JJ (1990). *Interviews with German Contributors to Aviation History.* Westport, CT: Meckler.

Filtzer, D (2002). *Soviet Workers and Late Stalinism: Labour and the Restoration of the Stalinist System after World War II.* Cambridge, England: Cambridge University Press.

Frey, BS (1997). On the relationship between intrinsic and extrinsic work motivation. *International Journal of Industrial Organization,* 15(4): 427–439.

Golley, J (1987). *Whittle: The True Story.* Shrewsbury, England: Airlife.

Golovanov, I. (1994). Korolev: Fakty i mify. Moscow: Nauka.

Gordon, Y and Dexter, A (1999). Beyond the frontiers: Soviet mixed-power fighters, 1939–46. In: *Wings of Fame,* Volume 15. London: Aerospace Publishing, pp. 148–157.

Grigor'ev, NV (1994). Aviatsionnoe motorostroenie v Velikoi Otechestvennoi voiny 1941–1945 gg. In: Biushgens GS (ed.), *Samoletostroenie v SSSR (1917–1945)* (Volume 2, 164–196). Moscow: TsGAI.

Harrison, M (1996). *Accounting for War: Soviet Production, Employment, and The Defence Burden, 1940–1945.* Cambridge, England: Cambridge University Press.

Harrison, M (1998). The Economics of World War II: An Overview. In: Harrison M (ed.) *The Economics of World War II: Six Great Powers in International Comparison*. Cambridge, England: Cambridge University Press, pp. 1–42.
Harrison, M (2000). New Postwar Branches (1): Rocketry. In: Barber J and Harrison M (eds.), The Soviet defence-industry complex from Stalin to Khrushchev. Basingstoke, England: Macmillan, pp. 118–149.
Harrison, M (2003a). The Political Economy of a Soviet Military R&D Failure: Steam Power for Aviation, 1932–1939. *Journal of Economic History*, 63(1): 178–212.
Harrison, M (2003b). Soviet Industry and the Red Army under Stalin: A Military-Industrial Complex? *Les Cahiers du Monde russe*, 44(2–3): 323–342.
Hayek, FA (1945). The Use of Knowledge in Society. *American Economic Review*, 35(4): 519–530.
Holloway, D (1982a). Innovation in the defence sector. In: Amann R and Cooper J (eds) *Industrial Innovation in the Soviet Union*. New Haven, CT: Yale University Press, pp. 276–367.
Holloway, D (1982b). Innovation in the defence sector: Battle tanks and ICBMs. In: Amann R and Cooper J (eds.), *Industrial Innovation in the Soviet Union*. New Haven, CT: Yale University Press, pp. 368–414.
Holloway, D (1994). *Stalin and the Bomb: The Soviet Union and Atomic Energy, 1939–1956*. New Haven, CT: Yale University Press.
Iakovlev, AS (2000). *Tsel' zhizni. Zapiski aviakonstruktora* (6th edn.). Moscow: Respublika.
Khlevniuk, OV (1996). *Politburo: Mekhanizmy politchecheskoi vlasti v 1930-e gody*. Moscow: Rosspen.
Khlevniuk, OV, Kvashonkin, AV, Kosheleva, LP and Rogovaia, LA (eds.) (1995). *Stalinskoe Politbiuro v 30-e gody. Sbornik dokumentov*. Moscow: AIRO-XX.
Le Grand, J (1991). Quasi-Markets and Social Policy. *Economic Journal*, 101(408): 1256–1267.
Le Grand, J and Bartlett, W (eds.) (1993). *Quasi-Markets and Social Policy*. Basingstoke, England: Macmillan.
Lebina, N (2000). The Defence-Industry Complex in Leningrad: The Post-war Uranium Industry. In: Barber J and Harrison M (eds.) *The Soviet Defence-Industry Complex from Stalin to Khrushchev*. Basingstoke, England: Macmillan, pp. 184–194.
Litwack, JM (1991). Discretionary Behaviour and Soviet Economic Reform. *Soviet Studies*, 43(2): 255–279.
Liul'ka, AM and Kuvshinnikov, SP (1981). K istorii sozdaniia pervogo sovetskogo turboreaktivnogo dvigatelia. *Voprosy istorii estestvoznania i tekhniki*, 2: 86–96.
MacKenzie, D (1996). *Knowing Machines*. Cambridge MA: MIT Press.
Markevich, A and Harrison, M (2006). Quality, Experience, and Monopoly: The Soviet Market for Weapons under Stalin. *Economic History Review*, 59(1): 113–142.
Medvedev, ZA (1978). *Soviet Science*. New York: W.W. Norton.

Mises, L von (1949/1998). *Human Action: A Treatise on Economics.* Auburn, AL: Ludwig von Mises Institute.
Mokyr, J (1990). *The Lever of Riches: Technological Creativity and Economic Progress.* Oxford: Oxford University Press.
Morishima, M (1984). *The Economics of Industrial Society.* Cambridge, England: Cambridge University Press.
Mukhin, M (2004). Employment in The Soviet Aircraft Industry, 1918–1940: Work Culture, Organization, and Incentives. PERSA Working Paper no. 36. Department of Economics, University of Warwick. Available at: http://www.warwick.ac.uk/go/persa.
Olson, M (1993). Dictatorship, Democracy, and Development. *American Political Science Review*, 87(3)4: 567–576.
Ordway, FI and Sharpe, MR (1979). *The Rocket Team.* London: Heinemann.
Ozerov, G (1973). *Tupolevskaia sharaga* (2nd ed.). Frankfurt am Main: Posev.
Perakh, M (1998). Laughing under the covers (Russian oral jokes). Available at: http://www.nctimes.net/'-'mark/htmjokes/.
Pikhoia, RG (1998). *Sovetskii soiuz. Istoriia vlasti. 1945–1991.* Moscow: RAGS pri Presidente RF.
Raushenbakh, BV (ed.) (1998). *S. P. Korolev i ego delo. Svet i teni v istorii kosmonavtika.* Moscow: Nauka.
Samuelson, L (1996). *Soviet Defence Industry Planning: Tukhachevskii and military-industrial mobilisation.* Stockholm: Stockholm School of Economics.
Samuelson, L (2000). *Plans for Stalin's war machine: Tukhachevskii and Military-Economic Planning, 1925–41.* London and Basingstoke: Macmillan.
Serov, G (1997). V nachale reaktivnoi ery. Samolety mira, 3–4, 2–7.
Shavrov, VB (1988). Istoriia konstruktsii samoletov v SSSR 1938 1950 gg (2nd ed.). Moscow: Mashinostroenie.
Shleifer, A. (1985). A Theory of Yardstick Competition. *RAND Journal of Economics*, 16(3): 319–327.
Siddiqi, AA (2000). *Challenge to Apollo: The Soviet Union and the Space Race, 1945–1974.* Washington, DC: NASA History Division (NASA SP-2000-4408).
Siddiqi, AA (2003). The Rockets' Red Glare: Technology, Conflict, and Terror in the Soviet Union. *Technology and Culture*, 44(3): 470–501.
Simonov, NS (2000). New Postwar Branches: The Nuclear Industry. In: Barber J and Harrison M (eds.), *The Soviet Defence-Industry Complex from Stalin to Khrushchev.* Basingstoke, England: Macmillan, pp. 150–172.
Smith, GG (no date). *Gas Turbines and Jet Propulsion for Aircraft.* London: Flight Publishing Co.
Sobolev, DA (1996). *Nemetskii sled v istorii Sovetskoi aviatsii. Ob uchastii nemetskikh spetsialistov v rasvitii aviastroenii v SSSR.* Moscow: Aviantik.
Starkov, B (2000). The Security Organs and the Defence-Industry Complex. In: Barber J and Harrison M (eds.), *The Soviet Defence-Industry Complex from Stalin to Khrushchev.* Basingstoke, England: Macmillan, pp. 246–268.
Stoecker, SW (1998). *Forging Stalin's Army: Marshal Tukhachevsky and the Politics of Military Innovation.* Boulder, CO: Westview.

Sultanov, IG (1998). Istoriia sozdaniia pervykh otechestvennykh turboreaktivnykh samoletov. Mosow: Vusovskaia kniga.

Voronkov, IS (1984). Teoriia i konstruktsiia: istoriia vzaimodeistviia pri sozdanii pervykh aviatsionnykh reaktivnykh dvigatelei. In: *Issledovaniia po istorii i teorii razvitiia aviatsionnoi i raketo-kosmocheskoi nauki i tekhniki*, volume 3. Moscow: Nauka, pp. 113–122.

Whittle, F (1953). *Jet: The Story of a Pioneer*. London: Frederick Muller.

Wible, JR (1998). The Economics of Science: Methodology and Epistemology as if Economics Really Mattered. London and New York: Routledge.

Williamson, OE (1996). *The Mechanisms of Governance*. Oxford: Oxford University Press.

Chapter 8

The Political Economy of a Soviet Military R&D Failure: Steam Power for Aviation, 1932–1939

Mark Harrison

The Soviet economy was managed by a vertical hierarchy in which agents supplied principals with flows of information from below and principals issued commands from above.[1] Commands flowed downward, but the more closely we study this system the more we find that self-interested agents rarely did exactly as they were told. They ignored some orders, exceeded others, and did many things that were not ordered at all. Information flowed upwards but the most difficult problem for those who received it was to verify exactly what people were really doing when they appeared to be obeying commands.

In this chapter I explore the operation of the Soviet command system and the interactions among agents within it. These interactions were horizontal as well as vertical. What I mean by "horizontal" and "vertical"

This chapter first appeared in the *Journal of Economic History*, 63(1): 178–212. © The Economic History Association 2003.

[1]I thank the Leverhulme Trust, the British Academy, and the University of Warwick Research and Innovations Fund for financial support of his research on "Invention, Imitation, and the Birth of Soviet Aerospace," the University of Warwick for study leave, and the staff of the State Archive of the Russian Federation, the Russian State Economics Archive, and the Russian State Military Archive for access to documents. He also thanks Jean Vezina for sharing his knowledge of steam aviation projects in other countries and documents from his personal collection, and for permission to cite from personal correspondence; Godfrey Carr for German translation services; and the participants in seminars at the Universities of Edinburgh, Glasgow, and Warwick, Jörg Baten, the late Joseph Berliner, Stephen Broadberry, Keith Dexter, Malcolm Hill, Peter Kneen, Valery Lazarev, Ivan Rodionov, Paul Stoneman, and the anonymous referees for advice and comments.

is this: relationships between those entitled to issue commands and those obligated to obey them occupied the vertical space, whereas horizontal relationships were formed among agents at the same vertical level. Each agent had to optimize in both dimensions simultaneously.

We can study the process through the prism of an R&D failure. This was the Soviet attempt to create a new aviation-engine technology using steam power. It was a serious venture that consumed significant resources for a few years and provides a relatively self–contained episode for investigating the incentives and strategies that drove the allocation of scientific and technical resources.

What is an R&D failure? All research may produce new knowledge of both explicit and tacit kinds. It also generates externalities that are potentially of benefit to society; even failure may have a value if we learn from it. Here I define failure narrowly within a principal–agent framework: an R&D project fails even though it yields a return to society, if it does not generate a surplus for the funding principal from the useful application that the principal intended. Thus I will classify Soviet investment in steam power for aviation as an R&D failure even though it may have enhanced human knowledge and generated all sorts of useful spinoffs.

The economic significance of R&D failures can be considered from three perspectives. R&D failure may be analysed within a profit-seeking framework as a cost to an economic principal, within a rent-seeking framework as consumption by an economic agent, and within a political economy framework as a channel through which a political principal may distribute rewards to loyal agents.

In considering R&D as a profit-seeking activity it is useful to distinguish the structure of competition between and within nations. In general R&D can be motivated by both "profit" and "competitive threat" (Beath *et al.*, 1995). From a national perspective each country pursued its defense objectives as a strategic competitor in relation to other nations. Military technologies provided the means to these objectives, and investments in military R&D made the technologies more efficient. By investing in technological enhancement, each nation could expect to lower the costs of realizing its strategic goals, controlling for the R&D efforts of other states. This is the profit motive. The second motive arose from knowing that its rivals were also engaging in military R&D: the nation that fell behind might fail to maintain its strategic position and be beaten, i.e., lose ownership rights over its assets to a competitor. This motive is the competitive threat.

The dimension of competitive threat gave the process of interwar military R&D its character of a winner-takes-all race among the great powers. At the same time, competitive threat did not overwhelm the profit motive. If it had, a nation that fell behind the leader would have lost any incentive to invest in trying to come in second. In military aviation engineering we see that each great power invested substantial resources in those areas where it was not the technological leader; it was an R&D success to come in first, but it was not a failure to come in second or third; it was only a failure to come in nowhere. One reason for this is that the significant breakthroughs came too late in the war to affect the outcome. Another reason is that some of the new knowledge produced by R&D did not become common knowledge because a part was kept secret, and some of the rest remained tacit (McKenzie, 1996: 215–216) and could only be acquired through "learning by doing" (Arrow, 1962). Thus, Germany was the first power to create jet aircraft, and was able to exploit a temporary advantage over its enemies as a result, but the British and Russians rationally maintained their efforts to come in next.

The structure of competition within each country created the same motivations in a different balance. Within the British, German, and Soviet markets for military invention individual agents also faced a profit motive and a competitive threat. Think of the returns to R&D success as partly financial and partly reputational. In terms of reputation the only thing that mattered was to come first: for example each country now remembers its own pioneer of the jet aviation engine, Frank Whittle, Hans von Ohain, and A. M. Liul'ka, whereas their rivals are completely forgotten. Much more than between countries, the competition within each country took the form of a race into which each rival was drawn by the chance to scoop the winnings before the others.

Other than in terms of reputation, none of the jet pioneers was allowed to make a profit from realizing their dream. In Britain the first contracts for serial production of jet engines were given to Rolls Royce, deliberately side-lining Whittle's company Power Jets, and in Germany to Junkers and BMW, favouring the rivals of Ohain's sponsor Otto Heinkel. For different reasons Soviet arrangements gave designers no expectation of a stake in the producer profits arising from their inventions. This might even have been a good thing: economic theory has shown that, when the pool of potential inventions is limited, winner-takes-all may lead rival agents to invest in R&D until all the potential gains have been competed away (Dasgupta and Stiglitz, 1980).

The approach outlined so far treats investments in military R&D as a cost to a profit-seeking principal. Because R&D outcomes were uncertain, there were many potential projects, each with a high probability of individual failure. Principals had to be willing to fund many projects in order to ensure that at least some successful projects would be included. Thus, failed inventions were part of the cost of success (Mokyr, 1990: 176–77).[2]

Within a rent-seeking framework, military R&D is not only an investment cost to the principal but also a source of the agent's consumption. Therefore, military R&D may become a focus for self-interested rent seekers. Several factors encourage it: the cloak of military secrecy; relatively soft budget constraints; intrinsic uncertainty about the timescale and expected value of returns to investment that impedes selection, including the rational expectation that many projects will fail; and large information biases that impede monitoring.[3] Under these circumstances R&D agents can be expected to invest resources in lobbying to win project funding, and some of the resources they invest will be diverted from previously awarded allocations to military R&D.

Thus R&D failures may have another significance. From the point of view of the principal, R&D failures were simply part of the cost of success: some experiments fail, and failed experiments are part of the necessary background against which success is achieved. In contrast, from the point of view of rent-seeking agents, unsuccessful projects provided consumption as effectively as successful ones. From this perspective R&D failures may be a gain to the agent although a loss to the principal. Why then should agents pursue success for the principal? Agents' indifference to failure might also be strengthened if reputational capital created by past success could not be translated into higher income.

Finally, within a political-economy framework, in the presence of rent seeking a political principal such as a dictator may deliberately design the allocation system to enable transactions in the political market place, for

---

[2]Sah and Stiglitz (1986) have suggested that, for given incentives and a given distribution of good and bad projects, a polyarchic market system in which screening decisions are decentralized and screening intensity is endogenous should possess an advantage relative to a unified hierarchy. Therefore, a prevalence of bad projects under a hierarchy may be a cost of maintaining that hierarchy. However, this hypothesis cannot be pursued through a single case study.

[3]Kornai (1979) originated the terminology of "hard" and "soft" budget constraints; he defined the firm's budget constraint as soft when the firm is managed on the basis of a rational expectation that the state will guarantee losses and that the firm's survival and growth will be uncorrelated with its intrinsic profitability.

example to distribute gifts to agents in return for loyalty (Wintrobe, 1998: 150–53). Thus, in a study of Soviet regional policy James Harris (1999) has shown that Stalin used investment allocations to reward loyal agents in the regions in his struggle with the opposition in the late 1920s; during the 1930s, however, his regional agents were called to account for their uses of these resources. Valery Lazarev and Paul R. Gregory (2003; see also Gregory and Lazarev, 2002) have shown in a detailed study of the Soviet allocation system for motor vehicles in the 1930s that the dictator maintained a stock of vehicles in reserve for use as rewards for loyal agents.

If the budgetary system is used to reward loyal agents, the effect must be that budgetary outlays will exceed an efficient level. The excess is the signal that loyalty is expected in return: if some waste did not result, those receiving the funding would have no reason to offer thanks to the government in exchange as any politician would rationally promise to undertake at least those expenditures that were efficient (Wintrobe, 1998: 31). If the system for military procurement and R&D were used in this way, then some R&D failures might be the intended outcome of a political exchange through which both agent and principal gained: the agent gained consumption and the principal gained loyalty, which is one source of political power.

In short, the incidence of R&D failures may reflect an economic experimentation process in which a certain proportion of failures is an unintended but necessary consequence, or it may reflect an economic process in which opportunistic agents extract rents from a funding principal, or it may reflect an intention on the part of a political principal to compensate agents directly for their loyalty. To discriminate among these hypotheses in the case of Soviet military R&D requires a close study of the allocation process, and this is one purpose of the analysis that follows.

I explore this process mainly through the documentary records of the Soviet defense industry held by the Russian State Economics Archive (RGAE), supplemented by those of the Red Army held by the Russian State Military Archive (RGVA), both in Moscow. What these records make possible is a detailed analysis of Soviet research and development in terms of the problem of the principal and the strategies adopted by both principals and agents. Before the archives were available, our interpretation of the management of defense resources was based on accounts that largely denied the existence of this problem. Anodyne official histories presented a version from the standpoint of the principal, say, the Politburo and defense industry leaders, which denied that the self-interest of agent or principal might diverge from that of society. The memoirs and biographies of the agents, the producers, designers, and scientists presented a more interesting and truthful account, but from a self-interested perspective that, while honestly reporting the tensions and disagreements among principals and agents as they actually occurred, tended to attribute such problems to the principals' low education, lack of trust, excessive interference, and oppressive behaviour towards those of superior culture and understanding, i.e. the agents (e.g. Holloway, 1984). Through the archives we can come to a more rounded understanding, observing the rules of the game and the strategies of the players from both sides.

This article is about invention, not innovation. A substantial scholarly literature (Berliner, 1976; Amann *et al.*, 1977; Amann and Cooper, 1982; 1986) has demonstrated that in Soviet industry innovation was sluggish because risks were high and rewards were meagre. Inventiveness, in contrast, was everywhere but, even when successful, it often did not benefit society. Joseph Berliner (1976: 506–09) distinguished between activities aimed at maximizing technological objectives, which he called "mission oriented", and activities aimed at maximizing an economic surplus. He suggested that, although Soviet R&D organizations were good at accomplishing technological missions regardless of cost, they were poor at organizing economic activities to the benefit of society. Here I hope to show that, in the Soviet economy, mission-oriented activity was also "economic activity" in the sense that everyone involved was looking for a payoff, and that the costs they were willing to incur or impose on others depended on the private stakes.[4]

Problems and Solutions

Despite a number of major postwar studies of the comparative development of Soviet technology in various fields (Sutton, 1968; 1971; 1973); Amann *et al.*, 1986; Hanson, 1981; Amann and Cooper, 1982, 1986; Albrecht, 1993), a scholarly history of Soviet aircraft propulsion has yet to be written. Before the revolution aviation engineering barely existed in Russia and in the early years of Soviet power it remained the "Achilles heel" of Soviet aviation

[4]Applied more generally this approach may eventually lead to new comparisons with military-industrial planning and procurement processes in market economies, including the United States, but the necessary analysis is still at a preliminary stage. See Gregory (2003) and Harrison (2003).

(Sobolev, 1996: 30). The establishment of a modern industry began in the mid-1920s with the importation and adaptation of French, German, and British engines to the conditions of Soviet industry. Thereafter the Soviet Union quickly established a substantial manufacturing capacity. By the years 1936–1938, the total annual output of five factories had reached more than 15,000 engines of the conventional reciprocating type with an airscrew propeller, including radial and in-line, air- and water-cooled, gasoline and diesel fuelled, and supercharged models (Kostyrchenko, 1992: 429). However, continuously challenged by the rapidity of technological advance in other countries, the Soviet industry remained dependent on imitation and adaptation of foreign designs and did not establish an autonomous capacity for development of conventional engines (Albrecht, 1993: 28–32).

In the interwar period, aircraft performance neared the limits of the piston and propeller engine (Grigor'ev, 1994: 189). The mechanical efficiency of propellers was found to fall away beyond a point as rotation speeds increased, with the result that propeller-driven aircraft could never approach supersonic speeds or stratospheric altitudes. At full power piston engines quickly shook and ground themselves to pieces; consequently they also required frequent and intensive maintenance and had short service lives. Such limitations prompted a search in several European countries for completely new types of aviation engine based on a continuous thermal cycle giving rise to a jet reaction. In Great Britain, Germany, and the Soviet Union much effort was invested in two alternatives: the rocket motor and the jet engine.[5]

The rocket principle had been well understood for hundreds of years, and European armies had used small, solid-fuelled rockets on the battlefield since the Napoleonic era. The new fuels and heat-resistant materials being developed in the early twentieth century promised significant applications for the rocket principle in aircraft propulsion. However, to create a primary rocket motor for aviation implied a design of unprecedented size and complexity by interwar standards, depending on more powerful liquid fuels and substantial further advances in material and fuel sciences and control systems. At the end of the development process lay apparatuses that could

[5]Soviet rocketry before and during World War II has been investigated by Ordway and Sharpe (1979); Holloway (1982a); Siddiqi (2000); and, with the benefit of new archival documentation, Harrison (2000). On the early development of the Soviet jet aviation engine, see Harrison (2005).

attain extraordinary speeds and limitless altitudes but consumed fuel at rates that limited powered flight to a few minutes' duration.

The concept of the jet engine was of more recent origin. It overcame the restricted flight duration of the rocket by breathing oxygen from the air, but this limited its use to atmospheric flight. Earlier designs could not provide a primary power plant because they could not be ignited unless the aircraft was already moving at flight speed. Later designs such as the turbojet eliminated this defect at the cost of added complexity including moving parts rotating at very high speeds and temperatures; these could not be brought to realization without still greater advances on interwar benchmarks in terms of heat-resistant alloys, fuels, and the control of combustion.

Hindsight tells us that by the end of World War II the turbojet would solve the problem of high-speed, high-altitude aviation. Liquid-fuelled rocketry, while not the answer for aviation, would solve other problems of strategic bombardment and space travel. During the interwar period this outcome was not yet clear and several avenues that would later be seen as stopgaps or failures were also explored. Among these were various attempts to build a propulsion unit based on the steam turbine.

The basic scheme is shown in Figure 1. It involved an oil-fired boiler to create steam; expansion of the steam in a turbine converted part of its thermal energy into kinetic energy, initially in the form of the torque required to rotate an airscrew propeller; the propeller drove an airstream backwards and the aircraft forwards. As the aircraft in flight could not replace its water, the steam was passed through a condenser and returned to the boiler in a closed circuit.

Figure 2 arranges the spectrum of modern aviation engines in two dimensions. Along the vertical axis the technologies of more recent

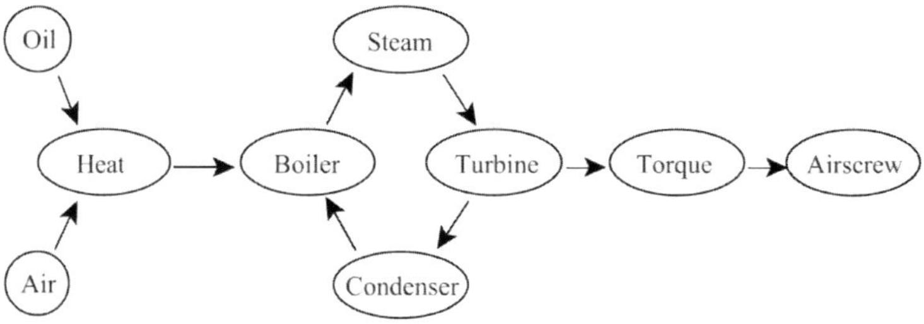

Fig. 1. The Steam Turbine in an Aviation Propulsion Unit.

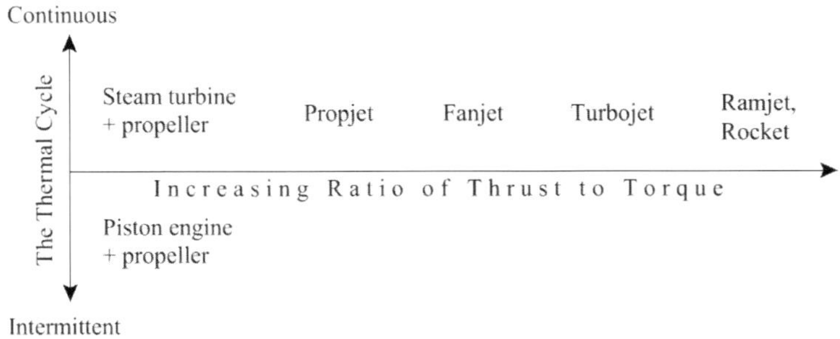

Fig. 2. Where the Steam Turbine Engine Fits.

development that make use of a continuous thermal cycle are demarcated from the traditional technology that relied on the intermittent action of reciprocating pistons. Along the horizontal axis the proportion of the energy that is delivered in the form of a jet stream or thrust, rather than of torque to a propeller or turbine, rises from zero to infinity.

The steam turbine lies at the most conservative extreme of the range of continuous thermal cycle engines, in that it aimed to deliver torque more efficiently and reliably rather than replace torque by thrust. This meant that the steam turbine, even if it worked, would never give access to supersonic speeds or stratospheric altitudes. Its main advantage lay in offering to replace the reciprocating engine with a familiar continuous-cycle technology that operated quietly and reliably for long periods at moderate temperatures and rates of rotation, with less destructive vibration and fire hazard, using materials that were already available and cheaper fuels (Knörnschild, 1941: 367).[6]

Only hindsight makes the concept of steam aircraft appear surprising. Between the development of the steam locomotive and the advent of the light gasoline engine there were many projects for steam-powered flight. Charles Parsons, who invented the steam turbine in 1884, successfully tested a model steam airplane in 1893 (Gibbs-Smith, 1970: 70; more generally, 59–68). In the interwar period steam turbines were widely used in naval propulsion, their other main use being in electric power generation. Historically the gas turbine came after the steam turbine and arose from the latter, so it was the steam turbine that was already the more proven

[6]I thank Jean Vezina for this reference and Godfrey Carr for his translation of it from German.

technology (Voronkov, 1984: 115). After World War I research on these lines in the United States was resumed by the Navy Department Bureau of Engineering in 1922 (Knörnschild, 1941: 369). Others followed not only in America but also in Britain, France, and Germany. In the 1930s, however, as will become apparent, it seems likely on present information that more aviation steam power projects were pursued in the Soviet Union than in the rest of the world put together.

If steam turbines replaced the reciprocating engine at sea, why did this not happen in the air as well? The main problem was that existing land- and ship-based applications of the steam turbine were based on ratios of power to weight and volume that were too low to take to the air. Aviation applications required not only efficient boilers but also bulky condensers to convert steam back to water; these could scarcely be accommodated within an airframe light enough to be lifted into the air by the power that the turbine would develop (Liul'ka and Kuvshinnikov, 1981: 88). Improving power relative to the mass of the engine was therefore the main focus of aviation-related development efforts. This meant primarily increasing the fuel efficiency of the boiler, and reducing the volume of the condenser and the mass of the circulating medium.

Scale and Scope

Scale

Table 1 provides an overview of the main Soviet R&D projects in aviation steam turbines from 1932 to 1939. It is compiled from plans, reports, and memoranda of the commissariats of defense, heavy industry, the defense industry, and the aircraft industry. The table shows that in the prewar years there were 11 major projects; taking into account the life of each venture, the total of activity amounted to 34 project-years. The eleven projects involved eleven research establishments, but the association between projects and establishments was less tidy than this would suggest since some designers had more than one institutional affiliation both through time and contemporaneously. There were five main funding interests: the Red Army air force, the civil aviation authority, the Academy of Sciences, the aircraft industry, and the electricity generation

Table 1. Major Soviet Projects for Aviation Steam Turbines, by Establishment, 1932 to 1939.

	1932	1933	1934	1935	1936	1937	1938	1939
KhAI	Tsvetkov 15,000 hp turbine	≫	≫					
	Tsvetkov "air-naval" PT-6 6,000 hp turbine	≫	≫	≫	PT-6 3000 hp turbine	≫	≫	
NII GVF	PT-3 "air-naval" 3000 hp turbine	≫	≫	≫				
VVA	Aksiutin PT-1 1,500/2,500 hp turbine (continued at SKB and EI)	≫	≫	≫	≫			
SKB	Sinev turbine					Sinev 1,600/2,500 hp turbine	≫	
SKB and EI						Aksiutin turbine from VVA	≫	
KB-2		Kozhevnikov 400 hp gas-steam turbine	≫	≫	≫			
VTI				"Air-naval" 1,000 hp turbine				

(*Continued*)

Table 1. (*Continued*)

	1932	1933	1934	1935	1936	1937	1938	1939
VTI and MAI						Przheslavskii binary-cycle 2,000 hp steam turbine	≫	
TsKTI						Hüttner turbine	VT-1, VTK-100 (100 hp), and VTK-3000 (3,000 hp) turbines	PT-1M 2000 hp and VTK-300 100 hp turbines
Zavod no. 18						Dybskii-Udod gas-steam turbine	≫	
TsIAM								1600 hp, single-, and binary-cycle turbines

Key to Design Establishments:
KhAI = Khar'kovskii Aviatsionnyi institut NKTP–NKOP–NKAP: Kharkov Aviation Institute of the People's Commissiat of Heavy Industry (later Defense Industry, later Aircraft Industry)
NII GVF = Nauchno–Issledovatel'skii institut Grazhdansko–Vozdushnogo Flota GUGVF SNK: Research Institute of the Civil Air Fleet of the Chief Administration of the Civil Air Fleet of the Council of People's Commissars

(*Continued*)

Table 1. (*Continued*)

VVA = Voenno–Vozdushnaia Akademiia RKKA: Air Force Academy of the Workers' and Peasants' Red Army
SKB, later SKB-1 = Spetsial'noe Konstruktorskoe biuro Pervogo Glavnogo upravleniia NKOP [Special-purpose Design Bureau of the First Chief Administration of the People's Commissariat of the Defense Industry]
EI = Energeticheskii institut AN SSSR [Energy Institute of the USSR Academy of Sciences]
KB-2 = Konstruktorskoe biuro no. 2 UVI RKKA [design bureau no. 2 of the Administration of Military Inventions of the Workers' and Peasants' Red Army, Leningrad]
VTI = Vsesoiuznyi Teplotekhnicheskii institut im. Dzerzhinskogo NKTP [Dzerzhinskii All-Union Thermal-Technical Institute of the Electricity Supply Industry Administration of the People's Commissariat of Heavy Industry]
MAI = Moskovskii Aviatsionnyi institut NKTP-NKOP-NKAP [Moscow Aviation Institute of the People's Commissariat of Heavy Industry, later the Defense Industry, later the Aircraft Industry]
TsKTI = Tsentral'nyi Kotlo-turbinnyi institut Energoproma NKTP [Central Boiler and Turbine Institute of the Electricity Supply Industry Administration of the People's Commissariat of Heavy Industry, Leningrad]
Zavod no. 18 = zavod no. 18 NKOP [factory no. 18 of the People's Commissariat of the Defense Industry, later the Aircraft Industry]
TsIAM = Tsentral'nyi Institut Aviationnogo Motorostroeniia NKTP-NKOP-NKAP [Central Institute for Aeroengine-Building of the People's Commissariat of Heavy Industry, later the Defense Industry, later the Aircraft Industry]

Sources: The supporting documentation comprises plans, reports, and memoranda of the People's Commissariats of Defense, Heavy Industry, the Defense Industry, and the Aircraft Industry: RGAE, 8044/1/994, ff. 21–23; 8328/1/696, f. 25; 8328/1/824, ff. 1–50; 8328/1/919, f. 84; 8328/1/992, ff. 6–7; 8328/1/996, ff. 16–18 and 22–23ob. RGVA, 4/14/2800, f. 4; 34272/1/167, ff. 23–24, 47–55, and 102–119. See also Rodionov (2002), 1932 under the month of September, 17 September, and 31 December; 1933 under 10 February and 8, 21, and 23 March; and 1934 under 3 and 14 July. Other steam power aeroengine projects not considered here include work on the Besler engine, discussed in the Appendix, at the Moscow Aviation Institute (MAI), and a plan, evidently soon abandoned, to build a steam-turbine supercharger for a reciprocating aeroengine in 1932–3 at the Khar'kov Aviation Institute (KhAI) and under the Red Army Bureau of New Designs (BNK); on the latter see Rodionov (2002), 1932 under September (Sovnarkom decree on the construction of steam turbine engines), and 1933 under 17 January (Unshlikht to the Council of Labor and Defense) and 23 March (Almazov to Gosplan).

industry.[7] Four designers in three urban centres accounted for three-fifths of the total activity: S. A. Aksiutin and N. M. Sinev in Moscow, P. L. Kozhevnikov in Leningrad, and V. T. Tsvetkov in Khar'kov.

The time profile of these projects is illustrated in Table 2 and Figure 3. Officially work began in 1932. From then to the end of 1937 ten projects were initiated. Four of these had been curtailed already by the end of 1936, so the number of major projects in progress peaked at the end of 1937 at six. Of the six, five were cancelled during 1938 and, although one more was begun during 1939, bringing the total to 11, all had been abandoned by 1940.

Did the Soviet effort represent a significant commitment of national resources? Based on the information available at present, the answer appears to be no. At the end of 1937, acting director Sinev of the special-purpose design bureau (SKB) of the commissariat of the defense industry put the total of sunk costs of the various steam turbine projects so far at 20 million rubles over five and a half years, i.e., not more than four million rubles per year.[8] On the basis of an annual average wage for that period of 3,000 rubles, four million rubles would also represent the direct-plus-indirect employment of up to 1,300 public-sector workers. Included in this total were a few scientists and designers who were more highly paid and were a particularly scarce resource, but even 100 such specialists would have represented no more than one per thousand of the Soviet Union's stock of "scientific workers" in 1940.[9]

Although insubstantial in absolute terms, the Soviet interwar effort in the direction of aviation steam power represented a measurable proportion of overall Soviet military R&D resources; these are known only for 1936 and 1937, and were as follows (in million rubles): the 1936 plan was 119, actual was 88; the 1937 plan was 136, actual was 67.[10] Thus four million

---

[7]The involvement of the electricity generation industry might be construed as an attempt to extract rents from new uses for an existing technology in which it had a vested interest. In a similar case Henning and Trace (1975: 385) attribute the British interwar slowness to adopt diesel engines for marine propulsion in preference to steam turbines partly to "the over-optimistic performance claims by builders of steam-turbine machinery." I thank Stephen Broadberry for this reference. In the Soviet case aviation specialists came to see the electricity supply engineers as hostile rivals (RGVA, 4/14/1925, ff. 150–52, 17 May 1938).

[8]RGAE, *fond* 8328, *opis'* 1, *delo* 992, folio 15 (hereafter 8328/1/992, f. 15) (19 December 1937).

[9]There were 98,300 "scientific workers" in the Soviet Union in 1940 according to Goskomstat (1987: 64).

[10]Harrison and Davies (1997: 387). These figures in turn represented less than two per cent of total military equipment orders.

Table 2. The Number of Major Soviet Projects for Aviation Engines based on Jet Propulsion and Turbines, 1 January 1932 to 30 June 1941.

	1932	1933	1934	1935	1936	1937	1938	1939	1940	1941
Steam Turbines										
1. In progress at start of year	0	3	4	4	3	2	6	1	—	—
2. Starting during year	4	1	0	1	0	4	0	1	—	—
3. Continuing during year	4	4	4	5	3	6	6	2	—	—
4. Discontinued by end of year	−1	0	0	−2	−1	0	−5	−2	—	—
5. In progress at end of year	3	4	4	3	2	6	1	0	—	—
Gas Turbines, Jets, and Rockets										
1. In progress at start of year	0	1	1	1	1	3	3	3	6	4
2. Starting during year	2	0	1	0	2	1	1	5	0	3
3. Continuing during year	2	1	2	1	3	4	4	8	6	7
4. Discontinued by end of year	−1	0	−1	0	0	−1	−1	−2	−2	0
5. In progress at end of year	1	1	1	1	3	3	3	6	4	7

Note: I count as one "major project" work that is continued in one establishment from year to year even though the particular objects may vary, and also work that is continued on a particular object from year to year by one designer even though the sponsoring establishment may change. Row 1 equals row 5 in the preceding year; row 5 is the sum of rows 4 and 3; row 3 is the sum of rows 2 and 1.

Sources: This table is adapted from Harrison (2005: 10–11); for steam turbines specifically, see Table 1.

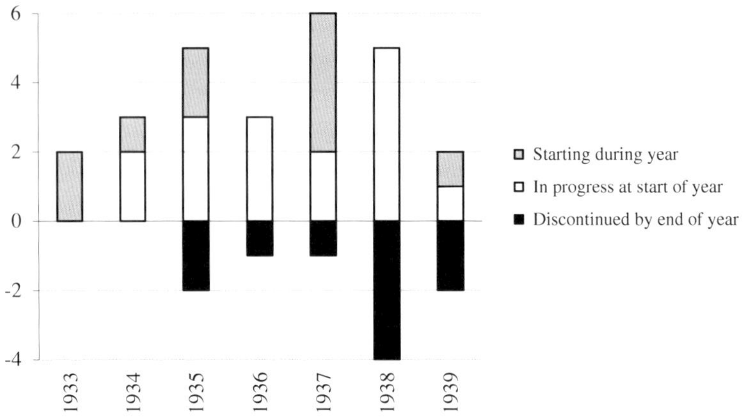

Fig. 3. Soviet Aviation-related R&D in Steam Turbines, 1932–1939 (Numbers of major projects).

Source: Calculated from Table 2.

rubles a year represented three per cent of resources budgeted for military R&D in these years, and five per cent of resources used.

What is also to the point is that Soviet funding of steam power plants for aviation probably exceeded that channelled into the much more famous projects for gas turbines, jet engines, and rockets. Table 2 shows that in every year between 1932 and 1938 at least as many and usually more steam turbine projects were in progress than the number of jet propulsion projects. Only in 1939 did efforts in jet propulsion begin to overshadow the steam turbine projects as the cull of the latter came to an end. Consequently, as Figure 4 shows, the cumulative investment of Soviet R&D resources in steam turbines, measured in major project-years, was overtaken by that in jet propulsion only in 1941.

How did the scale of Soviet research compare with efforts being made in other countries? Table 3 shows that there were eight other major steam-turbine ventures between the wars: three in the United States, three in Germany, and one each in Britain and France. The Besler engine is also worthy of note; although not a turbine, and unsuitable for scaling up to the much larger capacities sought by Soviet designers, it was the first and only steam engine to succeed in powering piloted flight. Besler, having acquired Doble Steam Motors in 1931, had developed a compact 150 horsepower steam reciprocating engine for use in rail and road transport. Installed in a light aircraft it took to the air on 12 April 1933, and this attracted

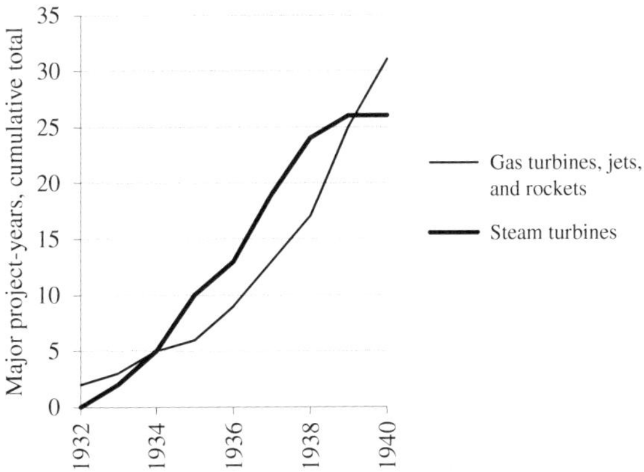

Fig. 4. Soviet Aviation-related R&D in Steam Turbines versus Gas Turbines, Jets, and Rockets, 1931–1941: Cumulative Investments in Major Project-Years.

Source: Calculated from Table 2.

the attention of the Soviet government; further information is given in the Appendix. Leaving the Besler engine to one side, it would appear that at this time there were fewer ventures in the whole of the rest of the world than in the Soviet Union.

Scope and Performance

Despite a common basic inspiration there was considerable technical variation among steam turbine designs. One variant with British, German, and Soviet derivatives was the Hüttner turbine; this replaced the conventional static boiler which blew hot gas past water pipes with a rotating boiler that moved U-shaped water pipes at high speed through the hot gas while the turbine was rotated in the opposite direction. Other refinements could include using a multi-stage turbine to convert more of the overall steam pressure drop into useful energy; expanding the boiler's exhaust gases in a gas turbine to supercharge the boiler itself or to supplement the torque delivered to the propeller; and using the heat otherwise lost from the condensing steam to expand the cooling air in such a way as to add to the overall thrust. The "gas-steam" apparatuses of Kozhevnikov and the Dybskii-Udod partnership evidently made use of the boiler's exhaust gases in an auxiliary turbine. Przheslavskii's "binary-cycle" turbine used a high-temperature liquid, probably mercury, in a preliminary stage to a water

Table 3. Designs for Aviation Steam Turbines in the Rest of the World, 1922–1942.

Corporation or Designer	Country	Begun or in progress	Capacity (hp)	Specific weight (kg/hp)	Specific fuel consumption (g/hp/hour)
Navy Department Bureau of Engineering	United States	1922	—	3.0	500[a]
Great Lakes Aircraft and General Electric Corporations	United States	1932 or earlier	2 × 1,150	1.44[b]	300[b]
Hüttner[c]	Germany	1937 or earlier	—	1.0	280–300
Vorkauf[c]	Germany	1937 or earlier	—[d]	—	—
Wagner	Germany	1937 or earlier	2 × 3,000	1.2	220
Aero-Turbines[c]	United Kingdom	1938	2,000	0.9	160
William Brobeck[e]	United States	1941 or earlier	1,000	0.9	—
Béchard[c]	France	1941 or earlier	—	—	—

Notes: [a] This figure is based on Knörnschild's statement that a specific fuel consumption of 250 grams per horse power per hour would be possible with a 100 per cent improvement in weight and consumption.
[b] These figures, calculated from imperial units given by Collina, are preferred to slightly lower figures of 1.33 kilograms per horsepower and 270 grams per horsepower per hour from Knörnschild.
[c] Hüttner and Béchard manufactured boilers that rotated in the opposite direction to the turbine; the Aero-Turbines design was based on a Hüttner turbine.
[d] A capacity figure of 650 kW is given by Knörnschild.
[e] William Brobeck was formerly chief engineer of the Besler Corporation.

Sources: The fullest survey is provided by Knörnschild (1941); the subsequent wartime survey by Smith (no date: 36–40), appears to take much of its information from Knörnschild. Collina (1932) had previously placed details of the Great Lakes Aircraft-GEC venture in the public domain; this is another reference for which I thank Jean Vezina. I omit from the table three other steam power projects as not turbines or not prime movers: Knörnschild counts the Besler engine of 1932, further described in the Appendix; Smith lists a Breguet Aviation project of 1942 to build a steam powered supercharger for a jet engine, and also a United Aircraft design in the name of the helicopter pioneer Igor Sikorsky and others to build a steam turbocompressor onto a conventional radial engine for added thrust. A Soviet memorandum of 28 February 1937 (RGAE, 8328/1/919, f. 77) lists seven foreign steam turbine ventures by AEG, Wagner, Hüttner, Vorkauf, Besler, Doble, and General Electric-Great Lakes Aircraft, but without giving any details; thus it correctly identifies four that were included subsequently in Knörnschild's survey, but wrongly counts Besler and Doble which was a single venture and not a turbine. Despite the evidence of this memorandum I omit AEG from the table since no details are provided, there is no corroboration in other sources, and the accuracy of the Soviet information about other foreign ventures appears to have been poor.

boiler so as to economize on the mass of the circulating medium and reduce specific fuel consumption.[11]

Soviet design capacities ranged from a few hundred horsepower to the 15,000 horsepower rating of Tsvetkov's first monster that was never built. By comparison, the largest reciprocating aviation engines in serial production in the Soviet Union and elsewhere at the end of the 1930s somewhat exceeded 1,000 horsepower.

The performance characteristics that mattered most were specific weight and specific fuel consumption, both measured relative to capacity. These are not always reported in the documentation; Tables 3 and 4 show available figures for Western and Soviet designs respectively. The typical Soviet design reported in Table 4 proposed a 1,000 to 3,000 horsepower engine with a specific weight of 1 kilogram per horsepower and specific fuel consumption around 300 grams per horsepower per hour of operation; some were better and some were substantially worse. This pattern was similar to the variation among the Western designs listed in Table 3.

The specific weight of the Soviet steam turbines was generally not as good as that of contemporary piston engines. Thus the design weight of one piston engine that was mass-produced in very large numbers in the mid-1930s, the Mikulin M-34, was 0.8 kilograms per horsepower.[12] Moreover, the specific weight of piston engines was falling. The M-34 and Klimov M105, the Wright Cyclone and its Soviet derivative the M-25, and the Rolls Royce Merlin, all dropped to 0.6 to 0.7 kilograms per horsepower in their last versions of the prewar period (Grigor'ev, 1994: 86, 90, and 93).

Specific fuel consumption can also be compared. In the mid-1930s the design consumption of the M-34 in its RN or FN version was 255 grams of aviation spirit per horsepower per hour.[13] The TB-7 heavy bomber of 1936, an aircraft that some designers intended for a steam turbine, provides a cross-check from practical aviation: equipped with four M-34FRN piston engines its operating fuel consumption was roughly 300 grams per horsepower per hour.[14] Tables 3 and 4 suggest that

[11]The principle of the mercury boiler, first applied in the United States by General Electric in 1914, is evaluated by Stodola (1927: 1313–16). On the advantages see Knörnschild (1941: 367).

[12]Rodionov (2002: 1934 under 3 July).

[13]Rodionov (2002: 1934 under 14 July and 1936 under 5 July).

[14]This figure is obtained as follows. The maximum range and speed of the TB-7 in 1936 were 3,000 kilometers and 403 kilometers per hour at an altitude of 8,000 meters, but no aircraft could achieve both at the same time; a cruising speed of 350 kph would suggest

Table 4. Soviet Designs for Aviation Steam Turbines: Technical Characteristics.

Design establishment	Designer	Model	Capacity (hp)	Weight (kg)	Specific weight (kg/hp)	Specific fuel consumption (g/hp/hour)
KhAI	Tsvetkov	—	15,000	16,500	—	—
KhAI	Tsvetkov	PT-6	6,000 reduced to 3,000	9,600 reduced to 3,000	—	—
NII GVF	—	PT-3	3,000	3,000	—	—
TsKTI	—	PT-1M	2,000	—	1.0	300
SKB	Sinev	—	1,600 increased to 2,500	—	2.5	350–400
VVA-SKB-EI	Aksiutin	PT-1	1,500 increased to 2,500	—	1.0	220
VTI	—	"Air–naval"	1,000	1,000	—	—
KB-2	Kozhevnikov	—	400	—	0.65[a]	—

Note: [a] This figure excludes the weight of a gas turbocharger.
Sources: As for Table 1.

some more fuel-efficient steam turbines were competitive with this. Direct comparison is complicated by the fact that piston engines burned high-octane petroleum rather than the low-grade fuel oil used by a steam boiler.[15] This meant that the steam turbine engines used a fuel that was much cheaper and more readily available than for existing engines, so regardless of specific fuel consumption they offered immediate gains in operating costs.

Ultimately, lower running costs did not compensate for the sheer mass of steam turbine engines, including the mass of fuel. The ruble cost of fuel required was a less important limit on performance than its weight. A TB-7 with its fuselage rearranged to accommodate boiler and turbine and its wings filled by condensers had little space remaining for fuel, crew, or payload.

Project Finance

Rationing, Rivalry, and Budget Constraints

What limited Soviet outlays on steam propulsion R&D? Stalin's Politburo set cash limits on ministerial budgets from year to year, but detailed allocation was delegated to officials at or below the level of minister. The main problem these officials faced was technological uncertainty: R&D agents presented them with many proposals for long-term projects, one of which would eventually solve the problem of aviation propulsion for the next half century, but no one knew which one. Rationing was an appropriate response. Funding was rationed across new projects, and was also rationed through time for established projects. As a result, R&D agents competed for both initial funding and subsequent refinancing. Making limited allocations to many projects permitted funding principals to exploit the rivalry of the agents and, by monitoring the progress of competing projects, to use each to provide information about the others and thus learn about their true worth. And by rationing funding through time principals could monitor the cumulative progress of each project towards its goal so that subsequent

a maximum flying time of 8 hours 35 minutes. Its four M-34FRN engines totalled 3,600 nominal horse power but for cruising speed say 3,200 hp, making a maximum flight total of just over 27,400 horsepower hours. The TB-7 carried 8,250 kilograms of fuel, making roughly 300 grams of fuel consumed per horsepower per hour. Technical details are from Gunston (1995: 280–81). Thanks to Keith Dexter for advice.

[15] According to Jasny (1952: 151–52), the official wholesale prices of fuel oil in Leningrad and of grade 2 motor petroleum in Moscow in 1937 were 155 and 900 rubles per ton respectively.

refinancing decisions could be taken in the light of more information than was available initially.

Such competition was not normal elsewhere in Soviet industry. In civilian branches funding principals and agents were more likely to form closed bilateral relationships. This did not rule out parallel projects or duplication of work, but it did ensure that organizations working on related issues did not necessarily compete with each other and might not even be aware of each other's work; thus producer interests took priority over the consumer (Berliner, 1976: 120–27). In the defense industry, in contrast, the government promoted the interests of a powerful consumer, the defense ministry, by assigning similar specifications to rival design bureaus; the latter were then forced to compete for their designs to be adopted. This practice was pioneered before the war in aircraft and aviation-engine development (Holloway, 1982b: 317–19). As will become apparent, however, the intensity of competition was varied through time. Indeed, after World War II it was suggested to Stalin that the Soviet Union was lagging in aircraft and aviation engines because relationships between the air force and the leading aircraft design bureaus had been or had become too cosy, enabling favoured designers to establish monopolies and relax the pace of development. This was one source of the "aviators' affair" of 1946 in which leaders of the air force and aircraft industry were arrested and imprisoned (Bystrova, 2000: 320–21).[16]

By rationing limited finance across competing projects and through time the principal could exploit agents' rivalry, but rationing also carried important costs. Rationing through time had the effect of softening budget constraints on individual projects because, in the presence of sunk costs, a project could attract refinancing even when it was known to be bad. Because some costs were already sunk, it could still be efficient for both the funding principal and the R&D agent to continue a project that the principal would have preferred not to finance in the first place.[17] Rationing across projects brought the danger of a fragmentation of effort in many rival projects, all underfunded. Both types of rationing invited agents to

[16]This was not the only source of the purge; according to Pikhoia (1998: 45–47), Stalin also designed it to undermine the position of the leader of the war effort Marshal G. K. Zhukov.

[17]Dewatripont and Maskin (1995) propose centralized credit and sunk costs as the general explanation of soft budget constraints and shortage phenomena in a socialist economy.

invest resources in lobbying to soften constraints and to use the argument that costs had already been sunk as an argument for project continuation.

Financial constraints on projects in progress therefore tended to become soft, and occasionally this became explicit. The technical council of the chief administration for the aircraft industry met in August 1936 to review the development of a number of steam turbine projects. The deputy minister for heavy industry Mikhail Kaganovich (brother of Lazar, the Politburo member and Stalin's deputy) presided. In the passage below he refers to his boss, minister for heavy industry G. K. "Sergo" Ordzhonikidze, a member of Stalin's inner team. One of the designers present reported: "they told me [that] to carry out the testing of the turbine in our factory funds of the order of a million rubles [were] needed". Kaganovich interrupted: "I can provide it". Kaganovich told another designer to proceed if he could to a flight test: "Whatever it costs I'll pay". He summed up[18]:

> I am not so poor, we have money in the sums that are needed, the boss is not hoarding it, comrade Ordzhonikidze says to take what you need.

What were the consequences of a soft budget constraint? The soft budget constraint might seem like a good thing for technical progress at first sight because the chances of successful invention would appear to increase with the number of generously funded projects. However, a soft budget constraint probably also encouraged adverse behavioral responses. Philip Hanson and Keith Pavitt (1987: 46) have noted a tendency of government-funded R&D to generate "too few technical alternatives"; this is because, when budget constraints are soft, R&D agents display insufficient care in coming to conclusions. They "do not apply the normal, prudent practice of commercial firms in carefully exploring technical alternatives in order to reduce technical and commercial uncertainties. Instead, they tend to follow a high risk strategy of premature commitment to the full scale commercial development of a particular technical configuration, before the reduction of key technical and commercial uncertainties".[19] In short, uncertain outcomes are not a good reason to throw money at a problem.

---

[18]RGAE, 8328/1/824, ff. 40, 51, and 52ob (22 August 1936).

[19]Why may hierarchies accept more bad projects than a market system? Hanson and Pavitt's reasoning proposes that a hierarchy generates specific incentives and is therefore made on different grounds from that of comparative "architecture" proposed by Sah and Stiglitz, "Architecture."

Rationing and soft budget constraints help explain an institutional cycle that accompanied Soviet progress in new aviation-engineering technologies in the interwar years (see also Harrison, 2005). The first phase was one of exploration. For the first time the principal defined the mission and agents formed a lobby. As a result cash was allocated and initial funding was dispersed among agents. With first results, existing projects were refinanced and further projects were designated. At a certain point the principal lost patience with rising expenses and lack of results, and declared a need to concentrate efforts and focus them more narrowly. For example, in August 1936 Kaganovich wrote to the Council for Labor and Defense with Ordzhonikidze's support to unify all work on steam and gas turbines under his chief administration for the aircraft industry.[20]

Exploration now gave way to rationalization. In the rationalization phase funding was removed from those projects judged less successful, which were terminated, and was concentrated on fewer projects reflecting more limited priorities. Agents responded both defensively and aggressively. Then the cycle was repeated because in the course of rationalization the principal made mistakes, curtailing some good projects as well as bad ones. Therefore, rationalization was often temporary because agents would eventually mount challenges to central priorities and organizational monopolies from below and exploration would begin again. This was a resilient cycle; it could only be curtailed by force.

The Financing Decision

On the demand side of the financing decision were the *funding principals*. The fund holder and the funding department were not necessarily the same. The *fund holder*, the legal owner of the R&D assets, was usually an industrial ministry, but the Red Army also maintained its own R&D establishments. The *funding department* paid for R&D services. Some centralized orders were paid directly out of the USSR state budget. In addition, budgetary institutions such as the defense commissariat were entitled to enter into decentralized contracts with industrial institutes and design bureaus for R&D services. Finally, the fund holder could also commission in-house research from its own establishments.

[20] Rodionov (2002: 1936 under 3 August). Evidently, this request failed.

Both fund holders and funding departments operated within a framework of strategic directives that were issued from time to time by high-level government committees: the Council for Labor and Defense or the executive subcommittee of the Council of People's Commissars responsible for defense matters. In practice, regardless of the formal issuing authority, Stalin personally made such decisions in secret consultation with a varying circle of individual Politburo members, usually after receiving representations from the funding department and fund holder (Rees, 2001).

Within this framework both funding departments and fund holders formulated operational plans. The most important planning horizon was annual. The Red Army had an annual plan for the development of military inventions some of which it funded directly through its own R&D establishments; it contracted the rest out to other organizations. Industrial ministries, including the branches of the defense industry, had their own R&D plans. This included the aircraft industry's annual plan for aviation-engine research and experimentation to be carried out by its own institutes and bureaus, part of which was made up by contracts accepted from the Red Army.

How did projects win a place in the plan? There was a variety of routes, but their common feature was that the initiative lay with the designer. This was not a process whereby an all-seeing and all-knowing planner identified needs from above, sought out designers, and put them together with resources to meet the needs identified. Rather, proposals came first from below. Established designers continually brought proposals for radical innovations to the attention of funding principals; it was their job to do so. For example, here is deputy minister for the aircraft industry M. M. Kaganovich again in August 1936[21]:

> Three years ago comrade Tsvetkov came to me and proposed making such a turbine, I went to the boss, the people's commissar signed a decree to the effect that, in urgent order, under personal responsibility, [inaudible] to build a turbine . . . .

Successful proposals required investments in lobbying. Such investments could bring not only success for individual proposals but also long-term privileged relationships with government officials responsible for funding. To win support for their projects and adoption of their

[21] RGAE, 8328/1/824, f. 35 (22 August 1936).

designs, designers had to be "heterogeneous engineers" capable of reshaping organizational as well as technological constraints.[22] To create a demand for new designs they had to build coalitions with soldiers or industrialists to overcome producer and consumer interests vested in markets for products that already existed.

Much has been written about Marshal M. N. Tukhachevskii, Red Army chief of armament from 1931 to 1936, as the military patron of jet propulsion in the Soviet Union between the wars. In November 1929 the post of chief of armament of the Red Army was created to help carry through its equipment modernization. The first chief of armament was Army Commander I. P. Uborevich, followed in 1931 by Army Commander, later Marshal, M. N. Tukhachevskii.[23] Among the departments reporting to the chief of armament was an administration for military inventions. In 1936 the post of chief of armament was abolished, its place taken by a chief administration for supply of weapons and equipment, and under the latter a department for inventions (Holloway, 1982b: 321).

An assiduous networker, Tukhachevskii used his oversight of military R&D to forge exclusive links with designers and acquire a monopoly of jet propulsion development for both artillery and aviation as both funding principal and fund holder. In this ambition he was never successful. It is notable that the first projects in steam propulsion were also sponsored by establishments of the Red Army, including KB-2 which was directly under Tukhachevskii's control as chief of armament at the time.

Which problem was more important for the principal: was it to *promote*, or to *limit* the number of projects involving steam propulsion? The funding principals struggled continually to limit and constrain initiatives and proposals from below, which were diverse and flowed from many sources, and were much more numerous than initiatives from above. As a result, R&D projects had a tendency to proliferate. Reports and resolutions prescribing the consolidation or cancellation of existing rival projects greatly outweighed the number of decisions authorizing new ones.

---

[22]On heterogeneous engineers see MacKenzie (1996: 13); and Harrison (2005) for further illustration.

[23]For reasons that are largely unrelated to this topic Tukhachevskii was arrested in May 1937 and, along with many other officers, subsequently executed as a traitor. On Tukhachevskii and jet propulsion see Holloway (1982a); Siddiqi (2000); and Harrison (2005). On Tukhachevskii and Red Army rearmament generally, see Samuelson (1996, 2000) and Stoecker (1998).

Refinancing

When projects are long term, projects in progress require periodic refinancing. Alternatively, they must be discontinued. By examining refinancing decisions affecting projects in progress we can learn more about the incentives facing designers and funding principals and the calculations they made.

Funding principals reacted to initiatives from below, and planning decisions tended to validate these initiatives. Consequently projects in progress were normally refinanced without an explicit decision to this effect being reported. The fact that a project had been previously approved so that initial funds had been disbursed and work begun was a sufficient reason, other things being equal, for funding to be continued.

This raises the possibility that principals were indifferent to R&D failure. Could it be that they distributed project funding to agents in return for political rather than technological payoffs? If so, one could expect the principal to have responded to the agent's failure by emphasizing shared objectives, the difficulties intrinsic to the task, the value to the principal of the agent's efforts as a foundation for future progress, the value to the agent of the principal's continued support, and the value to society of the experience so far accumulated and other positive externalities. But the evidence does not match this at all. An unusual insight is provided by the ministerial review of continuing steam turbine projects held in August 1936. Kaganovich's mood was one of intense frustration, not indifference to failure; he interrupted the designers repeatedly with heavy sarcasm[24]:

> With existing dimensions is it sensible or feasible to place such a plant in an aircraft? One turbine engineer suggested placing 5 turbines in an aircraft, but for this the aircraft must weigh 125 tons without additional payload. You could put a F[eliks] D[zerzhinskii] locomotive in an aircraft, but then the aircraft would weigh 2,000 tons. This is comrade B[inaudible]'s fantasy, he's got 245-meter wings and a 45-meter fuselage.
>
> [...] We're not talking about a boiler on a Tsvetkov locomotive. Whoever's first to give us a turbine, we'll take it and work with it and the result will be that the airscrew will turn on the ground, if we put an airscrew on a locomotive it'll also turn, but we need to put it in *an aircraft at altitude* [emphasis added].
>
> [...] Three years ago comrade Tsvetkov came to me and proposed making such a turbine, I went to the boss, the people's commissar signed a decree to the effect that, in urgent order, under personal responsibility, [inaudible] to make a turbine, [they] began to make it, and now he comes and says: "There's

[24] RGAE, 8328/1/824, ff. 12, 15, 35, 51, and 52 respectively (22 August 1936).

> a turbine but no boiler." That's how they move technology forward. It's as if we got pig-iron but no metal.
>
> [...] I said to comrade Aksiutin [...] I'll give you a TB-7 airplane, smash it to pieces if you want, but taxi it along, lift it up to 100 meters, and then it will be a deed of proof that a turbine lifts up. Whatever it costs I'll pay. But [...]
>
> [...] I can't sit for three years and see no results.

The designers' response was to plead for time to allow the technology to evolve. They promised to build smaller, more efficient boilers and condensers. The aircraft designers Petliakov and Lavochkin were present. It was obvious that the engines being designed would not fit an existing airframe, so Petliakov asked that the turbine engineers should give more consideration to aircraft design and Kaganovich made him responsible for liaison.

Money and time had been spent, and while there was the smallest possibility of a positive outcome Kaganovich was not going to give up. The costs already sunk meant that the steam turbine projects drifted on for two or three more years. During 1937 a turbine of the Khar'kov Aviation Institute was prepared for installation in a TB-7, but the attempt was recognized to have failed by the end of the year.[25] One defensive response to the lack of progress was diversification: in January 1938 SKB director Sinev referred his superiors to the value of potential spinoffs from his bureau's work on aviation steam turbines for naval and locomotive engineering.[26] Only one decision to terminate a project has been found: in July 1938 the Moscow Aviation Institute's design bureau was closed for failure to progress with the binary cycle turbine.[27] Other projects simply vanished one by one from plans and reports.

Competitive Threats

Takeovers and Mergers

An R&D project can be thought of as a long-lived capital asset. All economies need mechanisms for restructuring these assets and transferring

---

[25]Rodionov (2002: 1937 under annual prologue and epilogue).

[26]Rodionov (2002: 1938 under 15 January).

[27]Rodionov (2002: 1938 under 21 July). In the aircraft industry several design bureaus were closed during the Stalin years as a punishment for failure to create successful designs: Albrecht (1993: 136 and 215), lists those of Kalinin, Shcherbakov, Berezniak and Bolkhovitinov, and Gudkov. In some cases the chief designer was imprisoned (Gudkov) or executed (Kalinin).

ownership through time.[28] In the case of R&D projects this mechanism is created by their need for periodic refinancing, which has the necessary effect of creating a secondary asset market. Under Soviet law state ownership rights over R&D project were delegated to ministerial fund holders by whom such rights were not freely transferable. In reality there were substantial incentives for agents to mount takeover or merger bids for projects of other fund-holding departments.

One motive was profit: the predator could compare the value of a project in progress with the costs of taking it over. The value of a project lay in the sunk costs represented by its tangible and intangible assets. These costs had already been incurred at the expense of some other department to whom the new fund holder did not have to pay compensation. Takeovers were costly nevertheless. First, a bid required the payment of direct lobbying costs. Second, it required the expenditure of reputation; a successful bidder made promises for which he might later be held to account. Third, it weakened the ownership rights over economic assets on which all fund holders ultimately relied.

Another motive was competitive threat: it might be more dangerous to abstain from the secondary market than to enter it. For example, small establishments were continually at risk of being swallowed by larger ones. The command system favoured large projects because of their economies of scope: larger units required fewer lines of outside communication and were less reliant on outsiders for essential goods and services. The preference for scale was reflected in frequent calls to eliminate duplication of effort and "parallelism". Calls for rationalization and centralization were rarely if ever questioned; they were regarded as progressive almost beyond debate, especially when comparisons were made with the scale of R&D establishments in aviation engineering abroad.[29] Smaller units had to expand in order to hold off threats from larger rivals, and one method of expansion was through takeover. Consequently no one benefited from restraint, and relatively small organizations behaved as aggressively as larger ones if the occasion arose.

---

[28]Gregory and Lazarev (2002) provide a study of the Soviet economy's informal secondary market in another capital good, the motor car.

[29]RGAE, 8044/1/460, ff. 49–51 (31 December 1940): an explanatory memorandum by People's Commissar for the Aircraft Industry A. I. Shakhurin on the 1941 plan for aviation engineering research and experimentation.

The logic of the takeover bid was a restructuring of liabilities. Consider a failing project, i.e., one that had incurred significant sunk costs without giving results on schedule. Was the project intrinsically bad, or just badly funded or led? If the lack of results compared with the sunk costs could be ascribed to excessively dispersed funding or poor organization, then it was efficient to write off the sunk costs and refinance the project under new management. Such logic was strengthened when the scope of activity and the number of projects was on the increase because this also brought a rising number of potentially weak projects.

For example, in December 1937 SKB acting director Sinev submitted a memorandum listing six steam-turbine projects in progress in four different institutes and three different cities. Welcoming the piecemeal advances already made, he criticized their "cottage-industry" scope (kustarshchina). Claiming the support of his own team and the Khar'kov project leaders, he called for all the groups to be brought together in a single "unified production-experimentation base" in Moscow, with close links to the aircraft industry.[30]

Another channel for proposals for concentration at this time was the system of peer review. Thus, late in 1937 the gas turbine designer V. V. Uvarov of the All-Union Thermal-Technical Institute (VTI) was commissioned to report to the commissariat of the defense industry on the progress of the "gas-steam turbine" being developed at factory no. 18 by designers Dybskii and Udod. After commenting on the weaknesses that he had observed, Uvarov commented[31]:

> The continuation of work on the lines under investigation should be curtailed, the more so since work on steam and gas turbocompressors is already going on [elsewhere]. *These two lines* [of work] *completely cover the authors' design, and for this reason duplication will yield nothing new.*

Defensive Measures

One way in which R&D agents defended projects in progress against competitive threats was by creating and reinforcing monopolies in new explicit knowledge. Implicitly, designers did not trust or sufficiently value

---

[30]RGAE, 8328/1/992, ff. 14–18 (19 December 1937). By emphasizing future links with the aircraft industry Sinev aimed to limit the continuing involvement of the electricity supply engineers.

[31]RGAE, 8328/1/996, ff. 22–24 (1 January 1938): emphasis added.

existing rights of authorship under Soviet law. They sought to prevent rivals from grabbing new knowledge to underpin competing proposals for development funding. Such rival projects would have looked "good" because they would not have had to account for costs of experimentation already sunk.

At the August 1936 ministerial review of steam-turbine projects it became apparent to Kaganovich that the designers themselves, supported by their departmental superiors, created some barriers against the collaboration that he desired. Development work for the Aksiutin turbine was proceeding at the Leningrad Kirov factory (LKZ), but without results. Why had engineer Vinblad failed to make himself useful to Aksiutin on the LKZ site? Because no one would issue him with a pass. Why not? A participating engineer commented: "[...] because there was rivalry, a special proprietary interest [opeka] of each in this business. Each was trying to turn this business into one [associated with] his own name."[32] In response, Kaganovich was simultaneously reassuring and threatening[33]:

> I will take all measures to protect the authorship of one or another comrade at work. If it's Aksiutin's turbine so let it be, but if he's up to some fabrication, and not up to realizing a technical solution to the problem, and for this reason has kept Vinblad away from the installation for a full year, then that is an obvious criminal act and an obvious detriment to the value of the turbine for our work.
>
> In our country there are no secrets and the designer who holds on to big secrets and does not carry them out into life — in the capitalist world he would simply perish and in the socialist [world] he is simply good for nothing. That's why we will set in train all measures and powers to help you realise the ideas and creativity that you have performed, while you are guaranteed full protection of authorship.

Designers also defended themselves against hostile takeovers by lobbying. For example, in 1938 a new struggle arose for control over the development of aviation steam turbines: yes, this was still "work in progress", and the lack of results was being attributed not to an intrinsic badness of the project but to the dispersion of resources and duplication of effort. In June the commissariat of the defense industry submitted to Molotov its long-delayed draft plan for aviation-engine experimentation for that year. In a stunning vote of no confidence in its own designers, it proposed that by October all work on aviation steam turbines should be

[32]RGAE, 8328/1/824, f. 38 (22 August 1936).
[33]RGAE, 8328/1/824, ff. 35 and 52 (22 August 1936).

concentrated in the Central Boiler and Turbine Institute (TsKTI) belonging to the electricity generating industry in Leningrad and that the central government should issue 2.5 million rubles from its contingency fund to TsKTI to expand its plant and equipment for this purpose.[34]

Thrown onto the defensive, SKB director Sinev wrote to Molotov, the Kaganovich brothers (one the commissar for the defense industry, the other the responsible central committee secretary), and defense commissar Voroshilov to protest this recommendation.[35] Sinev made three charges against TsKTI: it lacked an "aviation culture"; it was ineffective even at its primary task, the design of steam turbines for power stations; and it was already "over-encumbered" (*gromozdkaia*). Again he proposed the formation of a new bureau in Moscow based on one from a range of existing aviation establishments.

This time the defense failed: it was referred to air force chief Loktionov, who rejected it and upheld the recommendation in favor of TsKTI.[36] And as Table 1 shows, 1938 saw the end of aviation steam turbines at SKB. On the other hand the victory of the electricity supply engineers of TsKTI was hollow, because steam aviation was going nowhere and all such projects had been closed down by the end of 1939. In the end, after spending tens of millions of rubles, it was recognized that these were just bad projects.[37]

Good and Bad Projects

The problem of the principal was how to select and monitor long-lived projects of uncertain worth. In the presence of sunk costs there was a tendency for both the funding principal and the R&D agent to be motivated to continue projects that the principal would have preferred not to finance in the first place. The result was that selection could become adverse: R&D agents had an incentive to understate needs and overstate expected returns so as to obtain the first instalment of funding. Once the first instalment was paid and had become a sunk cost, the payment of the next instalment became more likely. Moreover, if results fell short when refinancing became necessary, the designer could always shift blame to the funder because

[34]RGVA, 4/14/1925, ff. 232–48 (26 June 1938).
[35]RGVA, 4/14/1925, ff. 150–52 (17 May 1938).
[36]RGVA, 4/14/1925, f. 155 (19 May 1938).
[37]And Sinev was arguing for 10 million rubles *more*: RGAE, 8328/1/992, f. 15 (19 December 1937).

the first instalment of funding was always less than the amount originally proposed.

What factors determined whether a project was "good" or "bad" from the point of view of the national mission? This depended on four factors: the unknown state of nature, the level of funding, the organization of resources and teamwork, and the motivation of the design team. First, the state of nature determined whether or not the project was intrinsically bad, i.e., technologically infeasible given contemporary knowledge about the physical world. Second, even for an intrinsically good project the level of funding needed to be appropriate to the task. Third, the physical and human resources employed on the project required effective organization, including teamwork and leadership; a design team that lacked the right equipment or was poorly led would give poor results. Finally, success depended on motivation: what was good or bad depended on whether the state saw it the same way as the designer. Thus some inventors involved in jet propulsion R&D may have been motivated otherwise: to realize a dream, to build an empire, to live in style, or to live in peace. In 1937/38, official suspicions of "other" motivation were sometimes hardened into the designation "enemy of the people". It is not necessary to go to this extreme to understand that R&D agents' motivations were not aligned at all times with the preferences of the state.

When a project failed, did it matter whether it was intrinsically bad, or potentially good but poorly funded or organized? With funding rationed and entry controlled, the danger was that bad projects might drive out good ones. Therefore the funding authorities made great efforts to diagnose the causes of project failure to see if they could be rectified. However, it was also extremely difficult, and perhaps impossible, to do so without hindsight. Even with hindsight it is still very difficult, and for this reason I avoid comment on the intrinsic goodness or causes of failure of individual projects. Only classes of project can be evaluated in this way; for example, all the aviation steam turbine projects were intrinsically bad, but I do not know which ones were also poorly funded or poorly led.

The various research establishments reported regularly to higher authority on each project in progress. From time to time the same authorities launched special reviews, which ranged from round-table exchanges of specialist opinion concerning common difficulties shared by several projects to investigations of specific projects thought to be at risk of failing.

The difficulty of establishing the causes of project failure made it easy for designers to displace the blame for their own lack of success.

As has already been shown, designers sometimes faulted the funder for dispersing funding too widely, that is, sharing it with rivals: they argued that more time and preferential funding would turn their own project round. Designers also blamed producers for failure to share the motivation of the design, leading to incompetent or neglectful preparation of components and assemblies. For example, steam-turbine designer Aksiutin complained to Tukhachevskii in 1935 that the Leningrad Kirov factory (LKZ) was incapable of playing a constructive role because it was gripped by "a certain conservatism utterly alien to the aviation culture" and commented that LKZ had declined a contract to build an Uvarov turbine for VTI giving as its official reason that the turbine required "too many parts to be completed to 'aviation standards' that would be an embarrassment for the factory [*chto dlia zavoda zatrudnitel'no*]".

According to the recollection of the aircraft designer A. S. Iakovlev (2000: 501), Stalin himself reflected on the tendency of designers to displace the blame for their own lack of results:

> A designer is a creative worker. Like the painter of a picture or the writer of a literary work, the product of a designer's or scholar's creativity can be successful or unsuccessful. The only difference is that from a picture or verse you can tell the author's talent right away. [...] With a designer it's more complicated: his design can look very attractive on paper, but final success or failure is determined much later as a result of the work of a numerous collective and after the expenditure of substantial material means .... Most designers get carried away with themselves and are convinced of their own and no one else's righteousness; on the basis of an overdeveloped self-regard and the mistrust that is characteristic of every author they tend to attribute their own failures to prejudice against themselves and their creations.

Were the different intrinsic motivations of R&D agents a factor in project success and failure? Stalin understood that Soviet funding institutions offered a degree of protection for self-serving interests; this created a rationale for him, with his security chiefs N. I. Ezhov, then L. P. Beriia, periodically to mount cruel inquisitions into the souls of the scientists and engineers.

What was the logic behind the ending of research into aviation steam power in 1938/39? Since most of the decisions remain undocumented we have to guess. With hindsight, the curtailment of these fruitless investigations might be seen as timely. But in the context of the time such a judgement could be questioned: was it not premature when the underlying problem of a replacement for the reciprocating engine based on internal combustion had not yet been solved? The context of a fundamental problem still unsolved helps us understand why interest in steam power for aviation still persisted in other countries, as Table 3 shows.

It is possible to set the Soviet decision to write off steam-power research in a different context, that of Moscow's growing impatience with the work of military inventors combined with changing external threat perceptions. In 1937/38 the leading specialists in rocket propulsion were arrested and imprisoned or executed (Harrison, 2000: 128–30; Siddiqi, 2000: 10–11).[38] The punishment of individuals was followed by sanctions against organizations. In early 1939 the Red Army decided to cancel payment for a wide range of ventures under the nominal heading of ammunition R&D, including several concerned with aviation jet propulsion. This unilateral decision left some 40 million rubles' worth of project commitments, including the entire work of two research institutes, completely unfunded.[39]

The mood of frustration was evidently sharpened by the growing expectation of war in Europe, and a desire to redirect funding away from risky long-term projects with uncertain returns towards research that could be relied on to bring quick results with immediate military application when war broke out.

It would be useful to know whether the steam turbine engineers were penalized as individuals for the failure of their research, but the fates of those who never achieved prominence can be hard to discover. It is not clear how many of them survived the general bloodletting of the late 1930s. Besides, this was a time when success could attract a punishment nearly as easily as failure, and many were punished for no reason at all.[40] The MAI experimental design bureau was closed down but we do not know the fate of its designer Przheslavskii. We do know that SKB director Sinev went on to become deputy chief designer at the Leningrad Kirov factory (LKZ) in wartime, then director of the LKZ experimental design bureau, and a leader of the postwar uranium industry (Lebina, 2000: 188). The designer Aksiutin reappeared in Voronezh in 1947/48 as chief of the experimental design bureau of aviation-engine factory no. 154.[41]

The only known victim of repression is A. Iu. Vinblad, chief of the Leningrad Kirov factory special-purpose design bureau for turbine

---

[38] For an earlier account of the purge see Holloway (1982a: 387–88).

[39] RGAE, 8162/1/299, ff. 36–54 (March to April 1939). Subsequently the jet propulsion design bureau KB-7 was closed down.

[40] On the complex relationship between success, failure, and punishment in the Great Terror, see Manning (1993).

[41] RGAE, 8044/1/1637, f. 112 (12 June 1947) and 8044/1/1795, f. 109 (May 1948).

engineering and a participant in the August 1936 ministerial review of steam turbines. He was arrested in February 1940 and sentenced to eight years in a labor camp. In September 1942 he was transferred to a scientific penal colony where in May 1946 he was listed as working on a jet aviation engine.[42] Had he taken the blame for a steam power project failure? Vinblad was sentenced for espionage under Article 58(6) of the RSFSR penal code. Probably he had been careless with his paperwork, had travelled abroad, or had communicated with relatives abroad. Had he been punished for an R&D failure, he would more likely have been charged with economic sabotage under Article 58(7).[43]

Finally, given that the steam turbine projects were bad ones, and that a number of their leaders survived unharmed, what light is thrown on the possibility that principals deliberately tolerated or fostered them in order to share rents? There is no evidence that the funding of steam turbine projects was continued despite evidence of failure in order to promote vertical relationships of trust and loyalty. Their continued refinancing is sufficiently explained by sunk costs and the difficulty of diagnosing the causes of project failure. Evidence from other fields of Soviet aviation R&D (Harrison, 2005) also confirms that, when rent-seeking was identified, it was punished.

Conclusions

During the 1930s the Soviet Union probably invested more resources in steam power R&D than in jet propulsion or rocketry, and possibly as much

[42]GARF, 9401/2/136, f. 207 (18 May 1946). This person, called both Vinblad and Vinblat in the verbatim record of the August 1936 ministerial review, is Vinblad in the presently cited report by a senior MVD official, but Vinblat in an accompanying memorandum by minister S.N. Kruglov (ibid., f. 204). Both documents were addressed to Stalin personally. One hopes the latter was not in a mood to see sabotage in misspelling.

[43]Article 58, which dealt with "counter-revolutionary crimes," is reproduced in part by Conquest (1971: 741–46). Since Vinblad's sentence exceeded 3 years he must have been convicted of "transmission, theft or collection" of "an *especially guarded* state secret" (my emphasis) with a view to handing it to "foreign states, counter-revolutionary organisations, or private individuals," but not involving "*especially grievous* consequences to the interests of the USSR" which would have merited death by shooting. The reference to private individuals means that suspicious personal contacts could have constituted evidence sufficient for a conviction. In contrast the rocket aircraft designer S.P. Korolev, arrested in June 1938, was imprisoned under Article 58(7) (economic sabotage), 58(8) (terrorism), 58(11) (premeditation and conspiracy), and 58(17) (political sabotage); see Raushenbakh (1998): 64–66).

as all other countries put together. An investigation of the results suggests a number of findings of more general significance.

First, in the Soviet Union steam power R&D was carried out in the context of a vertically organized command system. But within this context there was a great deal of market-like activity on the supply side including horizontal rivalry and competitive rent seeking, a secondary market in R&D projects involving takeover and merger activity, and attempts to create and defend monopolies.

In the Soviet command system the designer took the initiative. There was no shortage of inventiveness and there were more proposals for radical innovations than the authorities were willing to fund. The main problem for the authorities was to regulate inventive activity, not promote it.

In the Soviet economy the scale of steam-power R&D was that of an artisan industry. The resources available for such research were extremely limited and funding was rationed. However, budget constraints on individual projects in progress tended to become soft. Once a project had been selected for funding it had a good chance of its funding being continued until wide limits on the funding principals' resources and patience were breached.

Designers who were successful in getting their proposals selected for initial funding and subsequent refinancing were "heterogeneous engineers". They invested resources in lobbying and political reputation to ensure that their projects were selected for funding and, once selected, to protect them against termination from above or takeover by rivals in the name of rationalization.

It was difficult or impossible for the Soviet authorities to tell good ideas from bad ones at the time. The difficulty reflected technological uncertainty and agents' unobserved characteristics. It remained difficult when projects in progress did not give results because ideas could fail for reasons unrelated to their intrinsic merits. In the presence of sunk costs, refinancing a project in progress was usually easier than terminating it. It is possible that adverse selection resulted.

Finally, the evidence of rent seeking is plentiful but there is no evidence that principals deliberately shared rents with their agents through the military R&D system so as to win their trust or reward their loyalty.

Appendix: The Besler Connection

A Travel Air 2000 biplane made the world's first piloted flight under steam power over Oakland, California, on 12 April 1933. The strangest feature of the flight was its relative silence; spectators on the ground could hear the pilot when he called to them from mid-air (FitzGerald, 1933).[44]

The aircraft, piloted by William Besler, had been fitted with a two-cylinder, 150 horsepower reciprocating engine. An important contribution to its design was made by Nathan C. Price, a former Doble Steam Motors engineer; When Doble went bankrupt in 1931 William Besler and his brother George bought its assets and patents and founded the Besler Corporation. At Besler, Price was working on high-pressure compact engines for rail and road transport; the purpose of the flight was to obtain publicity for this work. Following its unexpectedly favorable reception Price went to Boeing and worked on various aviation projects, but Boeing dropped the idea of a steam aviation engine in 1936 (Schlaifer, 1950: 448–49). Price later worked for Lockheed where his experience with developing compact burners for steam boilers helped to design Lockheed's first jet engine.[45]

On its own the Besler engine weighed 180lbs but the boiler and condenser took the weight up to 500lbs, some 200lbs more than the Curtiss OX-5 engine that it replaced.[46] The result was a specific weight of approximately 3.3lbs (1.5 kg) per horse power, a figure that was high relative to the internal combustion aviation engines of the time, and also some of the steam turbine designs listed in Tables 3 and 4, but the engine had not been designed for aviation use nor had it been adapted.

The advantages of the "Besler System" that were claimed at the time included the elimination of audible noise and destructive vibration; greater efficiency at low engine speeds and also at high altitudes where lower air temperatures assisted condensation; reduced likelihood of engine failure; reduced maintenance costs; reduced fuel costs, because fuel oil was used in place of gasoline; reduced fire hazard because the fuel was less volatile and operating temperatures were lower; and a lack of need for radio shielding (Martin, 1933; Nickerson, 1933).

Not surprisingly, Soviet aviation experts became aware of the Besler System. According to a report to defense minister Voroshilov and others dated 3 July 1934, the current plan for experimental aircraft and aviation engines included a steam engine with a capacity of 100 to 120 horse power

[44] I thank Jean Vezina for this and other references and information relating to the Besler experience.

[45] Personal communication from Jean Vezina (14 November 2001).

[46] Martin (1933) provides a technical description. A replica of the Besler steam engine is on display at the Smithsonian National Air and Space Museum in Washington, DC.

and weight of 120 to 150 kilograms. The chief administration for the aircraft industry had subcontracted the design work to a team at the Moscow Aviation Institute led by a professor Kvasnikov. The report states: "a steam automobile is being purchased abroad as a basis for designing a steam engine for an aircraft" (Rodionov, 2002: 1934 under 3 July). This appears to have been an engine commissioned from the Besler Corporation.[47]

There is no evidence that a Soviet aviation engine ever resulted. One reason may be that the Soviet authorities' primary motivation was to improve the power plant for heavy bombers. For capacities in excess of 1,000 horse power a turbine captures the energy released by the expansion of steam more efficiently than a piston. Thus, the steam reciprocating engine turned out to be unsuitable for scaling up to the needs of large aircraft, and interest today in the application of steam power to aviation remains confined to small reciprocating engines.

References

Official Papers

Russian Federation. Gosudarstvennyi Arkhiv Rossiiskoi Federatsii (GARF) [State Archive of the Russian Federation], Moscow.

Russian Federation. Rossiiskii Gosudarstvennyi Arkhiv Ekonomiki (RGAE) [Russian State Economics Archive], Moscow.

Russian Federation. Rossiiskii Gosudarstvennyi Voennyi Arkhiv (RGVA) [Russian State Military Archive], Moscow.

Publications

Albrecht, U (1993). *The Soviet Armaments Industry.* Chur (Switzerland): Harwood Academic Publishers.

Amann, R and Cooper, J (eds.) (1982). *Industrial Innovation in the Soviet Union.* New Haven, CT: Yale University Press.

Amann, R and Cooper, J (eds.) (1986). *Technical Progress and Soviet Economic Development.* Oxford: Blackwell, 1986.

[47] Later the Boeing executive Allan Bonnalie stated that the Besler brothers "got a contract from the Russian govt. to build an actual airplane power plant which they did." The transcript of this interview, undated but probably done not long after World War II, is in the collection of Jean Vezina. The latter adds, however: "According to Jim Crank, a retired Lockheed engineer and a renowned steam car collector who had personally known William Besler before his death in 1984, the engine delivered to the Soviet Union was in fact a Doble steam car chassis and not a dedicated airplane power plant. The Bonnalie statement was not first-hand since, at that time, the contract with the Boeing School of Aeronautics was completed and the deal with Russia was made at the Besler factory at Emeryville, California" (personal communication, 18 March 2002).

Amann, R, Cooper, J and Davies, RW (eds.) (1977). *The Technological Level of Soviet Industry*. New Haven, CT: Yale University Press.

Arrow, KJ (1962). The Economic Implications of Learning by Doing. *Review of Economic Studies*, 29(3): 155–73.

Beath, J, Katsoulacos, Y and Ulph, D (1995). Game-Theoretic Approaches to the Modelling of Technological Change. In: Stoneman PR (ed.), *Handbook of the Economics of Innovation and Technological Change*, Oxford: Blackwell, pp. 132–81.

Berliner, JS (1976). *The Innovation Decision in Soviet Industry*. Cambridge, MA: MIT Press.

Bystrova, I (2000). *Voenno-promyshlennyi kompleks SSSR v gody kholodnoi voiny. (Vtoraia polovina 40-kh — nachalo 60-kh godov)*. Moscow: Institut Rossiiskoi istorii Rossiiskoi Akademii nauk.

Collina, A (1932). A Steam Power Plant for Aircraft. *Aviation Engineering*, 7(6): 10–11.

Conquest, R (1971). *The Great Terror* (revised edn.). Harmondsworth (England): Pelican.

Dasgupta, P and Stiglitz, JE (1980). Industrial Structure and the Nature of Innovative Activity. *Economic Journal*, 90(358): 266–93.

Dewatripont, M and Maskin, E (1995). Credit and Efficiency in Centralized and Decentralized Economies. *Review of Economic Studies*, 62(4): 541–55.

FitzGerald, HJ (1933). World's First Steam-Driven Airplane. *Popular Science Monthly*, 123(1): 9–11 and 92.

Gibbs-Smith, CH (1970). *Aviation: An Historical Survey from its Origins to the End of World War II*. London: HMSO.

Goskomstat SSSR (1987). *Narodnoe khoziaistvo SSSR za 70 let. Iubileinyi statisticheskii sbornik*. Moscow: Finansy i Statistika.

Gregory, PR (2003). Soviet Defence Puzzles: Archives, Strategy, and Underfulfilment. *Europe-Asia Studies*, 55(3): 923–958.

Gregory, PR and Lazarev, V (2002). The Wheels of a Command Economy: Allocating Soviet Vehicles. *Economic History Review*, 55(2): 324–48.

Grigor'ev, NV (1994). Aviatsionnoe motorostroenie v Velikoi Otechestvennoi voiny 1941–1945 gg. In: Biushgens GS (ed.), *Samoletostroenie v SSSR. 1917–1945*, Volume 2, Moscow: TsAGI, pp. 164–96.

Gunston, B (1995). *The Osprey Encyclopedia of Russian Aircraft, 1875–1995*. London: Osprey Aerospace.

Hanson, P (1981). *Trade and Technology in Soviet-Western Relations*. London and Basingstoke, England: Macmillan.

Hanson, P and Pavitt, K (1987). *The Comparative Economics of Research, Development and Innovation in East and West: A Survey*. Chur (Switzerland): Harwood Academic Publishers.

Harris, JR (1999). *The Great Urals: Regionalism and the Evolution of the Soviet System*. Ithaca, NY, and London: Cornell University Press.

Harrison, M (2000). New Postwar Branches (1): Rocketry. In: Barber J and Harrison M (eds.), *The Soviet Defence-Industry Complex from Stalin to Khrushchev*. Basingstoke, England: Macmillan. pp. 118–49.

Harrison, M (2003). Soviet Industry and the Red Army Under Stalin: A Military-Industrial Complex? *Les cahiers du monde Russe*, 44(2–3): 323–42.

Harrison, M (2005). A Soviet Quasi-Market for Inventions: Jet Propulsion, 1932 to 1946. *Research in Economic History*, 23: 1–59.

Harrison, M and Davies, RW (1997). The Soviet Military-Economic Effort Under the Second Five-Year Plan (1933–1937). *Europe-Asia Studies*, 49(3): 369–406.

Henning, GR and Trace, K (1975). Britain and the Motorship: A Case of the Delayed Adoption of New Technology? *Journal of Economic History*, 35(2): 353–85.

Holloway, D (1982a). Battle Tanks and ICBMs. In: Amann R and Cooper J (eds.), *Industrial Innovation in the Soviet Union*. New Haven, CT: Yale University Press, 368–414.

Holloway, D (1982b). Innovation in the Defence Sector. In: R Amann and J Cooper (eds.), *Industrial Innovation in the Soviet Union*. New Haven, CT: Yale University Press, pp. 276–367.

Holloway, D (1984). *Stalin and the Bomb: the Soviet Union and Atomic Energy, 1939–1956*. New Haven, CT: Yale University Press.

Iakovlev, AS (2000). *Tsel' zhizni. Zapiski aviakonstruktora* (6th edn.), Moscow: Respublika.

Jasny, N (1952). *Soviet Prices of Producers' Goods*. Stanford, CA: Stanford University, Food Research Institute.

Knörnschild, E (1941). Dampftriebwerke für Flugzeuge. *Luftwissen*, 8(12): 366–73.

Kornai, J (1979). Resource-Constrained versus Demand-Constrained Systems. *Econometrica*, 47(4): 801–20.

Kostyrchenko, GV (1992). Organizatsiia aviatsionnogo krupnoseriinogo proizvodstva. In: Biushgens GS (ed.), *Samoletostroenie v SSSR. 1917–1945*, Volume 1. Moscow: TsAGI, pp. 413–35.

Lazarev, V and Gregory, PR (2003). Commissars and Cars: A Case Study in the Political Economy of Dictatorship. *Journal of Comparative Economics*, 31(1): 1–19.

Lebina, N (2000). The Defence-Industry Complex in Leningrad (2): The Postwar Uranium Industry. In: Barber J and Harrison M (eds.), *The Soviet Defence-Industry Complex from Stalin to Khrushchev*. Basingstoke (England): Macmillan, pp. 184–94.

Liul'ka, AM and Kuvshinnikov, SP (1981). K istorii sozdaniia pervogo sovetskogo turbo-reaktivnogo dvigatelia. *Voprosy istorii estestvoznania i tekhniki*, 2: 86–96.

MacKenzie, D (1996). *Knowing Machines*. Cambridge MA: MIT Press.

Manning, RT (1993). The Soviet Economic Crisis of 1936–1940 and the Great Purges. In: Getty JA and Manning RT (eds.), *Stalinist Terror: New Perspectives*, Cambridge: Cambridge University Press, 116–41.

Martin, M (1933). The Flying Tea-Kettle. *Popular Aviation*, 13(1), 8–10 and 73.

Mokyr, J (1990). *The Lever of Riches: Technological Creativity and Economic Progress*. Oxford: Oxford University Press.

Nickerson, AN (1933). More About Steam Planes. *Popular Aviation*, 13(2): 105–06 and 131.
Ordway, FI and Sharpe, MR (1979). *The Rocket Team*. London: Heinemann.
Pikhoia, RG (1998). *Sovetskii soiuz. Istoriia vlasti. 1945–1991*. Moscow: RAGS pri Presidente RF.
Raushenbakh, BV (ed.) (1998). *S. P. Korolev i ego delo. Svet i teni v istorii kosmonavtika*. Moscow: Nauka.
Rees, EA (2001). Leaders and Their Institutions. In: PR Gregory (ed.), *Behind the Façade of Stalin's Command Economy: Evidence from the Soviet State and Party Archives*. Stanford, CA: Hoover, pp. 35–60.
Rodionov, I (2002). *The Aviation and Aircraft Industry of the Soviet Union, 1916 to 1946*, Version 3.1. University of Warwick, Department of Economics. Available at http://warwick.ac.uk/aviaprom/.
Sah, RK and Stiglitz, JE (1986). The Architecture of Economic Systems: Hierarchies and Polyarchies. *American Economic Review*, 76(4): 716–27.
Samuelson, L (1996). *Soviet Defence Industry Planning: Tukhachevskii and Military-Industrial Mobilisation*. Stockholm: Stockholm School of Economics.
Samuelson, L (2000). *Plans for Stalin's War Machine: Tukhachevskii and Military-Economic Planning, 1925–1941*. London and Basingstoke: Macmillan.
Schlaifer, R (1950). Development of Aircraft Engines. In: Schlaifer R and Heron SD, *Two Studies of Relations Between Government and Business*. Boston, MA: Harvard University, Graduate School of Business Administration, Division of Research.
Siddiqi, AA (2000). *Challenge to Apollo: The Soviet Union and the Space Race, 1945–1974*. Washington, DC: NASA History Division, NASA SP-2000-4408.
Smith, GG (no date). *Gas Turbines and Jet Propulsion for Aircraft*. London: Flight Publishing Co.
Sobolev, DA (1996). *Nemetskii sled v istorii Sovetskoi aviatsii. Ob uchastii nemetskikh spetsialistov v razvitii aviastroeniia v SSSR*. Moscow: Aviantik.
Stodola, A (1927). *Steam and Gas Turbines*, New York: McGraw-Hill.
Stoecker, SW (1998). *Forging Stalin's Army: Marshal Tukhachevsky and the Politics of Military Innovation*. Boulder, CO: Westview.
Sutton, AC (1968). Western Technology and Soviet Development: 1917–1930 (Volume 1). Stanford, CA: Hoover Institution Press.
Sutton, AC (1971). Western Technology and Soviet Development: 1930–1945 (Volume 2). Stanford, CA: Hoover Institution Press.
Sutton, AC (1973). Western Technology and Soviet Development: 1945–1960 (Volume 3). Stanford, CA: Hoover Institution Press.
Voronkov, IuS (1984). Teoriia i konstruktsiia: istoriia vzaimodeistviia pri sozdanii pervykh aviatsionnykh reaktivnykh dvigatelei. In: *Issledovaniia po istorii i teorii razvitiia aviatsionnoi i raketo-kosmicheskoi nauki i tekhniki*, Volume 3. Moscow: Nauka, pp. 113–22.
Wintrobe, R (1998). *The Political Economy of Dictatorship*. Cambridge: Cambridge University Press.

Part 3

Communism and Coercion

Chapter 9

The Fundamental Problem of Command: Plan and Compliance in a Partially Centralized Economy

Mark Harrison

In *The Economics of Soviet Planning* Abram Bergson (1964: 327–9) distinguished between the economic merit and the rationality of socialism.[1] The merit of socialism was to be gauged by the efficiency with which it satisfied consumer welfare. Its rationality could be measured not only against consumer welfare but also in relation to "planners' preferences". Rationality in either sense, he suggested, is best measured by 'the degree to which conduct conforms to the material ends sought.... We are thus led to ask, in respect to Soviet conduct regarding resource use, the cardinal question that arises in respect to economic conduct generally: In terms of what material ends and to what extent is such behavior economically rational?'

Few will have changed their minds about the *merits* of socialism as a result of the opening of the archives. The main advances that new documentation has made possible relate to the *rationality* of the Soviet command system. We can learn in unprecedented detail how Soviet officials behaved in the privacy of their offices and why they behaved as they did.

This chapter was first published in 2005 in *Comparative Economic Studies* 47(2): 296–314. © The Association for Comparative Economic Studies 2005.

[1]This paper was presented to a conference on "Performance and Efficiency in the USSR: New Measurements and the Bergson Legacy," held in memory of Abram Bergson (1914–2003) at Harvard University, November, 23 and 24, 2003. I thank Leonid Borodkin, Paul Gregory, and Avner Greif for an inspirational conversation at Stanford University in July 2003, and Michael Ellman, Simon Ertz, Marshall Goldman, Paul Gregory, Gregory Grossman, David Lane, Peter Law, Valery Lazarev, Andrei Markevich, and Michael Waterson for comments and advice. I am responsible for errors.

Specifically, we find that officials at each level were continually issuing new decrees and instructions to those below them, who rarely did exactly as they were told. At the time the same officials were receiving reports and information from below, some distorted and the rest incomplete; their core problem was to verify and control what people were really doing when they claimed to be obeying orders. The fact that officials at each level of the command system were not merely passive instruments of higher authority is what makes its operation an interesting problem.

How did Soviet officials command people and allocate resources? How did Stalin and his successors get others to do their bidding? This paper is about the rational behaviour of a principal who gives an order and the agent who is expected to carry it out. In studying the behaviour that we observe historically it tries to answer Bergson's 'cardinal question': 'In terms of what material ends and to what extent is such behavior economically rational?'

According to Avner Greif (2000), any institution that persists must be understood as an equilibrium of individual motivations and constraints: each person participates in it because it is in their interest to do, conditional upon their expectation that others have made the same calculation. How is the equilibrium of individual motivations and constraints achieved in the case of market institutions? Greif describes the *fundamental problem of exchange* as a game that arises from the sequential nature of a transaction. Two players contemplate a contract to engage in mutually beneficial exchange. The first player must initiate the contract and advance the means to fulfil it to the second player, or alternatively abstain. The second must decide to cooperate by completing the contract or cheat by stealing the proceeds of the exchange, including the advance, at the expense of the first player. Without means of enforcement the best strategy of the second player is to cheat and that of the first is to abstain. Historically, therefore, the rise of market exchange required the development of a variety of private and public enforcement mechanisms that overcame this problem.

Hierarchies give rise to a corresponding *fundamental problem of command.* The origin of the problem is the same, the sequential nature of a transaction. A principal gives an order to an agent and advances the means to implement it. Why is an order necessary? Because the principal desires an allocation of resources that differs from the one that would result from the agent's pursuit of her own self-interest; if this were not the case then the desired allocation would result from the agent's action without having to be told. It follows that for the hierarchy to find an equilibrium the obedient agent must be induced to forego opportunities for private gain. This problem must be solved for a hierarchy to find an equilibrium.

Hierarchies have existed in many forms and for many reasons. One form is the modern corporation. The transaction-cost approach of Coase (1937) and Williamson (1975) suggests that corporations exist in order to realise potential efficiency gains. In neo-Marxian economics, however, they exist to exert power and monopolise resources (Hymer, 1960/1976; Marglin, 1974). In the first case the hierarchy grows because its superior efficiency gives it a competitive advantage, and society gains as a result. In the second case it grows by stifling competition and extracting a monopoly rent, while society suffers a welfare loss.

In this paper we will think about a particular kind of hierarchy, the command system. The command system is a hierarchy that is driven by monopolisation, not cost reduction: it attempts to monopolise a territory by force. Monopolisation is the strategy that Evgenii Preobrazhensky (1926/1965) advocated for the Soviet economy in the 1920s. We are also looking at something similar to Mancur Olson (1993) when he analyzed a proprietary dictator.

The problem that I describe is present in all complex hierarchical organisations, even where their existence is predicated on superior efficiency. This is because, even if an organisation exists solely to enable its members to exploit a mutual efficiency gain that is shared with society, any individual within the organisation can gain still more by shirking or cheating the other members. Such opportunities have always been present in command systems to the extent that their monopolizing drive was never completely successful and their control over resources remained partially, not fully centralized. We will use the Soviet economy as an example.

The Fundamental Problem of Command

In a market system each player's first option is to abstain. In a command system there is compulsion: to abstain is mutiny or desertion and these are punished. The agent can decide only to obey or disobey. Disobedience can take more than one form. Figure 1 illustrates the possible outcomes. The obedient agent receives the wage advanced w and returns the planned output x to the principal who gains a net surplus $x - w$. The disobedient agent may shirk, i.e. take the wage and ignore the command. The shirker's payoff is the wage plus the gain in reduced effort, $w + e$, which is always preferable to the wage alone, while the principal records a loss, $-w$.

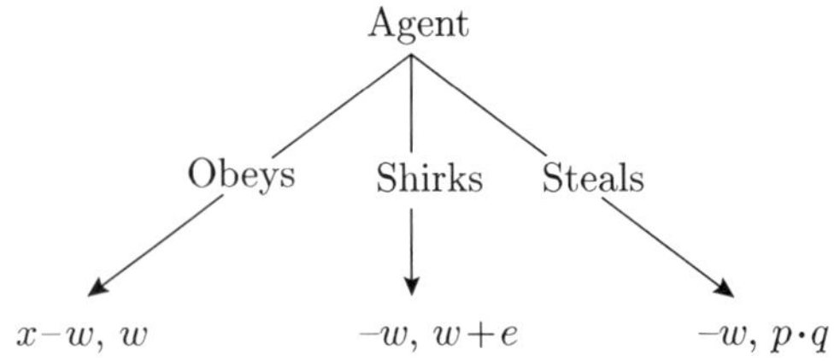

Fig. 1. The Fundamental Problem of Command.

When the economy is only partly centralized the agent has an outside option: steal the wage advanced and invest it with a private network of persons outside the hierarchy. There are profit opportunities that the command system has left unexploited either by intention, just as a monopolist deliberately restricts supply and capacity usage, or by mistake because of plan errors. Where these opportunities exist, an unofficial network can move in and reap some of the gains privately. The agent stands to gain $p \cdot q$ where q is the value of the agent's share in the opportunity and p is the probability that she can realise it. In fact, p measures horizontal trust or "honor among thieves". Where criminals mistrust each other crime cannot be organised and private networks will lack the scope to realise available opportunities. Also, where thieves lack honor the agent will not be able to enforce her share in the return from a private venture by private means; she will rationally expect others in the network to steal from her what she stole from the principal in the first place.

The one-shot game has bad and less bad news for the principal. The less bad news is that the agent will prefer compliance to stealing while horizontal mistrust is high. The bad news is that, even if the agent will not steal, she will always shirk.

The news is better when the game is repeated. Suppose the principal responds to one-time disobedience by terminating the agent's position and expelling her from the hierarchy, so that she has nothing left to take into the horizontal network in the next round; then obedience is the agent's only move that can allow her to take part repeatedly and this is the principal's best hope. Repeated infinitely, obedience will return $w/(1-\delta)$ to the agent where δ is the discount factor and $1 > \delta > 0$. It follows that a forward-looking agent will prefer compliance to shirking where $w \cdot \delta/(1-\delta) > e$, and the latter is more likely the less the agent discounts the future.

Similarly the agent will prefer compliance to theft if $w/(1-\delta) > p \cdot q$. To draw out the intuition, compliance is more likely the higher is the official wage compared with unofficial profit opportunities, but even in the presence of a clear efficiency deficit that makes tempting private opportunities the principal can still keep the agent compliant provided honor among thieves remains low and the agent places a low enough discount on the future.

The fundamental problem of command can be interpreted broadly. The principal can be anybody who has a non-market relationship with an agent, from a great ruler to a lowly office manager. The unofficial network can be anybody external to the agent's hierarchy who may share the gain arising from the principal's loss: a principal in another hierarchy, a family friend, a criminal gang, or a foreign power. The asset of which the principal is cheated need not be understood as a narrowly material object. For example, the problem of command also describes the situation of a proprietorial dictator (Olson, 1993) who aims to monopolise the loyalty of the citizens; in this context the citizen's loyalty is a reproducible, transferable asset that the citizen may invest in the dictator, withhold, or transfer to the opposition (Wintrobe, 1990). These correspond to the agent's choices to comply, shirk, or steal.

The Soviet Economy as a Command System

The fundamental problem of command may be fairly obvious as a theoretical idea and yet have little or nothing to do with real life in a command system. In order to show a connection between history and theory I will use the Soviet economy to provide some data.

Some details of the Soviet command system are relevant. As Bergson (1964: 15–48) described it most enterprises were state owned and most producers were agents of a government principal, usually a minister acting as the legal fund holder. Horizontal relationships of specialisation and exchange were organised by order from above through vertical hierarchies rather than directly between buyers and sellers on a voluntary basis. Vertical subordination permitted orders to flow downwards on the basis of the information that flowed upwards.

The Soviet hierarchy was formed by a dictator, Stalin, who was unchallenged by his closest associates from 1932 until his death in 1953 (Davies *et al.*, 2004). He ruled by decree: thus it was said that "the plan is the law" (e.g. Gregory, 2003: 164). He ruled by delegation, so under him many smaller dictators exercised unconstrained power within the specialised fiefs that he allocated to them. Power cascaded downward through branching networks of agents that collectively formed the so-called *nomenklatura*, but in practice the 'big' nomenklatura of a million posts was fractionalised because every little dictator nested his own little nomenklatura inside the big one and demanded its loyalty (Khlevniuk, 2003; Lewin, 2003).

Consequently, the Soviet state took the form of a nested dictatorship in which "each organisation duplicated the administrative control structures of its superior in the vertical chain of command" (Gregory and Markevich, 2002). The dominant administrative relationship at each level of the state was that of principal and agent, with each principal in turn acting as the agent of a higher principal until we reach the top where the great dictator ruled alone as *capo di capi*.

The Soviet command system was defined by hierarchical complexity: parallel hierarchies that were functionally differentiated at lower levels converged at the top in unified leading bodies such as the council of ministers and the politburo. To fulfil its function each specialised hierarchy had to trade horizontally with others. Thus the defense ministry purchased equipment and fuel from industry and food from agriculture; the ministry for engineering bought metals from the steel industry and power from the electricity generating industry. The coordination problem created by specialised hierarchies was solved only partly through centralization. Rather there was "centralized pluralism" (Nove, 1977: 60–4), meaning that much decentralized allocation went on behind a centralized façade (see also Zaleski, 1980; Markevich, 2003).

One reason for partial centralization was the impossibility of forcing full information sharing in hierarchies. The process of reaching economic equilibrium in markets requires agents to share information. This sharing is voluntary and partial: voluntary because self-interested, and partial because each agent does not need to know everything about all other agents for an equilibrium to result. According to Hayek (1945) economizing on information-sharing is one of the merits of the market system. In the Soviet command system, in contrast, the sharing of information was subject to coercion. Coercion arose because agents had incentives to conceal information rather than share it. The coercion was asymmetric: the principal had a right to know everything but the agent did not have a right to know anything. However, there were limits to the principal's coercion. The principal depended on the agent to select and aggregate information since he could not handle high-volume detail. The agent had self-interested motives to disclose information selectively and to distort it before disclosure. Thus, the principal could not trust what the agent told him, and did not know what the agent did not tell him, unless he sought additional information by other means at additional cost.

Faced with this information problem, central authorities delegated responsibility for the detail of inter-ministerial exchange to ministerial officials. Such exchanges were authorised in high-level plans and decrees only in broad outline; then the ministries themselves had to negotiate detailed contracts for specific commodities that implemented the plan (Gregory, 2003). In principle, the implementation of contracts rested on budget authorisation by the ministry of finance and credit authorisation by the state bank. In practice, because the plan dealt only with aggregates and the budget dealt only with rubles, there were wide spheres of informal market allocation where firms competed for real resources with each other and with households (Harrison and Kim, 2006). Markets were wider for general-purpose intermediate products and consumables than for more specialised products such as special-purpose machinery where supply or demand were more easily monopolised. Whether or not competition existed, weak oversight promoted ministerial discretion and led to protracted inter-ministerial bargaining and disputes (Kroll, 1986; Harrison and Simonov, 2000; Belova, 2005; Belova and Gregory, 2002).

It is well known that elements of market organisation persisted inside and alongside the command system. There were illegal markets for many industrial materials and consumer goods and services and legal markets for labor and labor-intensive foodstuffs (Katsenelinboigen, 1977). Historically these markets evolved in various ways. For the majority that escaped penal servitude the labor market remained relatively free but there were 15 years of much heavier regulation around World War II (Filtzer, 2002). At first the authorities tried to suppress the food market altogether, and they did achieve a broad state monopoly of grains and larger livestock, but the food shortages that resulted compelled them to concede legality to the market supply of backyard food products (Davies, 1994).

Monitoring and Enforcement

In the Soviet economy principals waged a continual struggle against shirking and theft. They also faced the fact that agents who intended to comply with orders did not always accept them unconditionally but sometimes tried to renegotiate them before complying. The authorities developed a wide range of mechanisms for monitoring and enforcement that secured the participation of the players in the command system. The outcome was an equilibrium of behaviours and expectations that secured the willing participation of all the players.

The need to solve the fundamental problem of command parsimoniously explains a number of Soviet economic institutions and suggests an agenda for more research. The development of several distinct mechanisms for monitoring and enforcement, each with different technologies, costs, and returns, implies that the dictator had scope to optimise on several margins at once. More research may lead us to a more unified explanation of the great shifts in Soviet reliance on propaganda and education, legal and relational enforcement, forced labor, secrecy, and military power, in terms of a general equilibrium rather than considering each in isolation.

Multi-Layer, Multi-Stage Monitoring

Information asymmetries pervaded the Soviet command system. Agents could exploit them at the expense of the principal, for example by persuading the principal that they had fulfilled an assignment when they had not (Harrison, 1998). If the principal could not tell whether he was receiving a yield of x, 0, or $-w$ he did not have much hope of solving the fundamental problem of command. Since the dictator could not trust agents in the direct line of command to tell the truth, he developed other means that met this need including separate hierarchies for planning and auditing (Gregory, 1990; Gregory, 2001; Belova and Gregory, 2002), financial and legal regulation (Gorlizki, 2002), contractual observance (Belova, 2005), and so on. Thus, many agencies monitored activities, tracked outcomes, and reported to the boss.

Himself inscrutable and devious, Stalin required complete openness in those he allowed to report to him. He reacted severely to the least suspicion of a self-interested lie, more so if it appeared to involve horizontal collusion (Gorlizki and Khlevniuk, 2004). Stalin's mistrust of his own monitors was rational. The problem that resulted was: *Quis custodiet ipsos custodes* [who watches the watchmen]? In the mid-1930s Stalin handed over the role of monitoring the monitors to the security "organs" with results that are well known.

The authorities intensified monitoring by advancing w in instalments and frequently checking progress. Breaking production and innovation cycles into many intermediate stages shortened the period over which an agent could behave badly before being expelled from the hierarchy.

A by-product was the creation of sunk costs that, in a context of credit and fiscal centralization, gave rise to Kornai's (1980) famous "soft" budget constraint (Dewatripont and Maskin, 1995).

Multi-layer, multi-stage monitoring was costly. The centralization of information that it required was only feasible within historical limits. Consider the information cost of monitoring a given outcome. Information costs fell rapidly in the nineteenth century and more rapidly in the twentieth century. However, the outcomes were changing too: the share of services in output was rising and there was growing variation in product and service quality (Harrison, 1998). Real output and value became harder to observe and accurately measure. As a result, the viability of centralized, multi-layer monitoring followed an inverse U curve: it was more effective between 1870 and 1970, less before and less after (Broadberry and Ghosal, 2004).

Enforcement Mechanisms

Internalisation and Promotion. One way to achieve a command equilibrium was to persuade the agent to desire the same outcomes as the principal, that is to internalise the latter's valuations. Through schooling and the media Soviet leaders invested in persuading citizens to respect authority and adopt patriotic and party-minded norms. They developed costly rites of passage into citizenship, party membership, and military service, and also used humiliation rituals such as the 'boards of shame' that stigmatised bad citizens. They induced many to internalise preferences for the state to own basic industries, guarantee employment, social care, and minimum incomes, and protect the collective before the individual. These preferences typically outlived the emigration of individuals (Silver, 1987) and the demise of the system (Blanchflower and Freeman, 1997).

The promotion of Soviet values may be thought of as the costly creation of a costly moral identity, that of a loyal and obedient *homo sovieticus*, to which the citizen could subscribe through loyal activity, acquire a stake, and so gain access to a stream of psychological rewards. Having subscribed to this identity she then found shirking or stealing to be more costly than before since it triggered the loss of identity. There were innumerable ethnic, departmental, and campaign-oriented variations on the general theme: for example, Siegelbaum (1988: 210–46) has described how the state promoted the identity of a Stakhanovite worker in the late 1930s. There was a risk that the variation could become more important than the theme and weaken loyalty to the dictator beyond a point; thus Stalin acquired a rational mistrust of the *ésprit de corps* of the armed forces and aimed to destroy it in 1937.

It was possible to induce the agent to adopt the values of the principal by offering promotion in return for loyal behaviour. Then the obedient agent could hope to become a principal. A problem was that the number of aspiring agents tended to exceed the number of retiring principals. Under Stalin this problem was solved by frequent purges combined with ministerial subdivisions that created new leading positions in great numbers; "I need not mention", Stalin (1940: 650) reported, 'that the division of organisations has made it possible to promote hundreds and thousands of new people to leading posts". But purges and reorganisations were also costly. Valery Lazarev (2005) suggests that the promotion mechanism eventually became unviable and threatened the command system with bankruptcy.

Artificial Punishment and Rewards. In the first turbulent years, the Stalinist leadership tried to run a high-employment, low-wage economy (Kuromiya, 1988) with predictable results: widespread shirking, demoralisation, and increasingly punitive reflexes. From 1931 onwards Soviet leaders tilted the balance of incentives towards compliance by supplementing w with side payments to induce high effort that Bergson (1964: 72–92) described. They also threatened to cut or confiscate incomes altogether as the penalty for shirking and other disloyalty. They expelled non-compliant agents from the hierarchy by resort to firing and forced labor.

The efficacy of this route was limited for two reasons. One is that rewards and punishments were hard to target on effort with much accuracy. Effort was hard to monitor while output became harder to measure. The output-effort relationship in the 1930s and 1940s was dominated by random shocks. As a result the application of penalties for supposed shirking was also nearly random (Filtzer, 2002). Finally, the penalties available were very costly. After Stalin the credibility of punishment went into a long decline (Harrison, 2002).

Another reason for limited effectiveness is that penalisation of some agents must be carried out by others who require a reward in return. Therefore, in any hierarchy the balance of rewards to penalties should rise as we ascend from one level to the next; a regressive distribution of managerial rewards for fulfilment is exactly what we see in the figures that Bergson (1964: 76) assembled. Given persistent labor shortage, however, it was often harder to find effective punishments for ordinary workers than for their managers. As a result, labor-management collusion pervaded the Soviet enterprise.

General Criminalisation, Selective Sanctions. Although the authorities could not eliminate the scope for the horizontal networks that threatened to corrupt official hierarchies, they could limit it. By criminalizing unauthorised transactions the authorities won a natural advantage in their struggle: they ruled out public enforcement of private deals so that agents who sought a return from diverting the principal's assets into the underground had to rely on relational enforcement alone. This nested the fundamental problem of exchange within the problem of command to the advantage of the principal.

Agents could enforce illicit transactions more easily when they were of low value, quickly sequenced, or based on close acquaintance through the 'economy of favours' (Ledeneva, 1998). But the mistrust associated with the fundamental problem of exchange impeded large, long-range, unique, or complex deals and so eased the problem of command.

The authorities discriminated between two kinds of corruption. Disloyal agents took bribes for personal enrichment. Others extracted cash, resources, and favours from loyal motivations, for use in fulfilling plan assignments. Those who intended to comply with commands often had little alternative but to deal corruptly with external networks in order to carry them out (Belova, 2001; Gregory, 2003: 164). Principals could control the balance between loyal and disloyal corruption by varying the tautness of plan assignments (Harrison and Kim, 2006).

When they detected illicit trade the authorities punished violations selectively, passing over the "loyal" but corrupt transactions that were designed to promote fulfilment of the plan (Gregory, 2003) and bearing down more heavily on those that enriched the agent personally (Belova, 2001). In turn, the risk premium that agents required for engaging in loyal trade should have been lower than for trade that lined their own pockets. By implication, the authorities discouraged disloyal networks while exploiting the networks of the loyal.

Heavy Defense Spending. This framework supports a simple interpretation of the domestic calculations behind foreign policy and military spending. Facing adversaries at home and abroad, Stalin did not fear them acting singly but he feared their combination (Khlevniuk, 1995; Simonov, 1996). The potential for external foes to exploit internal treachery is portrayed in the fundamental problem of command as the scope for an agent to trade with a network representing a foreign power. From this point of view the powerful armed forces supported by a large defense industry described in recent work (Simonov, 1996; Barber and

Harrison, 2000; Samuelson, 2000) drove down the chances of a deal between domestic and foreign agents by reducing both the reward that the foreign power would offer an agent and the probability that the agent could enforce the transaction privately and collect her reward from the foreign power.

By paying the soldiers and defense producers well Stalin was also able to raise w, the opportunity cost to his agent of collaboration with the enemy. Thus the scale of the defense effort and the privilege it gave to the servants of the military-industrial complex helped to solve the fundamental problem of command. This solution was very expensive, consuming a large share of national resources that was shrouded in state secrecy.

Secretiveness. The Soviet state carried secrecy to extraordinary lengths (Davies, 2001). Secrecy of economic information may have assisted principals in inhibiting the temptation to steal. A prerequisite for trade is the sharing of information: buyer and seller must be able to signal each other about ex ante supplies and demands (Hayek, 1945). When trade is between an agent in a hierarchy and an outsider, criminalizing the exchange of signals inhibits exchange by reducing information and increasing mistrust (Harrison, 2004). Unverifiable information reduced the probability p of honor among thieves. Alternatively, the agent must incur extra costs to restore trust and ensure the credibility of the signals she sends and receives. This cost arises even when illicit signalling is not detected or punished.

Secrecy was costly in terms of enforcement and efficiency. Enforcement required the costly tracking of secret documents and further efforts to monitor and investigate cases of disclosure. The punishment of disclosure led to the loss of human capital already sunk in agents who turned out to be disloyal. Efficiency costs arose because secrecy created barriers to the sharing of information that could have led to efficient reallocations; for example, principals fixed the overall allocation of resources in ignorance of specific facts, while agents made specific allocations while insufficiently informed of the wider context. Thus secrecy also increased the probability of planning mistakes that lowered x and raised q.

Summary

A range of enforcement mechanisms addressed the fundamental problem of command in the Soviet economy. Internalisation, promotion, and side payments and penalties controlled shirking. The criminalisation of disloyal networks, heavy defense spending, and secretiveness limited agents' trading at the principal's expense. But each was costly in different ways. The Soviet system combined them in a balance that shifted through time as principals and agents learned and the comparative costs of each mechanism changed. It seems that at the end of the 1980s the ensemble ceased to be viable and the system collapsed (Harrison, 2002). If the dictator had succeeded in optimizing simultaneously on each margin, one would expect that each mechanism failed at about the same time. This is roughly what is meant by the collapse of a system.

Command and Negotiation

In a speech to the sixteenth party congress in June 1930, Stalin (1955: 357) defended the recent radical increases in final targets for the first five-year plan on the basis that he recognised the possibility of plan mistakes. The five-year plan was, he suggested

> merely . . . a first approximation, which has to be made more precise, altered and perfected in conformity with the experience gained in the localities, with the experience gained in carrying out the plan. No five-year plan can take into account all the possibilities latent in the depths of our system and which reveal themselves only in the course of the work, in the course of carrying out the plan . . . Only bureaucrats can think that the work of planning ends with the drafting of a plan. The drafting of a plan is only the beginning of planning. Real guidance in planning develops only after the plan has been drafted, after it has been tested in the localities, in the course of carrying it out, correcting it and making it more precise.

At the time Stalin meant to encourage agents to make plans more ambitious, but in subsequent years he allowed his comments to be taken out of context and reinterpreted to sanction all kinds of plan modifications including the relaxation of targets when the plan turned out to be too ambitious (Harrison, 1985: 17).

In short, the ability to command well does not always accompany the power to command. So far I have assumed that (1) the principal was powerless to mitigate his plan mistakes that helped to form private profit opportunities and (2) the returns to different activities were common knowledge. However, it is normal that in hierarchies information is distributed unevenly. The principal has better general knowledge while the agent has the advantage in local knowledge. As a result, it was possible for a principal to make a mistake and an agent to know it. One way of raising returns in the command system was to let the agent bargain for an improved plan.

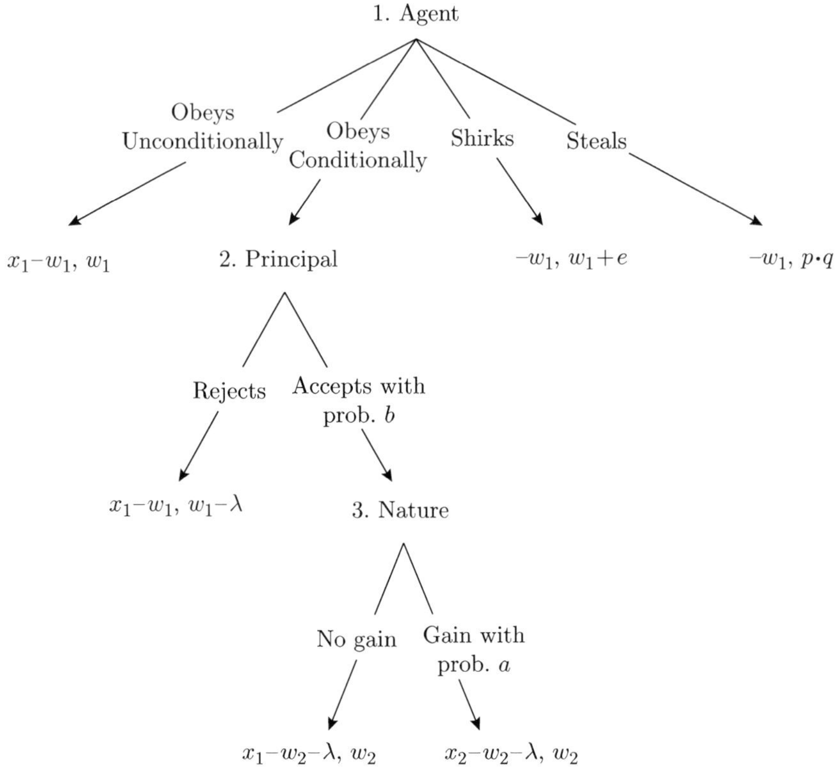

Fig. 2. Unconditional and Conditional Obedience.

The obedient agent could then choose to disobey, obey unconditionally, or bargain with the principal.

Vertical bargaining is shown in Figure 2. The agent offers to share information about plan mistakes and unexploited opportunities with the principal in return for a share of the principal's rent. Initially the principal advanced w_1 and expected to receive x_1. At stage 1 the agent chooses between disobedience and absolute and conditional obedience. Conditional obedience involves lobbying: the agent takes λ from her advance and invests it in making the case for a better project that can yield x_2 for the principal and w_2 for the agent. At stage 2 the principal decides whether or not to accept the agent's case. If yes, at stage 3 nature takes a hand: it deals $x_2 > x_1$ with probability a and x_1 otherwise so a is the probability that the original command was in error. The agent expects her lobby to succeed with probability b, which therefore measures the credibility of her case; if she succeeds, the agent's gain is that the principal will advance $w_2 > w_1$ to her and also reimburse λ, the cost of lobbying. If lobbying fails the agent is worse off: she will get only the original advance and must pay the lobbying cost from it.

If we take into account that bargaining takes time and that the players may have preferences over the time it takes to reach a solution, the resulting problem begins to resemble Rubinstein bargaining with an outside option (Rubinstein, 1982; Shaked and Sutton, 1984); the outside option is the return to covert disobedience. This case differs in that, while the bargaining costs of principal and agent are fixed, who bears the cost is part of the bargain.

In Figure 2, the fundamental problem of command is mitigated as long as both principal and agent gain from plan renegotiation over unconditional compliance. The agent gains provided $b \cdot (w_2 - w_1) > (1 - b) \cdot \lambda$, i.e. the expected wage gain adjusted by the credibility of her information about plan errors exceeds the lobbying cost taking into account the probability that she will have to pay it. Thus, a 'listening' principal can reduce the relative attractions of shirking or stealing. However, listening is costly. The principal will gain only if $a \cdot (x_2 - x_1) > (w_2 - w_1) + \lambda$, i.e. his expected efficiency gain exceeds the compensation the agent extracts from him as a return on her successful lobbying, plus the deadweight cost of renegotiation. The larger are the probability of a mistaken command and the scope for mistakes, the greater will be the principal's gain.

Plan bargaining went on in the Soviet economy at every level. In the Politburo ministers lobbied Stalin to improve their shares of budgetary or material resources (Gregory, 2001: 22–4). At lower levels Bergson (1964: 81–2) noted the tendency of firms to complain of burdensome quotas or demand increased supply allocations as a condition for fulfilling them (see also Kornai, 1992: 121–4; Gregory, 2003: 202–4). Of course the agent preferred to tell the principal that the plan was too ambitious while the principal preferred to be told that the plan was too modest. In cases such as the counter-plan campaign of the early 1930s or the Stakhanov campaign a few years later (Kuromiya, 1988; Davies and Khlevniuk, 2002) the authorities organised countervailing pressure. The information advantage of the agent was greatest at the frontiers of military science and technology (Holloway, 1982; Holloway, 1984). In research and development the inventors' enthusiasm gave rise to strenuous competitive lobbying to boost plans (Harrison, 2003). But when plans called for known advances to be applied to production the enthusiasm for higher plans disappeared (Berliner, 1976).

In the years of Stalin's dictatorship plan renegotiation remained informal. The post-Stalin economic reforms tended to institutionalise it. For example Kornai (1986: 1700) described Hungary in the 1980s as:

> a bargaining society, and the main direction is vertical, namely bargaining between the levels of the hierarchy, or between bureaucracy andfirm, not horizontal, between seller and buyer. All issues ... — entry, exit, appointment, output, input, price, wage, tax, subsidy, credit, and investment — are subject to meticulous negotiations, fights, lobbying.... The Hungarian literature calls this phenomenon 'regulator bargaining'; it has taken the place of 'plan bargaining' which had prevailed in the command economy.

What has escaped analysis is why such bargaining was increasingly in the principal's interest. Planning was based on imperfect information. Conservative planning "from the achieved level" (Birman, 1978), superimposed on growing complexity, gave rise to increasing misallocation. While faith in the system declined, agents' representations became increasingly persuasive and rent-sharing in return for information-sharing became increasingly formalised.

Conclusions

Abram Bergson sought to examine the rationality of the Soviet economic system according to various efficiency standards, including that of the system's directors. In terms of what material ends and to what extent was their behaviour rational? We have looked at Soviet institutions and behaviours from the point of view of a rational principal and agent in a command system that faces a problem. The fundamental problem of command is that of enforcement when agents may shirk or steal rather than comply. The equilibrium of the Soviet command system was conditional on finding solutions to this problem, and we have looked at some enforcement mechanisms that together may have provided the necessary conditions at the time but were also costly in different ways. This approach suggests that the system collapsed when the costs of enforcement could no longer be sustained.

References

Barber, J and Harrison, M (eds.) (2000). *The Soviet Defence-Industry Complex from Stalin to Khrushchev*. London and Basingstoke: Macmillan.

Belova, E (2001). Economic crime and punishment. In: Gregory PR (ed.) *Behind the façade of Stalin's command economy: Evidence from the state and party archives*. Stanford, CA: Hoover Institution Press, pp. 131–58.
Belova, E (2005). Legal contract enforcement in the Soviet economy. *Comparative Economic Studies*, 47(2): 387–401
Belova, E and Gregory, PR (2002). Dictators, loyal and opportunistic agents: the Soviet archives on creating the Soviet economic system. *Public Choice*, 113(3/4): 265–86.
Bergson, A (1964). *The Economics of Soviet Planning*. New Haven and London: Yale University Press.
Berliner, JS (1976). *The Innovation Decision in Soviet Industry*. Cambridge, MA: MIT Press.
Birman, I (1978). From the Achieved Level. *Soviet Studies*, 30(2): 153–172.
Blanchflower, DG and Freeman, RB (1997). The attitudinal legacy of communist labor relations. *Industrial and Labor Relations Review*, 50(3): 438–59.
Broadberry, SN and Ghosal, S (2004). *Technology, organisation, and productivity performance in services: lessons from Britain and the United States since 1870*. CEPR Discussion Paper no. 4428. London: Centre for Economic Policy Research.
Coase, RH (1937). The nature of the firm. *Economica*, 4: 386–405.
Davies, RW (1994) Changing economic systems: an overview. In: Davies RW, Harrison M and Wheatcroft SG (eds.) *The Economic Transformation of the Soviet Union, 1913–1945*. Cambridge University Press: Cambridge, pp. 1–23.
Davies, RW (2001). Making Economic Policy. In: Gregory PR (ed.) *Behind the Façade of Stalin's Command Economy: Evidence from the State and Party Archives*. Stanford, CA: Hoover Institution Press, pp. 61–80.
Davies, RW, Ilič, M and Khlevniuk, O (2004). The Politburo and Economic Policy-Making. In: Rees EA (ed.) *The Nature of Stalin's Dictatorship: The Politburo, 1924–1953*. Basingstoke: Palgrave, pp. 108–134.
Davies, RW and Khlevniuk, O (2002). Stakhanovism and the Soviet Economy. *Europe-Asia Studies*, 54(6): 867–905.
Dewatripont, M and Maskin, E (1995). Credit and Efficiency in Centralized and Decentralized Economies. *Review of Economic Studies*, 62(4): 541–55.
Filtzer, D (2002). *Soviet Workers and Late Stalinism: Labour and the Restoration of the Stalinist System after World War II*. Cambridge: Cambridge University Press.
Gorlizki, Y (2002). *Industry, Justice, and the Politics of Soviet Regulation*. Paper to the Summer Research Euro-Workshop, University of Warwick, Department of Economics, July 8–19.
Gorlizki, Y, and Khlevniuk, O (2004). *Cold Peace: Stalin and the Soviet Ruling Circle, 1945–1953*. New York: Oxford University Press.
Gregory, PR (1990). *Restructuring the Soviet Economic Bureaucracy*. Cambridge: Cambridge University Press.
Gregory, PR (2001). The Dictator's Orders. In: Gregory PR (ed.) *Behind the Façade of Stalin's Command Economy: Evidence from the State and Party Archives*. Stanford, CA: Hoover Institution Press, pp. 11–33.

Gregory, PR (2003). *The Political Economy of Stalinism: Evidence from the Soviet Secret Archives*. Cambridge: Cambridge University Press.

Gregory, PR and Markevich, A (2002). Creating Soviet industry: The House that Stalin Built. *Slavic Review*, 61(4): 787–814.

Greif, A (2000). The Fundamental Problem of Exchange: A Research Agenda in Historical Institutional Analysis. *European Review of Economic History*, 4: 251–84.

Harrison, M (1985). *Soviet Planning in Peace and War, 1938–1945*. Cambridge: Cambridge University Press.

Harrison, M (1998). Prices, Planners, and Producers: An Agency Problem in Soviet Industry, 1928–1950. *Journal of Economic History*, 58(4): 1032–62.

Harrison, M (2002). Coercion, Compliance, and the Collapse of the Soviet Command Economy. *Economic History Review*, 55(3): 397–433.

Harrison, M (2003). The Political Economy of a Soviet Military R&D Failure: Steam Power for Aviation, 1932–1939. *Journal of Economic History*, 63(1): 178–212.

Harrison, M (2004). *Why Secrets? The Uses of Secrecy in Stalin's Command Economy*. PERSA Working Paper no. 34. University of Warwick, Department of Economics.

Harrison, M and Kim, B-Y (2006). Plans, Prices, and Corruption: The Soviet Firm Under Partial Centralization, 1930 to 1990. *Journal of Economic History*, 66(1): 1–41.

Harrison, M and Simonov, N (2000). Voenpriemka: Prices, costs, and quality in defence industry. In: Harrison M and Barber J (eds.) *The Soviet Defence Industry Complex from Stalin to Khrushchev*. London and Basingstoke: Macmillan, pp. 223–45.

Hayek, FA (1945). The Use of Knowledge in Society. *American Economic Review*, 35(4): 519–30.

Holloway, D (1982). Innovation in the Defence Sector. In: Amann R and Cooper J (eds.). *Industrial Innovation in the Soviet Union*. New Haven, CT: Yale University Press, pp. 276–367.

Holloway, D (1984). Stalin and the Bomb: The Soviet Union and Atomic Energy, 1939–1956. New Haven, CT: Yale University Press.

Hymer, SH (1960/1976). The International operations of national firms: A study of direct foreign investment. Cambridge, MA: MIT Press.

Katsenelinboigen, A (1977). Coloured markets in the Soviet Union. *Soviet Studies*, 29(1): 62–85.

Khlevniuk, O (1995). The Objectives of the Great Terror, 1937–1938. In: Cooper J, Perrie M and Rees EA (eds.) *Soviet History, 1917–53: Essays in Honour of R. W. Davies*. London and Basingstoke: St Martin's Press, pp. 158–76.

Khlevniuk, O (2003). Sistema tsentr-regiony v 1930–1950-e gody. Predposylki politizatsii nomenklatury. *Les Cahiers du Monde russe*, 44(2/3): 253–268.

Kornai, J (1980). *The Economics of Shortage*. Amsterdam: North-Holland.

Kornai, J (1986). The Hungarian reform process: visions, hopes, and reality. *Journal of Economic Literature*, 24: 1687–1737.

Kornai, J (1992). *The Socialist System: The Political Economy of Communism.* Oxford: Clarendon Press.

Kroll, H (1986). Decentralization and Precontract Disputes in Soviet Industry. *Soviet Economy*, 2(1): 51–71.

Kuromiya, H (1988). Stalin's Industrial Revolution: Politics and Workers, 1928–1932. Cambridge: Cambridge University Press.

Lazarev, V (2005). Promotion Contracts and Support for the Soviet regime. *Comparative Economic Studies*, 47(2): 346–363.

Ledeneva, AV (1998). *Russia's Economy of Favours: Blat, Networking and Informal Exchange.* Cambridge: Cambridge University Press.

Lewin, M (2003). Rebuilding the Soviet Nomenklaturas, 1945–1948. *Les Cahiers du Monde russe*, 44(2/3): 219–252.

Marglin, S (1974). What do bosses do? The origins and functions of hierarchy in capitalist production. *Review of Radical Political Economics*, 6(2): 33–60.

Markevich, AM (2003). *Was the Soviet Economy Planned? Planning in the People's Commissariats in the 1930s.* PERSA Working Paper No. 25. University of Warwick, Department of Economics.

Nove, A (1977). *The Soviet Economic System.* London: Allen & Unwin.

Olson, M (1993). Dictatorship, Democracy, and Development. *American Political Science Review*, 87(3): 567–76.

Preobrazhensky, EA (1926/1965). *The New Economics.* Oxford: Oxford University Press.

Rubinstein, A (1982). Perfect Equilibrium in a Bargaining Model. *Econometrica*, 50(1): 97–110.

Samuelson, L (2000). *Plans for Stalin's War Machine: Tukhachevskii and Military-Economic Planning, 1925–41.* London and Basingstoke: Macmillan.

Shaked, A and Sutton, J (1984). Involuntary unemployment as a perfect equilibrium in a bargaining model. *Econometrica*, 52(6): 1351–1364.

Siegelbaum, LH (1988). *Stakhanovism and the politics of productivity in the USSR, 1935–1941.* Cambridge: Cambridge University Press.

Silver, BD (1987). Political Beliefs of The Soviet Citizen: Sources of Support for Regime Norms. In: Millar JR (ed.) *Politics, Work, and Daily Life in the USSR: A Survey of Former Soviet Citizens.* Cambridge: Cambridge University Press, pp. 100–41.

Simonov NS (1996). *Voenno-promyshlennyi kompleks SSSR v 1920–1950-e gody: tempy ekonomicheskogo rosta, struktura, organizatsiia proizvodstva i upravlenie.* Moscow: ROSSPEN.

Stalin, JV (1940). *Leninism.* London: Lawrence & Wishart.

Stalin, JV (1955). *Works, 12.* Moscow: Foreign Languages Publishing House.

Williamson, OE (1975). *Markets and Hierarchies: Analysis and Antitrust Implications.* New York: The Free Press.

Wintrobe, R (1990). The Tinpot and the Totalitarian: An Economic Theory of Dictatorship. *American Political Science Review*, 84(3): 849–72.

Zaleski, E (1980). *Stalinist Planning for Economic Growth, 1932–1952.* London and Basingstoke: Macmillan.

Chapter 10

Accumulation and Labor Coercion under Late Stalinism

Paul R. Gregory and Mark Harrison

The archives tell us that Stalin was obsessed with accumulation, which is hardly a surprise.[1] One of the great stylized facts about the Soviet economy under Stalin was the sharp increase in its accumulation rate, accompanied by the fierce repression of private consumption. Between 1928 and 1937 Stalin shifted nearly 30 per cent of Soviet GNP out of private consumption into investment and government purchases; the investment rate doubled (Bergson, 1961: 217, 237). More generally, investment rates in socialist countries were consistently higher than in capitalist countries of comparable size and income levels (Kuznets, 1963: 352–57). Olson (1993) argued that a self-interested proprietary dictator or sedentary bandit who settles on and monopolizes a territory would maximize his rent through time by taxing heavily and reinvesting the surplus.

The traditional literature said nothing about how Soviet investment and consumption rates were actually chosen for the simple reason that little was known about it. Bergson's authoritative *Economics of Soviet*

This chapter was first published by Moscow University Press in English (with a Russian title: "Nakoplenie i prinuzhdenie k trudu v period 'pozdnego' stalinizma") in 2006 in *Rossiia v kontekste mirovogo ekonomicheskogo razvitiia vo vtoroi polovine XX veka* (edited by V. A. Vinogradov and 14 others), pp. 308–324 (Moscow: Moscow University Press). © Russian Academy of Sciences Division of Historical and Philological Sciences; M. V. Lomonosov Moscow State University; N. P. Ogarev Historical and Sociological Institute of Moscow State University 2006.

[1]Paul Gregory is Cullen Professor of Economics of the University of Houston and a research fellow at the Hoover Institution on War, Revolution, and Peace, Stanford University. This paper was presented to a conference on "Russia in the Context of World Economic Development in the Second Half of the Twentieth Century" held in Moscow, 24 and 25 November 2004. The authors thank the organizers and participants for discussion.

Planning (1964) introduces investment first in Chapter 11 and largely puts aside the issue of its total volume, focusing instead on rational choice of individual projects. Bergson recognized (1964: 241) that desired investment depends "on the output mix to be produced from it"; this view percolated through the various editions of Gregory and Stuart (1974: 214–5). The most widely cited paper on investment (Grossman, 1953) deals exclusively with the choice to be made when a given increase in desired output could be obtained from alternative projects.

There is a considerable literature on how a socialist economy *ought* to determine aggregate investment; it stems from Marx's emphasis on accumulation as the source of economic growth and proceeds via the growth models of Preobrazhensky and Feldman in the early Soviet period (Erlich, 1960; Spulber, 1964) to the later models of Brus, Kalecki, and Dobb (Rutland 1985: 103–107). Explicitly or implicitly these models suggested that the optimal investment rate depended on society's discount rate, to which Preobrazhensky added the scope for redistribution away from "former" social classes.

No Politburo member was trained in economics, however. Their advisers in Gosplan were focused, at best, on balancing supplies and demands for broad commodity groups no more than one year ahead. When Stalin and the Politburo thought of the longer term, it was probably *not* in terms of intertemporal optimizing; their language consisted of "urgently", "with maximum speed", "by [a given] date", and so on. They had to balance multiple objectives that included holding power and eliminating enemies against economic goals. At best, we would expect that they resorted to rules of thumb. These rules should tell us not only what they hoped to achieve but also how they experienced and perceived the objective limits on their power to achieve it.

1. *The Politburo Accumulation Model*

At the core of the Politburo's strategy to "build socialism" in the first two five-year plans were massive programs for the hydroelectric dams, machinery complexes, vehicle works, blast furnaces, railways, and canals that were included on its "title lists" of approved projects. Their purpose was to embody the new society in cement and structural steel. How were these programs implemented and managed?

Although Politburo meetings for the 1930s left few formal minutes, his deputies wrote to Stalin from time to time to detail key Politburo meetings and seek approval of operational decrees (Gregory, 2004: 76–125). They reveal that the Politburo consistently set three targets in the 1930s: the *nominal* investment budget, grain collections, and foreign exchange. These three control variables all related to investment. The investment budget allotted funds to industrial and regional agencies for construction and machinery. Despite objections to the lack of material balances for investment goods, ministries and regional authorities were simply given "investment rubles", and no one appeared to know the real investment that resulted. Grain collections were designed to contribute to national savings through the excess of sale prices over purchase prices. Stalin personally directed foreign exchange to the import of capital goods rather than the luxury goods sometimes demanded by the Bolshevik elite.

If Stalin's goal was indeed to maximize investment, the archives provide two types of evidence that are, at first glance, confusing. First, Stalin was extremely concerned about consumption, particularly where it touched upon the productivity and morale of the industrial workers. Consumer supplies were one of the most frequent items on Politburo agendas; in Stalin's words the "provisioning of workers" was one of "the most contested issues" before the Politburo and trade was "the most complicated ministry" (quoted by Gregory, 2004: 93–94). Stalin interpreted declining labor productivity as a sign that workers "were not provisioned as well as last year," and personally ordered the delivery of consumer goods to cities where labor productivity was declining (Gregory, 2004: 76–109). The Politburo set itself up as the highest trading organization, deciding retail trade plans, prices, assortment, and even the opening of new stores. Second, we have the little-known empirical fact that the Politburo, which based the success of its program on capital construction, deliberately reduced nominal investment on two occasions, in 1933 and 1937 (Davies, 2001) citing the fear of inflation and rising construction costs.

This evidence can be interpreted in two ways. One is that that Stalin's preferences were unstable; fitful humanitarian impulses led him to direct more supplies to the workers from time to time; periodic fits of financial orthodoxy led him to regret overstraining the economy. Alternatively, Stalin's preferences were stable, and he reallocated consumption or reduced investment according to a consistent rule. Knowing of Stalin's capacity for calculation, patience, and self-control, we reject the first explanation and investigate further the second.

Figure 1 illustrates the model that the Politburo appears to have used to set investment and consumption. The figure mirrors the Marxian concept

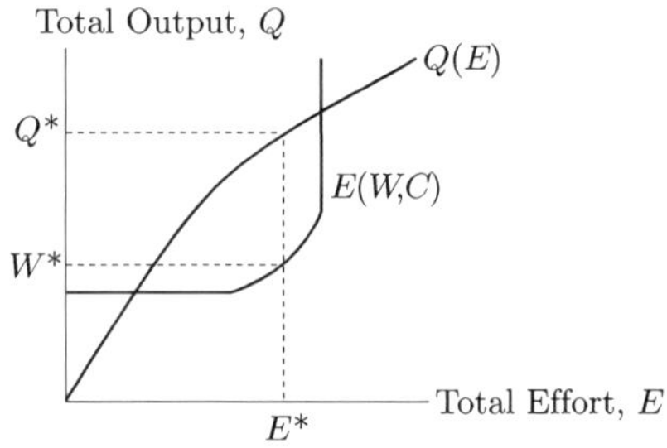

Fig. 1. The Politburo Model.

of the surplus product, the gap between output and consumption, as the outcome of a distributive struggle. The model has theoretical precursors in Schrettl (1982, 1984) and Podkaminer (1989), and is set out more fully by Gregory (2004: 76–109). It belongs to a general class of models in which a ruler's freedom of action is circumscribed by social "tolerance limits" (Kornai, 1980: 211–14) or a revolution or disorder constraint (Açemoglu and Robinson, 2000). By raising investment without limit the Politburo risked provoking the workers to go slow, strike, or rebel.

In the Soviet economy the demand for labor was always enough for full employment and besides all able-bodied persons were required to work by law (Granick, 1987), so the figure takes employment, N, as fixed exogenously; individual effort, e, was variable, so total effort, $E = e \cdot \overline{N}$, was variable although employment was not. Total output, Q, depended on total effort, E. Total effort varied with the real wage, w, as follows. The aggregate wage bill W, the consumer goods received by workers, is measured along the vertical axis in the same units as output, and is proportional to the real wage given that employment is fixed, i.e. $W = w \cdot \overline{N}$. There is a reservation wage, analogous to a tolerance limit or disorder constraint, below which effort is zero; there is also a "fair" wage at which effort is maximized.[2] As the economy moves from the fair wage to the reservation wage, effort declines as workers turn up late or drunk, or go slow or absent without leave; at the limit unrest simmers and threatens to boil over into local and general strikes and rebellion. Thus the effort curve intersects the horizontal

[2]The fair-wage concept in Figure 1 can be related to the observation of Akerlof (1984) and Akerlof and Yellen (1990) that workers in market economies give extra effort if they are paid a wage perceived to be fair, and withdraw it if the wage is reduced below the psychologically determined fair level. We suppose that effort has a maximum, so payments above the fair level do not elicit more effort.

axis at the reservation wage and becomes vertical at the fair wage. Effort also depends on the level of direct coercion, C, to which we turn in the next section. To maximize effort the dictator would pay the fair wage and get the maximum output, but this would not maximize the surplus. To maximize the surplus, $Q - W$, he would choose the intermediate wage, effort, and output levels denoted $W*, E*$, and $Q*$.

The effort curve makes the consequences of plan mistakes asymmetric: for the dictator, paying the workers too little is much worse than paying them too much. Starting from the investment optimizing position, too much consumption does at least raise effort and output, and this mitigates the fall in investment. A mistake of the same size in the opposite direction, however, not only cuts effort and output by more but also risks pushing the workers into outright confrontation with the state. An investment-maximizing dictator must tread a fine line between the pursuit of investment and the triggering of serious disorder.

As noted above, Stalin clearly understood the link between effort and consumption. Stalin managed worker morale and effort in two ways. When investment and consumption were about right in aggregate, detailed plan mistakes could still leave some workers with too much and others with too little. When this happened, it was imperative to reallocate supplies immediately to those left short, and Stalin personally ordered the reallocation of consumer goods to cities where labor productivity was declining. But there could also be aggregate plan mistakes; when too much investment threatened to disrupt the economy and provoke the workers, the Politburo preferred to sacrifice investment. This is how we interpret the unforced investment cutbacks of 1933 and 1937. Although Stalin did not express his concern for consumption directly, when he advocated "strengthening the ruble" as a justification for less investment (Gregory, 2004: 236–242) it was the same ruble that workers had to spend in retail markets; when he feared rising construction costs, these were fiscal costs that had to be covered through taxation or the inflation tax. In this sense Stalin's behaviour was stable and consistent, given the constraints under which he operated.

Figure 1 lays out other options and dangers. "Team Stalin" (Wheatcroft, 2004) could seek to manipulate the effort curve by offering ideological rewards in place of material payoffs. Idealists from Russia and abroad in fact assisted the first five year plan, motivated by the idea of building socialism for future generations. The attempt to transform *homo economicus* into *homo sovieticus* led, however, to a vicious circle of wage equalization, and declining productivity (Kuromiya, 1988; Davies, 1989; 1996). Subsequent mobilizations were limited to short lived campaigns such as for World War II and to cultivate the "virgin lands" of Kazakhstan and Siberia after the war. The most publicized mobilization campaign of the 1930s, the Stakhanovite movement, was driven by progressive piece rates that permitted participating workers to drive up their incomes by over fulfilling norms (Davies and Khlevnyuk, 2002); The archives suggest that Stalin abandoned it because it caused widespread upset to notions of a fair wage, and also threatened inflation (Gregory, 2004: 104–106).

Stalin could also elicit extra effort in the present by promising future promotion. The problem was that these promises eventually had to be made good. Lazarev (2005) has identified the creation of new posts that carried a wage premium to reward loyal supporters as an important feature of Stalin's policy. Relentlessly purging the potentially disloyal made it cheaper for him to keep his promises to the actually loyal.

In an extension of the model, the Politburo could bear down on consumption selectively. Rather than spread the burden equally, and invite widespread worker unrest, Stalin tried to confine the negative consequences to lower-priority workers by introducing ration entitlements differentiated by category and occupation according to the principle: "He who does not work *on industrialization* shall not eat" (cited by Gregory, 2004: 98 with emphasis added). It is notable that Stalin himself wrote the rules and procedures of the rationing system that remained in effect until 1935.

Scholars have taken different views of rationing. Osokina's (2001) archival study argues that Stalin positively wanted to drive down consumption and introduce a discriminatory ration. Hessler (2004) argues on the contrary that Stalin approved of non-rationed trade and a consumption oriented life-style for the workers but crises were so frequent that this goal was only achieved for short periods. Gregory (2004: 98–100) argues that Stalin saw rationing as a way to force accumulation without a loss of effort of high-priority workers, despite occasional pro-consumer statements.

After largely eliminating rationing in 1935, periodic crises — pre-war rearmament, wartime mobilization, and postwar famine — caused restorations of rationing. Davies and Khlevnyuk (1999) and Hessler (2004) have documented the drawbacks of widespread rationing for the regime: illegal markets spread rapidly and drew off the energy of the working population; officials charged with determining individual entitlements were paralysed by a flood of petitions and paperwork; local officials and managers charged with distributing goods in practice had to search, bargain, and trade, or even steal to trade on their own account. Productive effort and budget revenues fell away. Gregory (2004: 98–100) argues that Stalin abandoned rationing because the resulting distribution did not get consumer goods to priority workers.

It is not surprising, therefore, that the Politburo was typically planning to restore "normal" retail as soon as possible. Characteristically, as legal markets reappeared, Stalin sent the secret police into them to confiscate windfall gains of market traders, whom he regarded as class enemies (Hessler, 2004).

Figure 1 has a further ominous extension. The fair wage was set by a mass psychology that was unpredictable and hard to manipulate. If workers concluded from the propaganda of economic successes that they were being cheated the fair wage would rise, forcing the Politburo to cut investment back. Stalin used the vast informant network of the NKVD (interior ministry — *Narodnyi komissariat vnutrennykh del*) to monitor protests, strikes, anti-Soviet statements, and factory-wall graffiti, and eavesdrop on private conversations to gauge mass opinion (Berelovich, 2000). Stalin had obvious political motives to do this, but within our framework we conclude that wages and fairness lay at the cross-hairs of politics and economics.

2. *Coercion: Three Experiments*

In principle efficient penalties are cheaper than efficient rewards since "a promise is costly when it succeeds and a threat is costly when it fails" (Schelling, 1960: 177). In Figure 1, the fear of punishment may cause workers to lower their reservation wage without reducing effort; coercion may not make effort more productive but should make people willing to supply effort for less. As long as coercion displaces the effort curve downward while leaving the production curve undisturbed, the surplus is increased. In market economies, outside options, such as alternate employment, leisure, or unemployment, cannot easily be limited by force as the ubiquity of informal sector employment and illegal migration demonstrates. In the Stalinist state, however, the idea of controlling workers' alternatives by force proved attractive.

The archives show that Stalin believed that a wide range of problems could be solved by force. The language with which he addressed both administrative and economic problems was habitually violent. Faced with foot-dragging in the transport ministry in 1931, he wrote: "we must smash this gang... If you can manage without my help, smash the gang before it's too late". Considering the role of speculators in a legalized urban rural market, he wrote in 1932: "we must eradicate this scum... The OGPU [secret police] and its agencies must, without delay, start training its forces and studying the enemy" (Davies *et al.*, 2003: 95, 102, 165).

Some observers have concluded that coercion was the glue that held the Stalinist economy together. When coercion failed, or was abandoned, the economy fell apart (Kontorovich, 1986; Blanchard and Kremer, 1997; Harrison, 2002). Others have emphasized the huge social and economic costs of coercion (Conquest, 1987; Khlevnyuk, 2001; Davies and Wheatcroft, 2003). The balance of the argument is yet to be struck. Whether coercion paid for itself, and where the optimum lay, are unresolved issues.

Stalin conducted three notable experiments with the use of coercion to foster accumulation: the forced collectivization of the peasantry, the criminalization of workplace indiscipline, and the widespread use of forced labor. In each case the outlines have been known for decades, but the full consequences are only now coming to light and we review them below.

Stalin's belief in solving problems by force also contributed to the motivation behind the Great Terror of 1937/38, but in this case his purpose was not to foster accumulation. Manning (1993) speculated that Stalin deployed terror to solve widespread *economic* problems, but the archives have not yielded any substantial evidence to this effect (Davies, 2006). Rather the aim of the terror was political: to isolate and remove a potential "fifth column" from society that, Stalin believed, could endanger his regime in a time of rising international tension (Khlevniuk, 1995).

2.1. *Collectivizing peasants*

Politically, collectivization aimed to impose Soviet power in the countryside and eliminate the stratum of richer peasants, the kulaks. It was triggered, however, by a grain marketing crisis that reflected the peasants' perceived unwillingness to contribute sufficiently to investment-led industrialization. Collectivization began in earnest in December 1929, signalling Stalin's victory in the power struggle. The relatively open discussion in the late 1920s and other published sources supported a substantial literature on the collectivization decision, the most complete account of which is Davies (1980); Heinzen (2004) has now added a study of the agriculture ministry in the 1920s. Jasny (1949), Lewin (1968), and Strauss (1969) outlined some general results. On the basis of estimates by Soviet historians Millar (1974) and Ellman (1975) suggested that the investment surplus that Stalin hoped to gain from collective agriculture was probably not forthcoming. But the detailed consequences of collectivization and the mass deportation or detention of peasants who were excluded or resisted were concealed behind a thick veil of secrecy.

Tauger (1991), Davies *et al.* (1995), Wheatcroft and Davies (2002), and Davies and Wheatcroft (2003) have now reviewed the archives on the immediate aftermath of collectivization, including the famine of 1932–33. The collective farms enabled Moscow to replace local decision making with its own detailed plans, instructions, and rules. Stalin was focused on what he could control: sown acreage, and the state's share in what this acreage produced. But Stalin could not control the harvest. While yields collapsed, the share delivered to the state increased. Excessive procurements, bad weather, plan errors, and rural shortages combined to strip the countryside of grain; first the livestock were killed, then the farmers themselves starved. They were prevented from feeding themselves from their own harvests by severe punishments including death for petty theft from collective farm stocks. Davies and Wheatcroft dispel Conquest's (1987) notion that Stalin manufactured the famine to kill class enemies; rather they show the leadership subsequently trying to ameliorate the effects of its own bungling.

Famines usually occur in poor countries with limited statistical reporting. The Soviet archives have provided unusually good documentation of the two peacetime famines of 1932/33 and 1946/47. Ellman (2000) has applied the entitlement theory of Sen (1983) to the latter famine, which particularly affected the peasants of the Ukraine and other southern regions. Sen argued from historical experience that famines are more likely to arise from entitlement or distributional failures than production failures, and that the state should generally play a substantial role in cushioning sections of the population threatened by entitlement failures. Ellman makes two points from the evidence.

First, he argues, in the 1946/47 famine, production and entitlement failures interacted. The production failure was not such as to leave insufficient food available to keep everyone alive. In this sense the famine arose from a failure of entitlement. But the famine was triggered by a harvest failure; if the harvest had not failed, everyone would have had sufficient entitlements to stay alive. Therefore, the famine cannot be attributed to entitlement failures alone. Second, Ellman notes that in this famine the role of the state was essentially negative: it selected those who died by denying them entitlements. Therefore, it cannot be a general rule to concentrate grain stocks in the hands of the state in times of famine. The motivations of the state must also be considered. Under some circumstances, by implication, market allocation may do better than the state at keeping people alive. While Ellman does not extend his findings to the Soviet famine of 1932/33, to do so would seem reasonable to us; Davies and Wheatcroft (2003: 402) concur at least on the first issue, the interaction of entitlements with harvest failure.

2.2. *Regimenting state employees*

As the 1940s began, Stalin redirected coercion from specific class enemies to the entire public-sector work force. A battery of intimidating laws criminalized work place violations which had previously been managed by administrative sanctions within the enterprise. The laws themselves were not secret and were described in detail by Conquest (1967: 102–6) and Nove (1969: 260–265). Their consequences, however, remained obscure until the archives became available. Full texts of the relevant decrees and other official documents are published by Kozlov (2004).

The notorious law of June 26, 1940 lengthened the normal working day and week for all state employees. At the same time absenteeism, defined as any 20 minutes' unauthorized absence during the working day, or even idling on the job, also became a criminal offense; the penalty was set at up to six months' corrective labor at work with a 25 per cent reduction in pay. Three offenses in one month or four in two months counted as unauthorized quitting, punishable by two to four months' imprisonment. Enterprises managers were made criminally liable for failure to report worker violations. In August 1940, the minimum sentence for petty theft at work and "hooliganism" was set at one year's imprisonment. In December 1941, defense industry workers were conscripted at work under a quasi-military discipline that prescribed 5 to 8 years in a labor camp for unauthorized quitting. Further measures of April and May 1943 militarized transport workers and required unauthorized quitters to serve 3 to 10 years in labor camps. After the war a decree of June 1947 set the minimum sentence for any theft of state or socialized property at five to seven years imprisonment. Transport work was eventually demilitarized in March 1948 and work in the defense industry two months later; otherwise, these punitive laws remained on the books until Stalin died.

A report prepared as background for Khushchev's secret de-Stalinization speech of February 1956 (Kozlov, 2004, Volume 1: 603–607), shows that from 1940 through June 1955 the regular courts and military, transport, and labor camp tribunals sentenced 35.8 million persons in total for all offenses. Not allowing for repeat offenders, this would represent about one third of the adult population of roughly 100 million. Of the 35.8 million, 15.1 million were imprisoned and a quarter of a million were executed. The annual rate of imprisonment was one million or more in most years up to 1950, and more than half a million at the time of Stalin's death. These totals were dominated by workplace offenses. Between 1940 and 1952, there were 15 million convictions for absenteeism and unauthorized quitting, out of a total public sector workforce in 1940 of 31 million. There were 11 million sentences to corrective labor at work with loss of pay, more than 2.7 million prison terms of a few months, and nearly one million terms in labor camps of between 3 and 10 years. There were also 4.7 million convictions for theft of socialized property (Kozlov, 2004, Volume 1: statistical appendix; see also Filtzer, 2004: 159–67; Sokolov, 2003, 2005).

The archives show that Team Stalin, not an independent judiciary, determined the number of convictions. When the laws were enacted, Stalin had to intervene personally to force a reluctant judiciary to prosecute tardy or lazy workers (Sokolov, 2003).

Given the many factors that differentiated the Soviet judicial and penal system from others, it is hard to find appropriate standards for detailed comparison of workplace offending. It is simpler to go to the aggregate of all offenses. From 1940 to June 1955 the USSR imprisoned 15.1 million people or, on an annual basis, roughly one per cent of its adult population each year. In 1958, for comparison, the United States imprisoned not more than one fifth of one percent of its adult population.[3] Thus Stalin imprisoned his subjects at least five times more frequently than the United States.

2.3. *The labor camps*

Collectivization in the early 1930s and the repression of state employees in the 1940s created huge flows into labor camps. In the 1920s there was just one Soviet forced labor complex in the Arctic where mainly political detainees were held. The first major expansion came with collectivization

[3]See U.S. Department of Commerce (1960: 139, 141). U.S. statistics of the time did not give aggregate sentencing figures. Instead, we multiplied the number of known crimes committed (p. 139), by the proportion of offenses cleared by arrest and the proportion of arrests that resulted in convictions (p. 142). This calculation does not allow for multiple offending.

which threw hundreds of thousands of peasants into camps or special settlements; the Gulag, the interior ministry's chief administration of labor camps, was created in 1930 to handle the sudden inflow and labor camps spread across the remote interior and the far north and east to house, at their peak, more than 2.5 million inmates.

Although the Gulag was shrouded in intense secrecy, its human side leaked out through the recollections of former inmates. Estimates of its scale and scope, however, could be based on little more than guesswork until the first official figures were released in 1989. While contemporary estimates of the number of detainees ranged up to 20 million, we now know that there were approximately 2.3 million penal laborers at the outbreak of World War II and about 2.5 million on Stalin's death; similar numbers of deportees were also confined to labor settlements in the remote interior. The forced laborers were mostly engaged in forestry, mining, and construction, where they made up substantial shares of employment, but never more than about 3 per cent of the total workforce including farm workers, and less than this in terms of the value of national output. Bacon (1994) compares the figures reported in the archives with the range of estimates that went before. But these new facts must be set beside others including the very high rates of conviction and sentencing that we now know about: although the Gulag population was smaller than observers had sometimes guessed, it also had much higher turnover with very large numbers entering the system for relatively short periods before returning to society. Possibly, the higher estimates for numbers of forced laborers reflected the 20-million-plus cumulative flow rather than the much smaller stock at a point in time (Kozlov, 2004, Volume 4: 111, 134–136).

Internal Gulag documents confirm that political strategy, not economics, dictated the Gulag's development. The labor camps were populated as a consequence of collectivization, terror, draconic labor laws, and also the punishment of ethnic groups, prisoners of war, and others suspected of potential or actual wartime disloyalty. Before the archives there was speculation that in the course of the 1930s economics eventually took over from politics as a motivation for recruiting forced laborers, or alternatively that the NKVD became a lobby for forced labor (Swianiewicz 1965: 161–2). Neither of these has turned out to be the case, although our judgement has to be carefully shaded.

The Gulag had a consistent economic raison d'être: to explore and colonize regions that were resource-rich but inhospitable, since forced labor could be ordered around the country at will. Penal labor was to cover its own costs, which were to be held down to a low subsistence minimum (Khlevnyuk 2003). Repression of the forced laborers' wages combined with the enforcement of effort through close supervision and monitoring were supposed to promote low-cost accumulation. At the same time, no one was planning to lock everybody up for economic reasons.

At the level of the state, the continued primacy of politics over economics is shown by the fact that the NKVD-MVD did not lobby for expansion. The NKVD projected a shrinking number of Gulag inmates for the third five-year plan (1938–42), just as the first victims of Stalin's Great Terror began to flood in (Gregory, 2003: 4). After the war the NKVD again planned for contraction and politics again overturned this expectation (Khlevnyuk, 2003: 52). In the late 1940s, Gulag officials proposed to release all but the most dangerous prisoners from camps (Tikhonov, 2003), but this was unacceptable to Stalin. In 1953, within three months of Stalin's death in March, MVD chief Lavrenty Beriia had released one and a half million prisoners, 60 per cent of the Gulag's inmates. Tikhonov (2003) shows that this move had been prepared five years earlier. In fact, elements within the MVD were increasingly alarmed by its economic and social costs. The economic costs were reflected in its growing financial deficits; the social costs were measured by high rates of recidivism. By Stalin's death, as shown above, at least one quarter of Soviet adults had been through the Gulag for one reason or another. Although the camps were supposed to segregate hardened criminals from petty youth offenders, the camp population was a mixing bowl and recidivism soared. The high turnover spread the culture and *mores* of camp life throughout society.

Why did the Gulag fail? Research on the microeconomics of the Gulag is in its infancy but early case studies show a complex learning process. The Far Eastern camps (Nordlander, 2003) highlighted the perennial tension between the possibilities for exploiting the forced laborers and the requirements of maintaining them. The Karelian camps (Joyce, 2003) show the experimental process by which the authorities learnt the scope and limits of the exploitation of forced labor. The White Sea-Baltic Canal project (Morukov, 2003) was initially seen as a success and this stimulated illusory expectations for the future. The Gulag leaders willingly undertook the building of Noril'sk (Borodkin and Ertz, 2003) because they underestimated the risks and difficulties that would arise. The operation of Noril'sk (Ertz, 2003) helped to expose illusions about the ease with which the inmates could be manipulated and coerced into supplying effort.

By the postwar years Gulag officials believed they knew why the camps failed to generate a surplus. Labor productivity there was extremely low relative to that of free workers, while guarding detainees was very expensive; in 1950 there was one guard to ten inmates, leading to the widespread practice of "unguarded" prison contingents. Within the Gulag, prisoners formed protective networks among themselves and with the guards to cover for each other (Heinzen, 2005). Camp managers learned that an arsenal of punishments was not sufficient to motivate prisoners and that there were complicated trade-offs: prisoners placed on reduced rations for failing to meet work quotas were no longer able to work effectively. One of the most effective incentive systems, early release for exemplary work, deprived the Gulag of its best workers. Material incentives played an ever larger role in motivating penal labor (Borodkin and Ertz, 2003; 2005; Ertz, 2005). In the last years of the Gulag, there was a process of "conversion of slaves to serfs" (Khlevnyuk 2003: 57); the camps increasingly paid prisoners civilian wages and the distinctions between penal and free labor became blurred.

3. *Coercion Failure?*

The collective farms ceased to be an instrument of coercion in the mid-1960s when the farmers were placed on fixed wages like any other employee. The Gulag was emptied of mass prisoners between 1953 and 1957. The draconian labor laws of the 1940s were also rescinded in the mid-1950s. These relaxations have two possible interpretations: Either the post-Stalin leadership did not have the stomach for continued repression and abandoned a system that was working; or coercion was abandoned because it did not work. The new evidence supports the second interpretation, particularly the fact that the most coercive instrument, the Gulag, was dissolved by its own leaders, who included men such as Beriia renowned for their amorality.

In short, the initial belief that results could be obtained more cheaply by raising the penalty for failure than by increasing the reward for success appears to have been mistaken. Effective coercion requires that penalties be accurately assessed and targeted, and that the agents of repression are well informed about offenders and the costs of their crimes. It proved more difficult than they expected to target penalties accurately; as a result the relationship between true effort and punishment was "noisy." The devising of an efficient coercion system is then further complicated if workers and managers respond strategically to increase the noise.

The archives show that officials had little idea whether workers were exerting full effort or not; the law could do little more than ensure that they were physically at work and did not steal too much. Agricultural controllers could order the collective farms to sow more land but could not assess whether the land was being farmed efficiently (Davies and Wheatcroft, 2003). In industry, attempts to pin "normal" effort down to objective technological criteria proved fruitless; attempts to maintain effort norms when productivity should have been rising often gave rise to damaging social conflicts (Davies and Khlevnyuk 2002: 877; Filtzer, 2002: 232–41).

Because the state relied on indirect indicators of effort, mistakes crept into the effort/punishment relationship. The investigation of low effort could yield an error of Type I that punished the innocent, and a Type II error that acquitted the guilty. Errors of both types appear to have been prevalent. Numerous Type I errors are reflected in the very high rates of penalization that we have reported; these condemned hard workers along with neer-do-wells, drunks, and thieves. The law criminalized such a wide range of behaviors that their scope made virtually every worker liable to prosecution for something, including one-time and accidental violations: at this time, a broken-down commuter bus could make criminals of scores of hapless workers at a stroke. Rational managers might wish to select the truly guilty for prosecution, the problem workers and repeat offenders, but the laws subjected even petty offenses to harsh penalties and managers who did not seek them in every case were threatened with the same. As a result, the innocent were bundled through the courts and camps along with the guilty in extraordinarily large numbers.

Memoirs suggest that Team Stalin probably knew this but did not care. In March 1937 Stalin's chief instrument of the Great Terror Nikolai Ezhov told his officials to expect "some innocent victims... Better that ten innocent people should suffer than one spy get away. When you chop wood, chips fly" (Montefiore, 2003: 194). Stalin's prime minister, Molotov (interviewed by Chuev, 1991: 416) also cared more about condemning the guilty than acquitting the innocent, "never mind if extra heads fall."

Type II errors were also clearly numerous. The evidence for this is the surprising fact that, although penalization rates were very high, offending rates were even higher. Filtzer (2002: 167–8) used the records of eight production branch ministries in 1947 to show that almost one third of a million instances of "labor desertion" gave rise to fewer than 55,000 convictions for a conviction rate of 16 percent. In other words, a judicial system that made the chips fly somehow failed to chop the wood. The combination of severe penalization with low conviction probability for the guilty is consistent with high-cost policing and justice administration (Becker, 1968: 184); the high rate of conviction of innocents, however, is more properly seen as a cost of dictatorship (Djankov *et al.*, 2003), in the sense that the dictator's efforts to achieve a lower rate of offending than society was willing to tolerate had highly suboptimal results.

The effectiveness of the Politburo accumulation model rested on the dictator's ability to create a gap between the civilian wage as a "fair" return for effort, and low subsistence in the Gulag as the return to shirking, so that the difference between them was the intended punishment for shirking. The widespread mistakes that we have described subverted this strategy: Type I errors reduced the worker's expected return to effort while Type II errors raised the expected return to shirking.

Error rates were not exogenous. They were fashioned by the counter-actions of those threatened with punishment, who could take steps to reduce their risks. Workers and managers diverted effort from production into mutual insurance: since the threat was shared among them, they could agree to cover up each other's shortcomings. At the grass roots, Berliner (1952) showed how managers, accountants, and technical workers in enterprises covered up poor results. Expressions in everyday use like "pulling wool over [the authorities'] eyes" [ochkovtiratel'stvo], and *pripiski* and *tufta*, the civilian and Gulag slang for exaggerated or non-existent output, encapsulated the constant effort to mislead authority. The necessary collusion relied on horizontal trust accumulated through "acquaintance and relationships" [znakomstvo i sviazi]. The archives have added detail: Filtzer (2002) has shown that post-war managers tolerated lateness and absence to maintain goodwill, and colluded with workers against the centre to underreport violations. Managers did, however, pursue illegal quitters who undermined morale and the factory's capacity to fulfil the plan. It was the rural police and courts that then gave moral shelter to the young offenders who had run home. In all these ways, mutual insurance tended to cut the individual risk of punishment. The archives also show how collusion travelled vertically upward through the hierarchy. Harris (1999: 156–63) has described how the Urals regional leadership formed itself into a "protected, mutually reliant" clique that fended off criticism from below and investigation from above in the mid-1930s, but was eventually wiped out in the Great Terror. Belova (2001) has shown that high-level patrons could protect the most egregious embezzlers.

Faced with widespread enforcement failures at lower levels, officials responded in two ways. First, they aimed to undermine horizontal trust by rewarding whistle blowing and informing (Belova, 2001). Second, Stalin forced the legal system, local party offices, and the militia to increase arrest and conviction rates or suffer penalties themselves. The most common method of forcing repression was to distribute quotas by region and profession to officials at lower levels (e.g. Kozlov, 2004, Volume 1: 270–271). In the Great Terror of 1937/38 local officials had to work feverishly to achieve a set number of confessions per day (Vatlin, 2004). To fulfil such plans the police officials imputed individual guilt from increasingly trivial differences in behavior. Whether or not these measures reduced the Type II errors, they seem likely to have encouraged false denunciation and confession and so added to the errors of Type I.

Team Stalin hoped, we suppose, that increased coercion would induce agents to take a lower wage without withdrawing effort, making coercion effectively costless. The evidence suggests that in practice those threatened with punishment raised the chance of a Type II error through mutual insurance, but this diverted effort from production. At the same time successful insurance against Type II errors combined with the increased chance of a Type I error to blunt the effect of penalization on the reservation wage. Finally, the heavy direct costs of the repressive apparatus must have further reduced the net payoff from coercion.

At the moment we have only one empirical study of the aggregate effects of the terror on industrial production (Katz, 1975), which attributes the 1937 slowdown in industrial growth to adverse effects of political repression. New case studies show the disastrous effects of the Great Terror on important Gulag operations (Ertz, 2003). Finally, the archives have shown that Soviet leaders themselves saw a direct connection between the terror and falling productivity (Davies, 2006): the terror eliminated a generation of skilled and knowledgeable managers and technical staff, worsened labor discipline, and damaged effort.

Conclusion

The study of Stalin's economic decision making suggests how we can understand the motivation behind his application of coercion to the labor force: he expected it to increase the surplus of resources at his disposal. Closer analysis shows, however, that coercion did not yield the results expected. Harsh punishments for violations did not create obedient citizens or a law-abiding society. People who were threatened or frightened did not just do as they were told. Instead, they allocated resources to protecting themselves in ways that detracted from Stalin's objectives.

It is sometimes suggested that the Soviet economic system could not have been held together without coercion. Whether or not this is true in some ultimate sense, it seems clear to us that the level of coercion in the Stalinist economy up to the early 1950s was counterproductive. In fact, those who managed the instruments of coercion were fully aware of it at the time; when Stalin died, they could hardly wait to dismantle important components of the coercive system.

References

Açemoglu, D and Robinson J (2000). Political Losers as Barriers to Economic Development. *American Economic Review*, 90(2): 126–130.

Akerlof, GA (1984). Gift Exchange and Efficiency Wages: Four Views. *American Economic Review*, 74(2): 79–83.

Akerlof, GA and Yellen, JL. 1990. "The Fair Wage-Effort Hypothesis and Unemployment," Quarterly Journal of Economics, 105:2, pp. 255-283.

Bacon, ET (1994). *The Gulag at War: Stalin's Forced Labour System in the Light of the Archives*. London: Macmillan.

Becker, GS (1968). Crime and Punishment: An Economic Approach. *The Journal of Political Economy*, 76(2): 169–217.

Belova, E (2001). Economic Crime and Punishment. In: Gregory PR (ed.) *Behind the Façade of Stalin's Command Economy: Evidence from the State and Party Archives*. Stanford, CA: Hoover Institution Press, pp. 131–158.

Berelovich, A (2000). *Sovetskaia Derevnia Glazami VChK-OGPU-NKVD, 2: 1923–1929*. Moscow: Rosspen.

Bergson, A (1961). *The Real National Income of Soviet Russia since 1928*. Cambridge, MA: Harvard University Press.

Bergson, A (1964). *The Economics of Soviet Planning*. New Haven: Yale University Press.

Berliner, JS (1952). The Informal Organization of the Soviet Firm. *Quarterly Journal of Economics*, 66(3): 342–365.

Blanchard, O and Kremer, M (1997). Disorganization. *Quarterly Journal of Economics*, 112(4): 1091–1126.

Borodkin, L and Ertz, S (2003). Coercion versus Motivation: Forced Labor in Norilsk. In Gregory PR and Lazarev V (eds.) *The Economics of Forced Labor: The Soviet Gulag*. Stanford, CA: Hoover Institution Press, pp. 75–103.

Borodkin, L and Ertz, S (2005). Forced Labor and the Need for Motivation: Wages and Bonuses in the Stalinist Camp System. *Comparative Economic Studies*, 47(2): 418–436.

Chuev, FI (1991). *Sto sorok besed s Molotovym*. Moscow: Terra.

Conquest, R (1967). *Industrial Workers in the USSR*. London: The Bodley Head.

Conquest, R (1987). *The Harvest of Sorrow: Soviet Collectivization and the Terror-Famine*. Oxford: Oxford University Press.

Kozlov, VA (ed.) (2005). *Istoriia stalinskogo Gulaga*. 7 Volumes. Moscow and Stanford, CA: ROSSPEN and Hoover Institution Press.

Davies, RW (1980). *The Industrialisation of Soviet Russia (Volume I): The Socialist Offensive: the Collectivisation of Soviet Agriculture, 1929–1930*. Volume 1. Basingstoke: Macmillan.

Davies, RW (1989). *The Industrialisation of Soviet Russia (Volume III): The Soviet Economy in Turmoil, 1929–1930*. Volume 3. Basingstoke: Macmillan.

Davies, RW (1996). *The Industrialisation of Soviet Russia (Volume IV): Crisis and Progress in the Soviet Economy, 1931–1933*. Volume 4. Basingstoke: Macmillan.

Davies, RW (2001). *Why Was There a Soviet Investment Cycle in 1933–7?* PERSA Working Paper no. 16. University of Warwick, Department of Economics. Available online at: http://www.warwick.ac.uk/go/persa.

Davies, RW (2006). The Soviet Economy and the Launching of the Great Terror. In: Ilic M (ed.) *Stalin's Terror Revisited*. London and New York: Palgrave, pp. 11–37.

Davies, RW and Khlevnyuk, O (1999). The End of Rationing in the Soviet Union, 1934–1935. *Europe-Asia Studies*, 51(4): 557–610.

Davies, RW and Khlevnyu, O (2002). Stakhanovism and the Soviet Economy. *Europe-Asia Studies*, 54(6): 897–903.

Davies, RW and Wheatcroft, SG (2003). *The Industrialisation of Soviet Russia (Volume V): The Years of Hunger: Soviet Agriculture, 1931–1933*. Volume 5. Basingstoke: Palgrave.

Davies, RW, Tauger, MB and Wheatcroft, SG (1995). Stalin, Grain Stocks and the Famine of 1932–1933. *Slavic Review*, 54(3): 642–657.

Davies, RW, Khlevnyuk, OV, Rees, EA, Kosheleva, LP and Rogovaya LA (eds.) (2003). *The Stalin-Kaganovich Correspondence, 1931–36*. New Haven, CT: Yale University Press.

Djankov, S, Glaeser, E, la Porta R, Lopez-de-Silanes F and Shleifer A (2003). The New Comparative Economics. *Journal of Comparative Economics*, 31(4): 595–619.

Ellman, M (1975). Did the Agricultural Surplus Provide the Resources for the Increase in Investment in the USSR During the First Five Year Plan? *Economic Journal*, 85(340): 44–863.

Ellman, M (2000). The 1947 Soviet Famine and the Entitlement Approach to Famines. *Cambridge Journal of Economics*, 24(5): 603–630.

Erlich, A (1960). *The Soviet Industrialization Debate, 1924–1928*. Cambridge, MA: Harvard University Press.

Ertz, S (2003). Building Norilsk. In: Gregory PR and Lazarev V (eds.) *The Economics of Forced Labor: The Soviet Gulag*. Stanford, CA: Hoover Institution Press, pp. 127–50.

Ertz, S (2005). Trading Effort for Freedom: Workday Credits in the Soviet Camp System. *Comparative Economic Studies*, 47(2): 476–491.

Filtzer, D (2002). *Soviet Workers and Late Stalinism: Labour and the Restoration of the Stalinist System after World War II.* Cambridge: Cambridge University Press.

Granick, D (1987). *Job Rights in the Soviet Union: Their Consequences.* New York: Cambridge University Press.

Gregory, PR (2003). An Introduction to the Economics of the Gulag. In: Gregory PR and Lazarev V (eds.) *The Economics of Forced Labor: The Soviet Gulag.* Stanford, CA: Hoover Institution Press, pp. 1–21.

Gregory, PR (2004). *The Political Economy of Stalinism: Evidence from the Soviet Secret Archives.* New York: Cambridge.

Gregory, PR and Stuart, RC (1974). *Soviet Economic Structure and Performance* (1st edn.), New York: Harper and Row.

Grossman, G (1953). Scarce Capital and Soviet Doctrine. *Quarterly Journal of Economics*, 67(3): 311–343.

Harris, JR (1999). *The Great Urals: Regionalism and the Evolution of the Soviet System.* Ithaca: Cornell University Press.

Harrison, M (2002). Coercion, Compliance, and the Collapse of the Soviet Command Economy. *Economic History Review*, 55(3): 397–433.

Heinzen, JW (2004). *Inventing a Soviet Countryside: State Power and the Transformation of Rural Russia, 1917–1929.* Pittsburgh, PA: University of Pittsburgh Press.

Heinzen, JW (2005). Corruption in the Gulag: Dilemmas of Officials and Prisoners. *Comparative Economic Studies*, 47(2): 456–475.

Hessler, J (2004). *A Social History of Soviet Trade: Trade Policy, Retail Practices, and Consumption, 1917–1953.* Princeton, NJ: Princeton University Press.

Jasny, N (1949). *The Socialized Agriculture of the USSR: Plans and Performance.* Stanford, CA: Stanford University Press.

Joyce, C (2003). The Gulag in Karelia: 1929–1941. In: Gregory PR and Lazarev V (eds.) *The Economics of Forced Labor: The Soviet Gulag.* Stanford, CA: Hoover Institution Press, pp. 163–188.

Katz, BG (1975). Purges and Production: Soviet Economic Growth, 1928–1940. *Journal of Economic History,* 35(3): 567–590.

Khlevnyuk, OV (1995). The Objectives of the Great Terror, 1937–38. In: Cooper JM, Perrie M and Rees EA (eds.) *Soviet History, 1917–1953: Essays in Honour of R.W. Davies.* New York, NY: St Martin's Press, pp. 158–176.

Khlevnyuk, OV (2001). The Economy of the Gulag. In: Gregory PR (ed.) *Behind the Façade of Stalin's Command Economy: Evidence from the State and Party Archives.* Stanford, CA: Hoover Institution Press, pp. 111–130.

Khlevnyuk, OV (2003). The Economy of the OGPU, NKVD, and MVD of the USSR, 1930-1953: The Scale, Structure, and Trends of Development. In: Gregory PR and Lazarev V (eds.) *The Economics of Forced Labor: The Soviet Gulag.* Stanford, CA: Hoover Institution Press, pp. 43–66.

Kontorovich, V (1986). Discipline and Growth in the Soviet Economy. *Problems of Communism*, 34(6): 18–31.

Kornai, J(1980). *The Economics of Shortage*, 2 vols. Amsterdam: North-Holland.

Kuromiya, H (1988). *Stalin's Industrial Revolution: Politics and Workers, 1928–1932.* Cambridge: Cambridge University Press.

Kuznets, S (1963). A Comparative Appraisal. In: Bergson A and Kuznets S (eds.) *Economic Trends in the Soviet Union.* Cambridge, MA: Harvard University Press, pp. 338–339.

Lazarev, V (2005). Economics of One-Party State: Promotion Incentives and Support for the Soviet Regime. *Comparative Economic Studies,* 47(2): 346–363.

Lewin, M (1968). *Russian Peasants and Soviet Power: A Study of Collectivization. London*: Allen and Unwin.

Manning, RT (1993). The Soviet Economic Crisis of 1936–1940 and the Great Purges. In: Arch Getty J and Manning RT (eds.) *Stalinist Terror: New Perspectives.* Cambridge: Cambridge University Press, pp. 116–141.

Millar, JR (1974). Mass Collectivization and the Contribution of Soviet Agriculture to the First Five-Year Plan: A Review Article. *Slavic Review*, 33(4): 750–766.

Montefiore, SS (2003). *Stalin: The Court of the Red Tsar.* London: Weidenfeld and Nicolson.

Morukov, M (2003). The White Sea-Baltic Canal. In: Gregory PR and Lazarev V (eds.) *The Economics of Forced Labor: The Soviet Gulag.* Stanford, CA: Hoover Institution Press, pp. 151–162.

Nordlander, D (2003). Magadan and the Economic History of Dalstroi in the 1930s. In: Gregory PR and Lazarev V (eds.) *The Economics of Forced Labor: The Soviet Gulag.* Stanford, CA: Hoover Institution Press, pp. 105–25.

Nove, A (1969). *An Economic History of the USSR.* London: Allen and Unwin.

Olson, M (1993). Dictatorship, Democracy, and Development. *American Political Science Review*, 87(3): 567–576.

Osokina, E (2001). *Our Daily Bread: Socialist Distribution and the Art of Survival in Stalin's Russia, 1927–1941.* Armonk, NY: M.E. Sharpe.

Podkaminer, L (1989). Theory of Investment Cycles in the Centrally Planned Economies: An Economic Interpretation. *Forschungsberichte no. 158.* Wiener Institut für Internationaler Wirtschaftsvergleiche.

Rutland, P (1985). *The Myth of the Plan.* LaSalle, IL: Open Court.

Schelling, TC (1960). *The Strategy of Conflict.* Oxford: Oxford University Press.

Schrettl, W (1982). *Consumption, Effort, and Growth in Soviet-Type Economies: A Theoretical Analysis.* Ann Arbor, MI: University Microfilms International.

Schrettl, W (1984). Anspruchsdenken, Leistungsbereitschaft und Wirtschaftszyklen. In: Schuller A (ed.) *Wachstumsverlangsamung und Konjunkturzyklen in unterschiedlichen Wirtschaftssystemen.* Schriften des Vereins für Socialpolitik, Volume 142. Berlin: Duncker and Humblot., pp. 153–191.

Sen, AK (1983). *Poverty and Famines: An Essay on Entitlement and Deprivation.* Oxford: Oxford University Press.

Sokolov, AK (2003). Forced Labor in Soviet Industry: the End of the 1930s to the Mid-1950s: an Overview. In: Gregory PR and Lazarev V (eds.) *The Economics of Forced Labor: The Soviet Gulag*. Stanford, CA: Hoover Institution Press, pp. 23–42.

Sokolov, AK (2005). Before Stalinism: the Defence Industry of Soviet Russia in the 1920s. *Comparative Economic Studies*, 47(2): 437–455.

Spulber, N (1964). *Soviet Strategy for Economic Growth*. Bloomington, IN: Indiana University Press.

Strauss, E (1969). *Soviet Agriculture in Perspective: A Study of Its Successes and Failures*. London: Allen and Unwin.

Swianiewicz, S (1965). *Forced Labour and Economic Development: An Enquiry into the Experience of Soviet Industrialization*. London: RIIA and Oxford University Press.

Tauger, MB (1991). The 1932 Harvest and the Soviet Famine of 1932–1933. *Slavic Review*, 50(1): 70–89.

Tikhonov, A (2003). The End of the Gulag. In: Gregory PR and Lazarev V (eds.) *The Economics of Forced Labor: The Soviet Gulag*. Stanford, CA: Hoover Institution Press, pp. 67–74.

United States Department of Commerce (1960). *Statistical Abstract of the United States*. Washington, DC: U.S. Government Printing Office.

Vatlin, AI (2004). *Terror raionnogo masshtaba: "massovye operatsii" NKVD v Kuntsevskom raione Moskovskoi oblasti. 1937–1938 gg.* Moscow: ROSSPEN.

Wheatcroft, SG (2004). From Team Stalin to Degenerate Tyranny. In: Rees, EA (ed.) *The Nature of Stalin's Dictatorship: The Politburo, 1924–1953*. Basingstoke: Palgrave, pp. 79–107.

Wheatcroft, SG and Davies, RW (2002). The Soviet Famine of 1932–3 and the Crisis in Agriculture. In: Wheatcroft SG (ed.) *Challenging Traditional Views of Russian History*. Basingstoke: Palgrave, pp. 112–146.

Chapter 11

Economic Information in the Life and Death of the Soviet Command System

Mark Harrison

Without the most careful examination of all the statistical data that which we possess in far larger measure than at any other time and in any other country, without organising these data, without analysing them and generalising from them, no scholarly economic work is possible. It is a source of regret that the statistical data are still classified secret in the central statistical administration in comrade Starovskii's safes.[1]

What is the problem we wish to solve when we try to construct a rational economic order? [...] It is [...] a problem of the utilisation of knowledge not given to anyone in its totality (Hayek, 1945: 519–20).

This paper is about the value of information in a command system.[2] Paul David and Dominique Foray distinguish information from knowledge as follows: knowledge is the cognitive capability that empowers its owners to act, intellectually or practically, whereas information is merely the "structured and formatted data that remain passive and inert" until those empowered by knowledge use it (David and Foray, 2001). On board an

---

This paper was first published by Frank Cass in *Reinterpreting the End of the Cold War: Issues, Interpretations, Periodizations* (edited by Silvio Pons and Federico Romero), 93–115. © Silvio Pons and Federico Romero 2005.

[1]This is an except from the speech of A.I. Mikoian to the twentieth congress of the communist party of the Soviet Union held in Moscow in February 1956. Vladimir Nikonovich Starovskii was chief of TsSU (the central statistical administration of USSR Gosplan, later of the USSR council of ministers) continuously from 21 October 1940 to 6 August 1975.

[2]This paper was first presented to a conference on "The End of Soviet Communism in Europe" at the Gramsci Institute Foundation, Rome, 10 and 11 May 2002. I thank the organizers and participants for discussion, and also Stephen Broadberry, Yoram Gorlizki, Paul Gregory, Oleg Khlevniuk, Peter Law, Martin Lowry, and Valery Lazerev for advice and comments.

aircraft, the instruments and charts provide information whereas the pilot has the knowledge of how to fly the plane using the information. Thus information is much less than knowledge. It is only a small part of the stock of intangible capital that is used to fly the plane or manage society. But information is still very important. In the classic formulation of Kenneth Arrow, the value of information is that it reduces uncertainty (Arrow, 1984: 106–14). Flying in cloud or darkness, for example, the most knowledgeable pilot cannot pinpoint the plane's position and attitude relative to the ground without checking the dials on the instrument panel.

Three factors appear to make it timely to investigate the economics of information in command systems. One is that any study of information in command systems cannot be separated from the problem of official secrecy: why are there official secrets, and why does the coverage of official secrecy vary through time and across countries? While there is a rich literature on the economics of concealment, including commercial secrecy, the issue of official secrecy does not appear to have been addressed; therefore, a gap in our understanding exists and should be remedied. Another is that the historical literature on Soviet official secrecy is written almost exclusively in moral and psychological terms of the rulers' paranoia, their excessive or pathological secretiveness, and so on, which diverts us from the necessary attempt to understand secrecy as a rational–choice process. Finally, a number of recent studies from the former Soviet archives are providing us with a growing mass of "information about information" which could help us to solve these problems.

From the point of view of the economics of information, however, this mass is almost entirely unstructured, so there still is a big empirical task in front of us. It is not just that we do not have measures of information, but also that we do not know how it should be measured. There is an aggregate stock of information in the economy that depreciates continually as information becomes outdated or irrelevant. New information is continually added; some of it updates or replaces old information, and some of it is new. Because information is dated and technology is changing, information of different dates is heterogeneous. Information can be digital or analogue. Digital information can be measured in internet pages, disk files, bytes and bits. Analogue information that is written and bound can be measured in volumes, pages, and characters, but analogue information also exists in film, tape, newsprint, and manuscript. This technical heterogeneity makes the stock of information and its growth hard to measure without ambiguity. Further, the uses of information are many; usually we cannot even distinguish between information that serves the economy's productive needs and that which serves our human curiosity and is therefore a consumption good. The value of information depends on our knowledge, and new knowledge can have unexpected effects; it can make some kinds of information completely obsolete, and make others suddenly valuable, for example we are continually finding new ways of using historical information.

But for the most part we are not very good at measuring information. What we tend to measure is not information itself but the spread of information *technology*, for example computer units, telephone lines, pounds of newsprint, and so on. I will try to focus on the concept of stocks of economically valuable information of various kinds, but the reader should not expect quick results.

The paper is organised as follows. Part 1 argues that the growing role that information plays in modern market economies is not just a consequence of its falling cost. The growing role of information also results from rising incomes. The evidence shows that as economic development proceeds and incomes rise information is demanded in much larger quantities and may even display increasing returns. Part 2 contrasts the uses of information in command and market systems. I will focus particularly on intermediate uses of information, rather than information as a consumption good. I will start from the assumption that information is exchanged so as to permit transactions that create a surplus for someone, and the distribution of the surplus is likely to be very important, but I will not analyse the creation of the surplus itself. Part 3 considers some specific uses of information: information adds more value to transactions that take place increasingly as incomes rise because it supports reputations, permits customization, and provides yardsticks. In the Soviet economy such information was frequently not produced; if produced, it was often concealed in "comrade Starovskii's safes" or elsewhere. Often it was of poor quality; regardless of quality, it often suffered from low credibility outside the ruling circle. In short, the Soviet command system forced economic growth on the basis of a relatively low–value information stock. Part 4 considers the implications for Soviet postwar economic growth and slowdown, the collapse of the command system, and the persistence of a low level of output after the collapse. Part 5 concludes.

1. *Falling Costs, Rising Demands*

We are sometimes told that we are living in a "new economy". The basis of the new economy is said to be the rapidly falling costs and diffusion of new technologies for information storage, handling, and transmission from the desktop computer and mobile phone to the internet. Other aspects of the new economy, whether or not it has truly accelerated growth, are said to include declining old industries, the growth of new industries producing machinery and specifically IT equipment, the rise of the information and financial services sectors, and globalization.[3]

I single out the rise of services because the products of the services sector are relatively information–intensive. Information is a final product of the news, entertainment, and publishing media. Agencies engaged in business consulting, market research, credit evaluation, and economic forecasting consume it and supply it to other business users. Information is essential to the provision of personal services that range from education, health care, and the long–term care of sick, disabled, and old people, to housing, labor market and financial services. Community services such as defense, policing, the administration of justice, and public accounting rely in obvious ways on the flow of information.

The rise of services has recently accelerated but has long historical roots. This is one piece of evidence that supports the idea that the new economy may not be so new and has probably been emerging for centuries. Table 1 shows that the employment share of services in three of the most

Table 1. Employment in Distribution, Finance, and Services, 1870 to 1990: the USA, UK, and Germany, Selected Years (per cent of total employment).

	United States	United Kingdom	Germany
1870	18.3	27.0	16.0
1910	26.2	32.3	19.5
1930	33.1	35.2	22.3
Increase, 1870 to 1930	+14.8	+8.2	+6.3
1950	40.0	31.7	21.1
1990	62.2	57.0	36.1
Increase, 1930 to 1990	+29.1	+21.8	+13.8

Source: Broadberry (1997: 64).

[3]For a survey of uses, leading and misleading, of the idea of a "new economy", see Madrick (2001).

developed industrial economies has been growing since 1870. Its growth has been continuous in the United States, whereas in Britain and Germany the industrial efforts of World War II interrupted the trend. But if we take 1930 as an intermediate benchmark the table also shows that the gain in the employment share of the services sector over the last sixty years in each country was at least twice that of the first sixty years, and this supports the idea of an underlying acceleration of the restructuring of production towards services since World War II.

No single factor has been associated with the rise of the "new economy" more than diffusion of the personal computer. Within twenty years the United States has gone from negligible availability to a position where there is one PC for every employable adult. Other leading economies are not far behind.[4]

Current rates of decline in information and communication costs are dramatic by any standard. For example, in the three decades from 1930 to 1960 the cost of a three–minute transatlantic phone call fell by 5.4 per cent a year, but this rate increased to 8.4 per cent annually in the next three decades from 1960 to 1990. In the three decades from 1960 to 1990 the US Department of Commerce computer price deflator fell at 14.9 per cent a year, and at 27.6 per cent a year in the 1980s alone.[5] More recent estimates show further acceleration, with performance–adjusted personal computer prices falling at 30 to 40 per cent annually in the 1990s (Landefield and Grimm, 2000: 19). It is a striking irony that the years of accelerating global decline in information costs are also roughly the period over which the Soviet economy decayed and then collapse.

There may be historical parallels in past diffusion curves of radio and television, telegraph and postal services and mail order, and newsprint and books. The cost of the printed word followed a similar curve in the years after the introduction of the printing press. Before printing, a single book might represent months or years of a scholar's income. Only the church or the crown could maintain a library (Bell, 1936: 331; Cipolla, 1956: 57–63). With the introduction of printing to Italy in 1465, the price of books is said to have fallen by 80 per cent in just three years. This decline was repeated in the years that followed because falling prices then widened the market and enabled a rapid increase in printing runs; by the early sixteenth century

[4]World Bank (2001). In the United States, in 1999, there were 511 PCs in use per thousand of the population, followed by Australia (469) and Switzerland (462).

[5]Calculated from figures cited by the International Monetary Fund (1997: 46).

editions of 3000 copies were not uncommon. After this time, however, book prices stabilised for a lengthy period.[6]

The decline in information costs during such episodes may be astonishing and spectacular, but it is only part of the story. Conventional analysis of the information age focuses on falling costs because our attention is naturally drawn to the historical episodes of dramatic cost decline, one of which we are living through now. Moreover it cannot be denied that falling costs are very, very important. However, the relationship between information and economic growth has roots that are not only longer in historical terms than are often imagined, but also deeper than a narrow focus on falling costs would imply. We are not just observing a slide down a fixed demand curve that widens the market as costs fall. The uses of information have been driven by rising incomes as well as by falling information costs. Income growth is slow and gradual. However, there is unambiguous evidence that information becomes more useful as incomes rise and would be demanded in greater quantities even if costs did not change.

Cross–country comparisons at a point in time show clearly how the demand for information handling capacity rises with income when costs are given. This relationship is well established and robust, and it can be observed in quite widely separated historical periods. As far back as the 1930s incomes were correlated with the availability of information and communications facilities across countries just as strongly as with ingots of steel or kilowatts of electric power (Harrison, 1994: 249–51). Lenin and Stalin thought they were living in the age of steel and electricity, but they were wrong: a new era was already overtaking them. Similar evidence from the present day confirms that the age of information is not driven by falling costs alone.[7] For given technologies available at given costs, relatively wealthy countries have found it advantageous to invest much more heavily in information capacity than poor countries. This may be

---

[6]I thank Martin Lowry for this information, based on his own research and reference to Giovanni Andrea Bussi, *Prefazioni alle edizioni di Sweynheym e Pannartz prototipografi Romani, a cura di Massimo Miglio* (Edizioni il Polifilo, 1978), and Anna Esposito, Anna Modigliani, and Paola Scarcia Piacentini, "Il costo del libro", in *Scrittura, biblioteche e stampa a Roma nel Quattrocento, Atti del Seminario 6–8 Maggio 1982, a c. di M. Miglio* (3 vols, Citta del Vaticano 1983), Volume II, pp. 323–553.

[7]For data on the present–day cross–country distribution of personal computers, televisions, telephone mainlines, and mobile phones, strongly correlated with incomes, see World Bank (2001).

because information is a luxury good so that rich people are willing to spend proportionately more to obtain it, or because information has higher returns in the activities in which rich countries specialise.

It seems likely that information displays diminishing returns when other factors are controlled. However, network externalities clearly ensure increasing returns over a certain range to items of information technology *equipment* such as video and fax machines, telephones, and PCs (Katz and Shapiro, 1985). Further, while returns to information may strictly diminish at the margin for a given activity, it is possible and even likely that average returns tend to be higher in those activities such as services that grow most rapidly as incomes rise. Hence, the value of information may rise with incomes and the level of economic development.

2. *Hierarchy and Secrecy*

In order to understand the scope and purposes of information stocks in the Soviet economy, it is necessary to note the differences between command and market systems. In a market that is competitive the important relationships are horizontal: buyers and sellers meet and compete on approximately equal terms. Market information is available to all or, if costly, is itself a commodity that can be supplied and demanded. A hierarchy is a vertical network of principals and agents (Wintrobe, 1998: 212–13). In the Soviet command economy most enterprises were state owned and most producers were agents of a government principal, usually a minister, with the ministry being the legal fund holder. Horizontal relationships of supply and demand were organised by order from above through vertical hierarchies rather than directly between buyers and sellers on a voluntary basis. Vertical subordination was based on the principle that orders flowed downwards while information flowed upwards and did not leak downwards or sideways.

An essential feature of the Soviet command system is that its hierarchies were complex: there were many parallel hierarchies with functions that were necessarily specialised and often overlapping, that converged at the top in a unified leading body such as the council of ministers and the politburo. And of course there were minor parallel hierarchies within every major one. None of these hierarchies could fulfill its functions in isolation, without horizontal exchanges that required contact with others. For example the defense ministry had to purchase equipment and fuel from industry and food from agriculture; the ministry for engineering had to buy metals from the steel industry and power from the electricity generating industry. These exchanges were first of all authorised in very broad outline in high–level plans and decrees, and then the ministries themselves had to negotiate detailed contracts for specific commodities that implemented the plan.[8] Implementing these contracts rested on budget authorisation by the ministry of finance and credit authorisation by the state bank. Meanwhile the criminal and security police, planners, and party activists selectively monitored activities and transactions and tracked their consequences.

Within these complex hierarchies principals were faced with two kinds of opportunistic action on the part of agents that might restrict or dissipate the principal's rents. One was the agent's ability to influence the principal arising from the agent's control of the upward flow of information. The other was the agent's ability to collude with contacts in parallel hierarchies in unauthorised horizontal transactions that might profit the agent to the loss of the principal.

The practice of secrecy was a mechanism that effectively supported vertical structures at the expense of horizontal ones. Secrecy itself had two aspects. One aspect was the strict rules limiting the downward transmission of information. For example, between 1930 and 1941 the government and its main economic committee made more than 32,000 decrees but less than 4,000 of these were openly published, and more than 5,000 received the top security classification which meant that they remained known only to a few top officials (Davies, 2001: 63). Secrecy extended both to decisions and to the decision making process itself. The principles of "conspirativeness" (*konspiratsiia*), approved by the party politburo in the late 1920s, were aimed at limiting knowledge of the business of the politburo and central committee to the narrowest possible set of participants and, in the process of transmitting decisions downwards, to deny information to lower levels about the sources and context of higher–level decisions. Moreover, on 5 March 1931 the politburo resolved "categorically to forbid people with the right of acquaintance with the decisions of the c[entral] c[ommittee], when passing instructions onward in the apparatus, to refer to the fact that these

[8]Our knowledge of inter–ministerial contracting in the Soviet economy was virtually non–existent before the archives. See Harrison and Simonov (2000), Belova and Gregory (2002), and Belova (2005).

instructions are decisions of the c[entral] c[ommittee]" (Khlevniuk *et al.* 1995: 85, 73–82).

Such secrecy cannot only have been designed to prevent breaches of national security or to prevent society from holding the government and ruling party to public account. It was also designed to influence the behaviour of those within the state, but below the apex of power. If officials and activists at the middle level of the nomenklatura knew nothing of the decisions being made above them and of the superior bodies making them, they could be prevented from learning how to shape and direct information and lobby superiors so as to influence decisions in their own private interests.

Another aspect of secrecy was the construction of elaborate firewalls within the state to inhibit horizontal transfers of information among the parallel hierarchies. Again, these firewalls could hardly have been justified on grounds of national security or even by a desire to protect the authority of the state as a whole over society. Even within the government information was shared on the basis of need, not right to know, and the need to know was defined within limits that appear sometimes extraordinarily narrow. For example, in August 1948, the deputy chief of the interior ministry (MVD) Ivan Serov wrote to his boss Lavrentii Beriia that in the process of drafting the 1949 budget the ministry of finance was demanding to be informed of numbers in prisons, labor camps, and prisoner of war camps "and their physical condition", numbers of internal security troops, and figures for gold output and the gold content of ores; these figures were required to budget for the required outlays by the MVD on wages and subsistence. Serov warned: "Provision of these figures will lead to familiarisation with especially important information on the part of a wide circle of staff of the USSR Ministry of Finance, the State Bank and the Industrial Bank". An accompanying memorandum advised Beriia that in past years such figures were loaned temporarily to the finance ministry to be processed by no more than two or three highly trusted workers, then returned; it noted that the ministries of the armed forces and state security provided the finance ministry only with financial summaries, not head counts; and it proposed that from now on the MVD do the same.[9]

[9]Gosudarstvennyi Arkhiv Rossiiskoi Federatsii, Moscow, *fond* 5446, *op.* 50a, *delo* 4043, folios 1–2. On Serov's memorandum is noted by hand: "Comrades Popov and Serov: consider and resolve. L. Beriia". I thank Oleg Khlevniuk for this reference. For a similar example from the 1930s, see Harrison (2001:96).

Such firewalls were often buttressed by a low–trust environment that helped to align agents' incentives with those of their principals. For example, the archives show that in the 1930s industrial producers of military products frequently refused to release information about the production cost of weapons to the defense ministry, which was the purchasing department. They justified this on the grounds that equipment costs were a military secret that the defense ministry was not entitled to know, but of course this was a crude excuse: they did so simply because it promoted their horizontal bargaining power (Harrison and Simonov, 2000: 232–5; Barber *et al.*, 2000: 19–23).

A number of interlocking mechanisms thus inhibited horizontal transactions. Unauthorised contracts among lower level agents to engage in horizontal trades were prohibited and were not legally enforceable. In addition vertical networks of patronage and protection encouraged agents to invest in relationships of trust and dependence with superiors rather than with their opposite numbers in parallel hierarchies.

By protecting their information from horizontal spillage principals successfully strengthened their vertical hierarchies. However, the reinforcement of hierarchy was achieved at a cost, and this cost was paid in the diminished quantity and quality of information that flowed upwards from agent to principal. Quantities of information that flowed upward were limited by the common interest of both principal and agent. The upper reaches of the hierarchies had a very limited information handling capacity. This was only partly because of technological limits, although it should not be forgotten that until the end of the Soviet regime many officials had nothing more sophisticated on their desks than an abacus. In addition, the politburo and ministers positively preferred to rely on small expert staffs; evidently this made it easier to share motivation and sustain unified leadership (Gregory, 2001: 14–16). There were many kinds of information that the commanders of the system simply did not want to know; they did not want to be bothered with every petty setback or failure, preferring to place responsibility on those at lower levels to sort things out without the assistance of higher authority. But it also suited those at lower levels to preserve their own freedom of action by keeping their superiors under informed (Belova and Gregory, 2002).

Vertical relations also worked to spoil the quality of information. Because profitable opportunities for horizontal exchanges that undermined the plan were everywhere, principals needed good information about what their agents were doing. For the same reason, however, agents had strong incentives to conceal or lie about their activities. And for another reason too: even if the returns to concealed horizontal exchanges were sometimes low relative to the high risks and penalties involved, the most profitable alternative for the agent was not necessarily to obey orders, because carrying out instructions also required effort, and doing nothing required just as much concealment as doing something that was actively illegal. Thus, although downward and horizontal transfers of information were successfully impeded, the flows of information vertically upward became sluggish and were frequently distorted.[10]

While principals appear to have discouraged inferiors from lobbying in general, under specific circumstances they encouraged it because it helped to overcome the reluctance of agents to volunteer information. In a context of competition for resources and favours, rival agents were all too willing to supply principals with information of two kinds: about the relative worth of competing spending projects that were available for the principal to select, and about the relative loyalty of competing agents. This strategy was followed, for example, to create a "market for inventions" in the defense industry (Harrison, 2005; 2003a). The information resulting could be plentiful, though recognisably biased. Its supply depended on enforcing rivalry among agents at lower levels through a policy of "divide and rule"; if they formed horizontal links and learned to collude, the flow of information was worsened in both quantity and quality (Harrison, 2003b).

In summary, the role of information was quite different in the command system compared with a market setting. Specifically, principals had strong incentives to collect true information but these incentives were substantially weakened by the high costs of collecting and handling it. Agents often had only weak incentives to collect information or provide true information to principals, and strong incentives either to withhold or distort it. The equilibrium outcome was characterised by limited information stocks of poor quality.

---

[10]As a result the detailed information that reached the summit was sometimes not trusted as a basis for planning decisions and supra–ministerial agencies incurred costs in gathering data independently to duplicate and cross–check that supplied through the ministerial hierarchy. See Gregory (1990: 37–8).

3. *Market Information in a Command System*

The tradition of Austrian economics emphasises the role of markets as information systems in which prices enable reallocation by acting as signals (Hayek, 1945). Indeed this provided the argument by which that tradition sought to establish the superiority of the capitalist economic system over socialism. The focus of this paper is a little different: it deals with various ways in which the exchange of non–price signals can add value to transactions. Without this information the transactions that take place would either be of lower value or would not take place at all. I distinguish three ways in which information adds value to transactions: it supports producers' and consumers' reputations, permits customization of products, and provides yardsticks that are useful to all agents in comparative evaluation of products, producers, and investments. Reputation and customization depend on the diffusion of specific information, that is information about individual consumers and producers. Yardsticks on the other hand are enabled by "general knowledge", that is knowledge about the economy in various aggregate dimensions.

3.1. *Brands, Advertising, and Reputation*

In markets for goods and services where there are many brands and many sellers, advertising informs consumers of the availability and price of new products and reduces their costs of searching (Stigler, 1961). Where the quality of a product matters and consumers cannot easily tell good products from bad ones before purchase, producers can command a premium if they invest in a good reputation.[11] However, for the same reasons that Soviet producers did not need to know their markets, they had little reason to acquire reputation with the final users of their products. For related reasons producers did not advertise: Most Soviet consumers took what they could get without having to choose among brands or to seek out the lowest price. It was enough to be able to buy at all.

In a seller's market the information of most value to consumers was to know for certain that a given commodity would be available in given quantity at a given time and place. Knowledge of availability was more

[11] According to Stigler (196: 224), "Reputation commands a price [...] because it economizes on search". For elaboration, see Shapiro (1983).

important than knowledge of quality or even price. Market research that predicted availability would have served consumers, but was of no interest to producers. Without it, such market intelligence was circulated on the twin basis of rumour and privilege, and was traded within the "economy of favours" (Ledeneva, 1998).

It would be wrong to conclude that business reputation did not exist at all in the command system. Reputations were attached to both products and agents. Particular branded products were widely known for high quality: for example *Stolichnaia* vodka, Red October chocolate confectionery, the Bolshoi ballet, and MiG aircraft. This reputation was based primarily on consumers' experience, but the payoff came from reporting this experience within the vertical hierarchy. The market reputation brought no return; it was the reputation in the hierarchy that won rewards for the producers. These rewards were extremely varied. Organisational team benefits took the form of priority access to financial and material resources for production and the establishment of privileged retail, housing, and welfare clubs for employees. Individual benefits ranged from cash premia and state decorations awarded to leading individuals to the chance to travel abroad and retain some foreign earnings.

What factors formed the business reputation of an agent? Again reputation could be formed on both horizontal and vertical lines; however, whereas the market reputation of a product reinforced its reputation in the hierarchy, the reputation of an agent was formed in a more complicated way.

Agents' reputation with superiors was formed by both productivity and loyalty, but the weighting of these two factors was variable and the size of the return was uncertain as well. For example, a study of Soviet regional policy has shown that Stalin used investment allocations to reward loyal agents in the regions in his struggle with the opposition in the late 1920s; during the 1930s, however, his regional agents were called to account for their wasteful use of these resources (Harris, 1999). A detailed study of the Soviet allocation system for motor vehicles in the 1930s also shows that the dictator held a stock of vehicles in reserve for use as rewards for loyal agents (Lazarev and Gregory, 2003). In aviation research and development, in contrast, although Stalin had clear favourites and proposals required investments in lobbying to gain initial finance, a reputation for loyalty was no protection against deprivation of funding and sometimes also punishment in the event of productive failure (Harrison, 2003a; 2005).

Thus, it appears that under Stalin loyalty alone was not enough. Perhaps, however, the slowdown of elite circulation that followed the dictator's death permitted an increase in the importance of loyalty relative to that of productivity, with adverse consequences for the health of the command system.

In unauthorised horizontal contacts the ability to complete transactions depended on a business reputation for plain dealing and keeping one's word. Eugenia Belova has described the unauthorised "relational" contracting system that arose where contracts were costly to enforce or unenforceable by other means (Belova, 2005). Relational contracts typically rested on a handshake and were enabled by experience of personal contact and friendship; only these could overcome the culture of low trust (Gregory, 1990: 91–2).

An agent could exploit a good horizontal reputation to support a vertical reputation. Vertical reputation rested on fulfilling the plan, but the plan was highly aggregated and did not foresee many specific details of the inter–ministerial transactions necessary to achieve it. The system of inter–ministerial contracting that underpinned the aggregate plan was also highly incomplete. In order to supplement the plan, and eventually fulfil it, agents were frequently compelled to make unauthorised horizontal contacts and deals. Thus a horizontal reputation for honesty among equals could support a vertical reputation for serving superiors. This created a problem for principals, however, who could not easily distinguish those unofficial deals that agents made so as to fulfil the plan from those that helped agents to cheat the state and line their own pockets. Perhaps an agent's horizontal reputation could be "too good": to be well thought of by everyone could too easily be a cover for embezzlement or worse.

Similarly, a vertical reputation that was "too good" could damage horizontal reputation. Eugenia Belova finds evidence that a bad vertical reputation could signal a readiness to do unofficial business on horizontal lines. While relational contracting should have become more efficient as personal networks increased in scope, she argues, it was not in the interest of principals to allow such horizontal networks to expand without limit. As a result horizontal reputation could only accumulate privately or within small groups (Belova, 2005).

In market economies a further dimension of reputation is information about specific consumers. Consumer reputation is important primarily for consumer credit. Banks, loan societies, and credit card agencies prefer to lend only to individuals with desirable characteristics such as fixed places of residence, collateral assets, and good repayment records. In the Soviet command system these things were mostly unimportant. Without a freehold property market no one wanted to acquire a housing mortgage. Consumer debt did not arise either; in a shortage economy most consumers had plenty of liquidity in the form of either cash assets or waiting time.

Individuals' status as producers and their political reputation with superiors carried more weight than any financial credit rating. Another way of putting this would be to say that the currency in which people were rated for credit worthiness and were able to accumulate debts was measured in loyalty and favours. This "currency" was traded privately subject to special rules of reciprocity and was not convertible into rubles. The information on the basis of which this trade took place was correspondingly private and not available for dissemination to others.

3.2. *Customization*

In the early twentieth century, mass production brought the cost of consumer durables down to the point where they could become items of mass consumption. The result was a high degree of market penetration by relatively uniform, standardised products. Henry Ford said of his family automobiles: "You can have any colour you want, as long as it's black".[12] The same went for radio, telephone, and television sets. At lower incomes price tended naturally to be more important than variety, although when controlling for income consumers in different countries still differed somewhat in their willingness to trade off variety or quality (Broadberry, 1997: 77–89).

In market economies customised products are able to command a premium over standardised, mass–produced commodities. But customization implies a great increase in the exchange of information required to complete a transaction. In the era of mass production the consumer bought a uniform product in the combination of attributes determined by the producer. Today flexible production means that, by recombining

[12]This is what Henry Ford is said to have said in 1929, or in 1933, or at some other date, of the Model A, the Model T, or some other Ford vehicle, according to hundreds of internet references registered by www.google.com.

their attributes in different ways, suppliers can customise goods and services to the needs of individual purchasers without loss of economies of scale.

Mass customization, sometimes also called "mass personalisation," is enabled by the increased information that producers can obtain about consumers. In wealthy countries today big companies carry out market research, or employ specialised research agencies, to develop huge databases that store household– or individual–level data about characteristics and preferences. For more complex products consumers themselves provide the information about their preferences "just in time." For example a buyer ordering a computer over the internet specifies the processor make and speed, chip and disk memory, additional drives, modem capacity, display type, sound card, multi–media facilities, and so on; this information is then transmitted directly to the assembly line programme.

In the Soviet economy information about product prices and characteristics was not readily available in such detail. Indeed it may not even have been produced. For example the Soviet statistical archives appear to hold price and quantity information about every product ever produced in a state–owned enterprise, but even for the most complex and costly machinery this information does not appear to extend to more than two or three product characteristics.[13] By comparison the marketing and mail order catalogues available to most American consumers from the early twentieth century onward convey many times this quantity of information per product. The information provided was not only enough to sell the products but also to support subsequent studies of hedonic pricing of durable goods that have routinely used such catalogues for their primary data (Gordon, 1990: 417–89). In contrast the information held in the Soviet archives was collected specifically to support planning targets based on quality–adjusted pricing but did not in fact do so; on the contrary, it permitted concealed inflation (Harrison, 1998a; 2000a).

Similarly, Soviet producers did not need or carry out market research into consumer characteristics. Mass production displaced the artisan system of producer–driven customization in the 1930s and 1940s (Harrison, 2000b). Thereafter, the Soviet economy remained wedded to mass production. Standardised clothing, household durables, and civilian and military machinery

[13]Rossiiskii Gosudarstvennyi Arkhiv Ekonomiki, *fond* 1562, contains the files of the USSR central statistical administration (TsSU SSSR) that have been declassified up to 1963.

supplied an undifferentiated market. The state procured and distributed output, breaking the link between supplier and final purchaser. The seller's market left producers with no incentive to become informed about the market, and no means of doing so because the state insulated them from market responses. The state monopoly of foreign trade cut producers off from the export market even more thoroughly than from the home market, and exporters were given no special incentives to tailor production to the requirements of foreign buyers (Holzman, 1974). Only the defense ministry had the power to enforce customization upon producers through its institutionalised presence in defense industry and power of veto in the procurement process (Harrison and Simonov, 2000).

In practice there were two ways in which Soviet products and services could command a real premium in return for variation of attributes. The first was provided by the command economy, which allocated an implicit premium to new and regraded products by pricing them favourably in relation to existing products. This was not a deliberate policy but arose as follows: plan prices were based on product costs so as not to reward cost cutting at the expense of product quality, but the authorities could not process complex information about product quality so as to distinguish cost increases associated with the customization of products to the market from those designed to attract a rent at the expense of society. Thus "simulated" innovation paid more than true innovation, and producers obtained rewards for product variations that increased costs rather than quality (Berliner, 1976: 375–80; Harrison, 1998a). The outcome was customization of products to the plan, not to the final consumer.

The other way in which products and services could command a return on varied attributes was through illegal trade and side payments. In these respects the Soviet economy returned the concept of customization to an artisan framework of semi–legal or illegal self–employment or unregulated small–scale trade. For example, in the market for commodities western imports were one source of customised products that could command a premium. To the consumer the main positive attribute of western clothing or household equipment was simply that these were different from the standardised Soviet article. The premium payable was reduced, however, because access to imported commodities depended on privilege more than purchasing power. Thus the possibility of buying denim jeans from tourists was restricted mainly to those with a Moscow or Leningrad residence permit. Similarly, personal services from housing maintenance to medical care were often customised to the individual consumer through bribery and side payments, but in the "economy of favours" being able to trade a privilege could count more than purchasing power.

3.3. *"General Knowledge" and Yardsticks*

In market economies specific transactions are enabled by combining different kinds of information. Information that is specific to the transaction ("What am I buying? How much does it cost? How much do I need it?") is essential but often specific information alone is not enough. To complete the transaction general knowledge is required as well. Are similar products available elsewhere for less? What other products are available? For durable goods time will also enter the equation: will the price fall? Will something substantially better come along in a year or two? Will needs change? Typically problems that involve time, such as investment choices or providing for retirement, are especially demanding of general knowledge. Producers must forecast aggregate trends as well as those specific to their own market. Individuals must predict their lifetime capacity to save and the lifetime return to saving in the context of macroeconomic and demographic trends.

In all these cases we use general knowledge of trends in the economy as a whole to provide yardsticks against which we can judge individual firms and products.[14] For example we may judge product prices against the retail price index, share performance against the Wall Street index, or fixed–interest financial products against the Bank of England's discount rate.

Yardsticks also play an essential role in informing policy choices and choices in political markets. We rank police forces by crime and detection rates and use this information to judge chief police officers. We judge health administrators by waiting lists and operation success rates in hospital league tables, teachers by pupils' exam performance in league tables of schools, and politicians by the economy's place in league tables of unemployment, inflation, real growth, and environmental quality. An

[14]Shleifer (1985) proposed "yardstick competition" as a means of regulating franchised monopolies. Franchised monopolies typically have little incentive to reduce costs. However, if the price that the regulated firm receives is made to depend on the costs of identical firms, then in equilibrium each firm will choose a socially efficient level of cost reduction. Yardstick competition is recognisable in a Soviet context as the idea behind "socialist emulation", with the difference that Soviet competition were aimed not at cutting costs but at increasing gross output.

external yardstick for the Soviet economy in this sense was sometimes the economic performance of the west, as when Stalin set the goal to "make good the distance we are lagging behind the advanced capitalist countries" or "to outstrip the principal capitalist countries economically".[15] Soviet leaders also benchmarked their own technologies against specific western products and processes; Stalin himself sometimes withheld support from new ideas until they had been tested by Western experience (Holloway, 1994: 147).

Such use of general knowledge can result in intense pressure on the independence and objectivity of those who produce it: business consortia, academic organisations, and government bureaucracies.[16] The pressure is facilitated in so far as each has a natural monopoly in the supply of information about themselves that cannot easily be checked. Therefore the structure of the market for statistics has an important influence on their quality. To assure this quality usually requires a combination of transparent sources and methods, regulation through audit, and constitutional guarantees of independence from operational concerns.

In the Soviet economy much information of a potentially "yardstick" character was not required because restrictions on agents' freedom of action would have prevented them from benefiting from it; even if required, such yardsticks were frequently not supplied, and remained locked in "comrade Starovskii's safes" or elsewhere. Such yardsticks, even if supplied, were often of unreliable quality. The quality of statistical information was usually unchecked: the government statistical service was allowed to exercise a monopoly over statistical work, and was itself a loyal instrument of the political class.[17] Whether or not they were reliable, the general public often found government statistics not to be credible, as in the saying "*Izvestiia* ('The News') isn't the truth (*ne pravda*) and *Pravda* ('The Truth') isn't news (*ne izvestiia*)". Interestingly, government statistics found credibility only within the closed world of the ruling circle; for example, the leaders

---

[15] "The Tasks of Business Executives" (4 February 1931) and "Report on the Work of the Central Committee" to the eighteenth party congress (10 March 1939), both reproduced in Stalin (1940), quoting from pages 367 and 634 respectively. For a survey and history of Soviet expert attempts to benchmark the Soviet economy against western yardsticks, see Kudrov (1997).

[16] In the terms of Arrow (1984: 142), yardstick information is indivisible and inappropriable. It is a public good by definition and this explains why it is not typically supplied on a private basis.

[17] Wheatcroft and Davies (1994). On planners and statisticians as "honest brokers" within the Stalinist dictatorship, see Belova and Gregory (2002).

relied on the official growth rates when no one else did. This may be seen as an important and interesting paradox, sometimes described in terms of an out-of-touch elite believing its own propaganda that the rest of the world could see as transparent lies.

Let us examine specifically two aspects of this, the changing propensity to conceal useful information, and the credibility gap between rulers and society. To begin with, consider the availability of information that would have usefully contributed to general knowledge, for example crime rates, rates of alcoholism, disease and death rates, harvest levels, the money supply, and real growth and inflation rates were all on the secret list at one time or another. Evidently, such concealment inhibited the establishment of yardsticks by which the performance of government and the economy could be evaluated; in this way it diminished accountability. It also resulted in two kinds of waste. First, in the absence of yardsticks some people made costly mistakes; for example, many Russians counted on the command system to provide them with pensions in retirement and saved less for their old age than they might have otherwise, or saved in the wrong form, in rubles that are now worthless; as a result they are now unable to retire, or live in poverty. Second, those for whom yardstick information mattered had to use resources unnecessarily in order to duplicate it, for example disseminating information about society by costly samizdat.[18]

Soviet yardsticks were not all concealed consistently. The propensity for suppression followed a protracted cycle from relative openness in the 1920s to the almost total statistical blackout of 1937 to 1956, followed by a return to greater openness in the 1960s and 1970s, always with qualifications and partial retreats until the flood of revelations associated with *glasnost'*.

Why were such facts revealed in some periods but not in others? There appear to have been three reasons. First, without providing general information from time to time the regime could not credibly claim responsibility for economic advances and thus hope to win the population's loyalty.[19] Thus, a regime choice that shifted the underpinnings of power away from repression towards loyalty could tip the balance in favour of a policy of revelation that would allow the establishment of performance

[18] Arrow (1984: 143), specifies a third cost of secrecy when secrets can be traded privately, the destruction of markets for sharing risks. In the Soviet case such markets were precluded anyway.

[19] For the underlying theoretical argument, see Wintrobe (1990: 857, 864).

yardsticks. Second, for the same reason at times of regime change it often suited a new leader to reveal the failings of the old ones in order to blame them and avoid carrying responsibility for their failures. Third, at times when the regime chose to seek wider involvement of the intelligentsia in solving economic and cultural problems greater openness became necessary in order to foster critical discussion and evaluation of existing policies. All three motives were evident in the spring of 1956 in Mikoian's public plea to "comrade Starovskii" to open his safes.

In one respect a command system that practised a degree of statistical openness might be regarded as providing better information than a market economy. The information that is needed to make optimal production and consumption decisions includes information about the future prices of commodities but in market economies such futures markets mostly do not exist. The command system did not provide information about future prices, but it did at least provide guidelines about future quantities in the shape of detailed plans for future national economic development.[20] In practice, however, this information was of limited value since plans were not implemented in any detail. In the words of Eugene Zaleski the Stalinist plan was a "vision of growth [...] The Soviet experiment shows that, in order to exert a real influence, the vision does not have to be very accurate" (Zaleski, 1971: 297).

In periods of greater openness a problem was that much of the information revealed was itself of very low quality and unreliable as a basis for making decisions. The very fact that Soviet statistics were designed to be used as success indicators led to their distortion in ways that have been well known for many years.[21] Some observers have attributed very large consequences to low-quality macroeconomic yardsticks in Soviet history. For example, Grigorii Khanin has argued that exaggerated claims to success made the Soviet regime complacent about economic stability, encouraged an official belief in the ability of society to shoulder fresh burdens, and inhibited necessary adaptations to changing economic requirements in the

[20] "Even as a graduate student", writes Arrow (1984: 160), "I was somewhat surprised at the emphasis on static allocative efficiency by market socialists, when the nonexistence of markets for future goods under capitalism seemed to me a much more obvious target".

[21] Nove (1958); Treml and Hardt (1972); Wheatcroft and Davies (1994). "Goodhart's law" (see Goodhart, 1984: 96) states: "any observed statistical regularity will tend to collapse once pressure is placed upon it for control purposes", in the Soviet case, for example, the relationship between observed and true real output.

1970s and 1980s (Khanin, 1991: 41–51). From the standpoint of official data, the worsening of Soviet economic conditions remained largely invisible; serious long-term problems were admitted, but no crisis. Meanwhile, unofficial experience told all too clearly of the sharply worsening quality and availability of goods and services. Soviet leaders' reliance on official yardsticks, which blanked out the true picture and shielded them from the necessity to act, explains the uncomprehending complacency. The upsurge of economic discontent, followed by a turn to terminal disintegration, simply took them by surprise.

Khanin has shown that at least some who participated in the making of policy understood the defective character of their information and tried to improve it (Khanin, 1991: 51–102). They were opposed, however, by strongly entrenched interests that were vested in distortion and concealment. With hindsight we see that the "errors" that resulted were a permanent feature of the Soviet command system: distorted statistics suited the policy biases that led to famine and demographic disaster in the 1930s, and to less dramatic but still costly losses in the post-war period.

Given this, why do we find that Soviet statistics commanded more credibility in the Kremlin than in the street? An answer may be found in the literature on signalling: diversity of preferences between sender and receiver may limit the information that can be credibly conveyed (Crawford and Sobel, 1982). Consider the possible range of preferences for high effort and high accumulation on one hand, versus low effort and consumption on the other. Stalin's brutal treatment of statisticians in the 1930s may be interpreted as a process of aligning their preferences with his, which were for high effort and accumulation. Once accomplished, this ensured that the statistics they produced were believed in the Kremlin. But the dictator's preferences diverged from those of society. This can explain why officially "honest brokers" were believed officially while seen as dishonest in unofficial circles. It also explains the sudden brutality with which Stalin reacted when for any reason he began to doubt their loyalty, which was essential to his mechanisms of power.[22]

[22]Stalin's treatment of planners and statisticians in the 1930s is discussed by Belova and Gregory (2002). In 1949, Stalin lost confidence in his formerly much favoured chief of the state planning commission N.A. Voznesenskii when the latter was alleged to have engaged in covering up a plan failure. This was the start of a process that ended in the latter's trial and execution; for relevant documents, see Khlevniuk *et al.* (2002: 274–85).

4. *Soviet Growth, Transition, and Information*

At the twentieth party congress in 1956, Mikoian claimed that the Soviet state possessed statistical data "in far larger measure than at any other time and in any other country". Can he have told the truth?

In a superficial sense, possibly, yes. Our growing acquaintance with the archives is showing us that the statistical resources of the Soviet state were very, very large. However, the scale of these resources should be discounted by two factors: first, the sheer size of the Soviet economy and, second, the fact that the Soviet state combined government with functions undertaken in market economies by the private sector. When this is done the balance of advantage becomes less clear. Moreover, most Soviet information that was collected was restricted to a narrow sphere of circulation, and most of it was unreliable. Many kinds of information were not collected at all or, if collected, were not disseminated. For most of the time most officials had little idea about the true value of the physical and human resources at their disposal, the true productivity of the producers that they commanded or the true degree of satisfaction of the firms and households that they supplied. In short, the Soviet command system forced economic growth on the basis of an information stock of relatively low quality and low value.

Soviet and Russian post-war economic growth displays four features that demand explanation: rapid growth from the 1920s through the 1950s; the slowdown that began in the 1960s and set in more strongly in the 1970s; the collapse at the end of the 1980s, and the persistence of a low level of real output in the 1990s and since then. The Soviet economy's restricted information capacity can contribute to explaining all four. At the same time, given the present state of our knowledge, its explanatory power looks low. This is because all four problems already have too many possible solutions.[23] It does not seem particularly useful to propose an extra solution to each problem when the data available do not allow us to discriminate among the solutions that have been advanced already. However, it may help to outline some possible avenues for further investigation.

Given low–value information stocks, how did the Soviet economy grow rapidly up to and through the 1950s? Soviet economic growth was based on the rapid expansion of output of standardised goods and services (but

[23]For a survey and investigation of competing explanations of Soviet economic growth and post-war slowdown, see Harrison (1998b); for various explanations of Soviet economic collapse and a proposed solution, see Harrison (2002).

mostly *goods*) with low information requirements. Let me emphasise: it is not part of my argument that this growth was an illusion. The Soviet economy did grow. It grew in real terms and by several times over several decades. On the best measures available, between the 1920s and the 1980s Soviet real incomes per head increased by a factor of approximately five times (Maddison, 1995). Real consumption grew by less, however, and the extra welfare gained from the growing supply of consumer goods and services was clearly lessened by shortages and other restrictions on variety and choice, and by social and intertemporal inequalities. Moreover, although the level of the information stock may have been low relative to real output, it was presumably possible to widen this stock as output grew so that information shortages did not constrain growth. The outcome was that by the 1980s the Soviet Union had achieved an income level many times higher than could have been predicted from its endowment with information handling equipment measured by computing power.[24]

New research on long-run trends in the organisation of market economies suggests that the conditions under which the Soviet economy could grow like this were temporary. The century from 1870 to 1970 was characterised by two special conditions in which large hierarchical organisations could flourish: industry was providing a rising scale of production, while the modern office had reduced costs of information and monitoring to a level that was, as yet, "neither prohibitive nor trivial". Since then, however, further declines in information costs have favoured a switch away from hierarchies back to horizontally organised networks based on trust, reputation, and customised production.[25] In short, the Soviet model of rapid industrialisation that relied on standardisation and sparse information was favoured by special conditions of the last century that no longer exist.

The character of Soviet information stocks may help explain Soviet postwar productivity slowdown. The evidence of the Soviet real growth series is that there was a sharp deceleration in the mid-1970s that cannot be explained by any exogenous shock (Harrison 1998b). This was just the

[24]See Iacopetta (2001) for this finding and a new investigation and analysis of the Soviet postwar failure to adopt new information technologies. Probably the gap that Iacopetta found between incomes and computing power was eliminated in the 1990s by the collapse of Russian incomes. The classic general investigation of the poor Soviet innovation record is Berliner (1976).

[25]Broadberry and Ghosal (2002), and Lamoreaux *et al.* (2003) from which the quoted words are taken. I thank Stephen Broadberry for many discussions on these issues.

time when in other countries information costs began to fall much more rapidly and at the same time demand shifted more decisively towards more information–intensive products and services. Globally perhaps the return to information was increasing relative to other resources such as labor and other kinds of capital where returns were falling. Suppose that Soviet information channels could not be deepened to supply the information that would have maintained the returns to other factors in the Soviet economy. Then, a slowdown of aggregate and productivity growth was the inevitable result.

Consider more specifically how information relates to knowledge. Knowledge in the sense of "how to" and "can do" is useless without information. Decade after decade the Soviet education system increased the numbers of knowledgeable people whose desire for information was perhaps increasingly frustrated.[26] As a result the returns to investments in human capital fell or proved unexpectedly low.

The low value of Soviet information stocks may also add to our understanding of how the Soviet command system eventually collapsed. According to Mancur Olson a dictator of the stationary–bandit type will invest in public goods such as protection of the physical and human assets under his control up to the point where his share in their return to society equals the cost to him of doing so (Olson, 1993). Public information is a public good but if information is costly and the return on publication accrues mainly to society the dictator will keep it to himself or not provide it. The dictator administers his assets through agents. Each agent will stay loyal to the dictator provided his share in the dictator's expected rents from the assets he administers exceeds the expected value of the asset if he stole it. Among these assets must be counted the dictator's secrets. These secrets had a value to the dictator if kept, and a value in the economic or political market place if disclosed. One aspect of the process that began with *glasnost'* and ended in the dismantling of the Soviet state was the moment when agents began to realise that the market value of state secrets long preserved under their control exceeded their value if left in concealment. At

[26]The idea that the spread of education would eventually force a more open society in the USSR is not new; see for example Deutscher (1967: 59–60): "The force of the revolutionary tradition has been great enough to compel the bureaucracy to give the workers much more education than has been required on narrow economic grounds, and perhaps more than is safe for the privileged groups. It may be argued that the bureaucracy is thus breeding its own grave–diggers".

this point stealing information became one more dimension of what Steven Solnick has called "stealing the state" (Solnick, 1998).

Finally, the inheritance of a low–value information stock may help to explain the persistence of a low level of output after the transition to a market system. The important role of intangible "social capital" in long run economic development has long been accepted by economic historians (Abramowitz, 1986). More recently the poor economic performance of the former Soviet republics since the collapse of the command system has been attributed to low social capital in the form of an institutional quality deficit (World Bank, 1996). Earlier optimism about the growth prospects of transitional economies is necessarily tempered when this deficit is fully taken into account (Crafts and Kaiser, 2002). The command system impeded the accumulation of social capital but could force economic growth by relying on coercion instead. Now that the command system has gone, a market economy will not prosper in Russia until the necessary accumulation of social capital has taken place, a process that could take decades or centuries.

Social capital is conventionally measured along several dimensions: education and knowledge, the rule of law and property rights, a civil society, and so on. What makes these things "social" rather than private in nature is that their benefits spill over the narrow limits of private profit and loss. In this sense information stocks are also a part of social capital. The Russian economy today is suffering from a shortage of the information stocks that make transactions valuable and make markets work. Accumulating the missing information stocks will take time, and will not even begin without progress in the complementary dimensions of social capital that are already recognised in the literature.

5. Conclusions

The economics of information analyses some of the ways in which information adds value to transactions in market economies. While markets may thrive on information, hierarchies may choke on it. In the Soviet command economy valuable information was frequently not produced; if produced, it was often concealed; whether concealed or not, it was often of poor quality; regardless of quality, it often suffered from low credibility outside the ruling circle. In short, the Soviet command system forced economic growth on the basis of a relatively low–value information stock. This may help explain aspects of Soviet postwar economic growth and slowdown, the collapse of the command system, and the persistence of low output since the collapse. At the moment, however, such suggestions are no more than speculations because we have virtually no empirical measures of trends in the quantity or value of the command system's information stocks, flows, or transfers.

References

Abramovitz, M (1986). Catching Up, Forging Ahead, and Falling Behind. *Journal of Economic History*, 46(2): 385–406.

Arrow, KJ (1984). *Collected Papers of Kenneth J. Arrow, Volume 4: The Economics of Information.* Oxford: Blackwell.

Barber, J, Harrison, M, Simonov, N and Starkov, B (2000). The Structure and Development of the Soviet Defence–Industry Complex. In: Barber J and Harrison M (eds.) *The Soviet Defence–Industry Complex from Stalin to Khrushchev.* London and Basingstoke: Macmillan, pp. 3–32.

Bell, HE (1936). The Price of Books in Medieval England. *Library (1936) s4-XVII*, (3): 312–332.

Belova, E (2005). Legal Contract Enforcement in the Soviet Economy. *Comparative Economic Studies*, 47(2): 387–401.

Belova, E and Paul, RG (2002). Dictators, Loyal and Opportunistic Agents: the Soviet Archives on Creating the Soviet Economic System. *Public Choice*, 113(3–4): 265–286.

Berliner, JS (1976). *The Innovation Decision in Soviet Industry.* Cambridge, MA: MIT Press.

Broadberry, SN (1997). *The Productivity Race: British Manufacturing in International Perspective, 1850–1990.* Cambridge: Cambridge University Press.

Broadberry, SN and Sayantan, G (2002). From the Counting House to the Modern Office: Explaining Anglo-American Productivity Differences in Services, 1870 1990. *Journal of Economic History*, 62(4): 967–998.

Cipolla, CM (1956). *Money, Prices, and Civilization in the Mediterranean World: Fifth to Seventeenth Century*, Princeton NJ: Princeton University Press.

Crafts, N and Kaiser, K (2002). Long-Term Growth Prospects in Transition Economies: A Reappraisal. *Structural Change and Economic Dynamics*, 15(1): 101–118.

Crawford, VP and Sobel, J (1982). Strategic Information Transmission. *Econometrica*, 50(6): 1431–51.

David, PA and Foray, D (2001). An Introduction to the Economy of the Knowledge Society. *International Social Science Journal*, 54(171): 9–23.

Deutscher, I (1967). *The Unfinished Revolution: Russia, 1917–1967*, Oxford: Oxford University Press.

Deutscher, I (1967). *Monetary Theory and Practice: The UK Experience*, London: Macmillan.

Gregory, PR (1990). *Restructuring the Soviet Economic Bureaucracy*. Cambridge: Cambridge University Press.
Gordon, RJ (1990). *The Measurement of Durable Goods Prices*. Chicago: University of Chicago Press.
Harris, JR (1999). *The Great Urals: Regionalism and the Evolution of the Soviet System*. Ithaca, NY, and London: Cornell University Press.
Harrison, M (1994). GDPs of the USSR and Eastern Europe: Towards an Interwar Comparison. *Europe–Asia Studies*, 46: 243–59.
Harrison, M (1994). Prices, Planners, and Producers: an Agency Problem in Soviet Industry, 1928–1950. *Journal of Economic History*, 58(4): 1032–62.
Harrison, M (1998b). Trends in Soviet Labour Productivity, 1928–1985: War, Postwar Recovery, and Slowdown. *European Review of Economic History*, 2(2): 171–200.
Harrison, M (2000a). Soviet Industrial Production, 1928 to 1955: Real Growth and Hidden Inflation. *Journal of Comparative Economics*, 28(1): 134–55.
Harrison, M (2000b). Wartime Mobilisation: A German Comparison. In: Harrison M and Barber J (eds.) *The Soviet Defence Industry Complex from Stalin to Khrushchev*. London and Basingstoke: Macmillan, pp. 99–117.
Harrison, M (2001). Providing for Defense. In: Gregory PR (ed.) *Behind the Façade of Stalin's Command Economy*. Stanford, CA: Hoover Institution Press, pp. 81–110.
Harrison, M (2002). Coercion, Compliance, and the Collapse of the Soviet Command Economy. *Economic History Review*, 55(3): 397–433.
Harrison, M (2003a). The Political Economy of a Soviet Military R&D Failure: Steam Power for Aviation, 1932 to 1939. *Journal of Economic History*, 63(1): 178–212.
Harrison, M (2003b). Soviet Industry and the Red Army under Stalin: A Military-Industrial Complex? *Les Cahiers du Monde russe*, 44(2-3): 323–42.
Harrison, M (2005). A Soviet Quasi-Market for Inventions: Jet Propulsion, 1932 to 1946. *Research in Economic History*, 23: 1–59.
Harrison, M and Simonov N (2000). Voenpriemka: Prices, Costs, and Quality in Defence Industry. In: Harrison M and Barber J (eds.) *Soviet Defence Industry Complex from Stalin to Khrushchev*. London and Basingstoke: Macmillan, pp. 223–45.
Hayek, FA (1945). The Use of Knowledge in Society. *American Economic Review*, 35(4): 519–30.
Hayek, FA (1945). *Stalin and the Bomb: the Soviet Union and Atomic Energy, 1939–1956*. New Haven, CT: Yale University Press.
Hayek, FA (1945). *Foreign Trade under Central Planning*. Cambridge, MA: Harvard University Press.
Hayek, FA (1945). Dissemination of Technology in Market and Planned Economies. *The B.E. Journal of Macroeconomics*, 4(1): 1–32.
International Monetary Fund (1997). *World Economic Outlook*. Washington DC: IMF.

Katz, ML and Shapiro C (1985). Network Externalities, Competition, and Compatibility. *American Economic Review*, 75(3): 424–40.

Khanin, GI (1991). *Dinamika ekonomicheskogo razvitiia SSSR*, Novosibirsk: Nauka.

Khlevniuk, OV, Kvashonkin, AV, Kosheleva, LP and Rogovaia, LA (eds.) (1995). *Stalinskoe Politbiuro v 30–e gody. Sbornik dokumentov.* Moscow: AIRO–XX.

Khlevniuk, OV, Gorlizki Y, Kosheleva, LP, Miniuk, AI, Prozumenshchikov, MI, Rogovaia, LA and Somonova, SV (eds.) (2002). *Politbiuro TsK VPK(b) i Sovet Ministrov SSSR. 1945–1953.* Moscow: ROSSPen.

Kudrov, VM (1997). *Sovetskaia ekonomika v retrospektive. Opyt pereosmysleniia*, Moscow: Nauka.

Lamoreaux, Naomi R., Daniel M. G. Raff, and Peter Temin (2003), "Beyond Markets and Hierarchies: Towards a New Synthesis of American Business History," American Historical Review 108: 404–33.

Lamoreaux, NR, Daniel, MGR and Temin, P (2003). A Note on the Impact of Hedonics and Computers on Real GDP, *The Free Library* (December 1), http://www.thefreelibrary.com/A Note on the Impact of Hedonics and Computers on Real GDP.-a069662305.

Lazarev, V and Paul, RG (2003). Commissars and Cars: A Case Study in the Political Economy of Dictatorship. *Journal of Comparative Economics*, 31(1): 1–19.

Ledeneva, AV (1998). *Russia's Economy of Favours: Blat, Networking and Informal Exchange*. Cambridge: Cambridge University Press.

Maddison, A (1995). *Monitoring the World Economy, 1820–1992.* Paris: OECD.

Madrick, J (2001). *The Business Media and the New Economy.* Harvard University, John F. Kennedy School of Government, Joan Shorenstein Centre Research Paper R–24.

Madrick, J (2001). The Problem of 'Success Indicators' in Soviet Industry. *Economica*, 25(97): 1–13.

Madrick, J (2001). Dictatorship, Democracy, and Development. *American Political Science Review*, 87(3): 567–76.

Mikoian, AI (1956). *Speech by A. I. Mikoyan at the 20th Congress of the Communist Party of the Soviet Union, February 16, 1956.* London: Soviet News.

Shapiro, C (1983). Premiums for High Quality Products as Returns to Reputations. *Quarterly Journal of Economics*, 98(4): 659–80.

Shleifer, A (1985). A Theory of Yardstick Competition. *RAND Journal of Economics*, 16(3): 319–27.

Solnick, SL (1998). *Stealing the State: Control and Collapse in Soviet Institutions.* Cambridge, MA: Harvard University Press.

Stalin, J (1940). *Leninism*, London: Lawrence and Wishart.

Stigler, GJ (1961). The Economics of Information. *Journal of Political Economy*, 69(3): 213–225.

Treml, VG and John PH (eds.) (1972). *Soviet Economic Statistics.* Durham, NC: Duke University Press.

Wheatcroft, SG and Davies RW (1994). The Crooked Mirror of Soviet Economics Statistics. In: Davies RW, Harrison M, and Wheatcroft SG (eds.) *The Economic Transformation of the Soviet Union, 1913–1945*. Cambridge: Cambridge University Press, pp. 24–37.

Wintrobe, R (1990). The Tinpot and the Totalitarian: an Economic Theory of Dictatorship. *American Political Science Review*, 84(3): 849–72.

Wintrobe, R (1998). *The Political Economy of Dictatorship*. Cambridge: Cambridge University Press.

World Bank (1996). *From Plan to Market: World Development Report, 1996*. Washington DC: The World Bank.

World Bank (2001). *World Development Indicators*. Washington, DC: The World Bank.

Zaleski, E (1971). *Planning for Economic Growth in the Soviet Union, 1918–1932*. Chapel Hill, NC: University of North Carolina Press.

Chapter 12

Coercion, Compliance, and the Collapse of the Soviet Command Economy

Mark Harrison

The Soviet economy began to collapse in 1990.[1] As Figure 1 suggests, the suddenness with which it did so can scarcely be overstated. After rapid, but turbulent, economic growth under Stalin the post-war years saw nearly half a century of rarely interrupted growth. Then, Soviet real incomes fell by one third in four years. After that they fell more slowly, and during most of the 1990s incomes in Russia remained two fifths below the peak of 1989. In the middle years of the decade two fifths of the population of the Russian Federation, nearly 60 million people, lived below the official poverty line compared with two percent in the late 1980s (Ellman, 2000a: 126).

Existing explanations of this collapse are not entirely convincing. Among them are the propositions that the Soviet economy collapsed because it was fundamentally unstable, or alternatively because it was destabilised by one or another from a range of possible contingent factors, for example a growing gap between consumer aspirations and resources,

This chapter was first published in the *Economic History Review*, 55(3): 397–433. © The Economic History Society 2002.

[1]Earlier drafts of this paper were presented to seminars at the University of Warwick and the London School of Economics, to the second Oxford–Houston conference on "Initial Conditions and Russia's Transitional Economy", University of Houston (April 2001), and in nontechnical versions as an inaugural lecture at the University of Warwick, to the Soviet Industrialisation Project Seminar of the University of Birmingham, and to the Centre for Economic History Seminar of Moscow State University. I thank all the participants for their comments, especially Leonid Borodkin, Stephen Broadberry, Nick Crafts, Bob Davies, Nick Feltovich, Don Filtzer, Peter Law, Valery Lazarev, and Stephen White. Thanks also to Michael Ellman, Gregory Grossman, Philip Hanson, Byung–Yeon Kim, Grigorii Khanin, Vladimir Kontorovich, Kalin Nikolov, Robert Skidelsky, and the anonymous referees for advice.

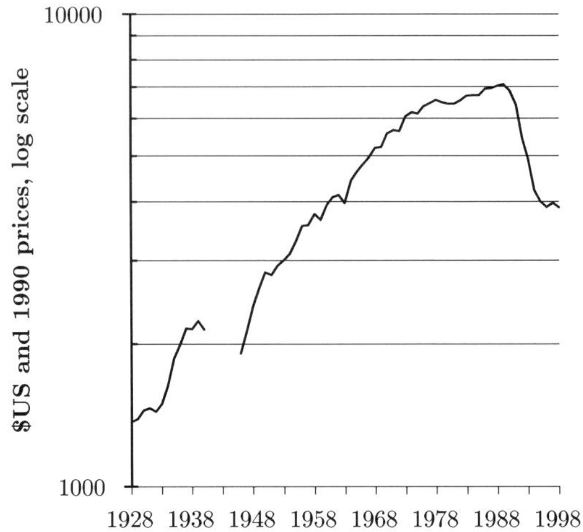

Fig. 1. Soviet and Former Soviet Real GDP per Head, 1928 to 1998.
Source: Maddison (1995: 200; 2001: 278).

the spread of rent-seeking and corruption, a loss of fiscal control, a loss of inter-industry coordination, and the "Gorbachev factor."

A widespread and influential view is that *the Soviet economy collapsed because it was predestined to do so.* "Essentialists" argue that Soviet society was fundamentally abnormal: stability requires normality, and normality requires consent, but the Soviet reliance on repression crowded out consent.[2] They maintain that the essence of the Soviet system made its eventual collapse inevitable and predictable; some of them claim to have predicted it.[3]

[2]For examples, see McNeill (1998), Rutland (1998), Brzeski (1999), Malia (1999), Pipes (1999); "essentialists" were first so called by Dallin (1992).

[3]Pryce-Jones (1995: 19–21), and Rutland (1998), have proposed alternative lists of scholars who predicted the collapse. Few of them were economists: Pryce-Jones mentions only one, the late Peter Wiles, and Rutland none. As an economist, Birman has a good case for inclusion; see Birman (1980) and Ticktin (1998). Khanin (1991) argued that the roots of the Soviet economy's ongoing self-destruction lay buried in past decades, and warned also that a sudden liberalisation would end in catastrophe; in Khanin (1992), he correctly predicted a collapse more severe and more permanent than the Great Depression of 1929–32, but blamed it on atrocious policy mistakes rather than systemic faults. Schroeder (1995: 223–4) lists a number of western economists who insisted over the years on the necessity of political and ownership change for significant improvement in Soviet economic performance but these were prescriptions rather than predictions. At the time most western economists would have agreed with Millar (1987: 183): "It would be wise to discount predictions of imminent Soviet economic collapse".

The essentialist argument appears strong. The Soviet system was repressive, and it did collapse. Yet the link from repression to collapse is usually asserted without specifics, and the mechanism is hard to substantiate. The contemporaneous evidence is that Soviet repression did not crowd out consent; on the contrary, there is much evidence of popular support for post-war Soviet institutions. Overt opposition was limited to intellectual dissent and emigration. The dissident movement's narrow social base suggests that in the 1970s most people were not interested (Churchward, 1975; Lane, 1976). Surveys of the Brezhnev-era emigration found that even émigrés remained loyal to a number of basic Soviet values (Silver, 1987). The Gorbachev era provides more substantial evidence of prevailing attitudes and aspirations. Surveys show that most people chose the extent of their participation in state and party institutions; the more they participated, the more influential they felt over outcomes (Bahry and Silver, 1990). They saw themselves as having more freedoms, with less censorship and less need for self-censorship, than many Americans and most black Americans (Gibson, 1993). While significant majorities favoured *perestroika* and a market economy in principle, most continued to support state ownership of heavy industry and state guarantees of basic incomes and jobs; they did not want such practical outcomes of a market economy as free prices, unemployment, or rich people (White, 1990; Finifter and Mickiewicz, 1992; Bahry, 1993; Fleron, 1996; Blanchflower and Freeman, 1997).

The evidence of the period refutes the idea that the Soviet economy was waiting to collapse. Time series for 1928 to 1987 show that Soviet productivity was growing.[4] It rose along a trend that was stable: the economy returned to it when subjected to a disturbance, although the latter were frequent and sometimes substantial. Trend growth became significantly slower in the mid-1970s, but remained positive. The welfare gains realised were large: between 1928 and 1987 GDP per head rose by a factor of five. Real consumption grew by less, and the extra welfare gained from the growing supply of consumer goods and services was lessened by shortages and other restrictions on variety and choice, and by social and intertemporal inequalities. Relative incomes and levels of job satisfaction and general happiness remained low by western standards (Millar and Clayton, 1987; Blanchflower and Freeman, 1997). Returns to Soviet accumulation were diminishing, and diminished more sharply

[4]On Soviet growth, see Ofer (1987), Bergson (1989), Easterly and Fischer (1995), Harrison (1998a), and Maddison (2001). On measurement, see Harrison (2002).

than they should have, especially after the mid-1970s, but they remained positive.[5] The Soviet economy was slowly catching up with the advanced capitalist countries until the mid-1970s, although it was still far from overtaking them. After that, its underlying growth declined further and became too slow to enable the Soviet economy ever to overtake its rivals, but growth did not fall to zero.

Thus, whether or not it was in fact unstable, the Soviet economy grew without any sign of instability for many years. Indeed, one needs to go back only to World War II to find the Soviet economy displaying much greater resilience under pressure than several more developed market economies.

A popular alternative view is that the postwar Soviet system, while not unstable at first, eventually fell victim to *a growing gap between consumer aspirations and resources*. It is often suggested that pressure on living standards from the heavy Soviet defense burden worsened this imbalance, so that military-economic rivalry with the United States added to instability. But it is hard to see such factors as decisive on their own. If the existence of a gap between households' disposable incomes and consumer aspirations was a sufficient condition for destabilisation, few economies would be left standing. And the evidence that the heavy defense burden contributed to a widening gap by damaging the growth rather than the level of living standards is not impressive (Easterly and Fischer, 1995: 348, 361). Indeed it is sometimes overlooked that the Soviet defense industry was itself an important source of growth.

Another popular conjecture is that the Soviet economy eventually collapsed because of the spread of *rent-seeking* and *corruption*. This suggests that the Soviet system failed not because of its repressive nature but because repression failed: that is, the Soviet regime could not stop special interest groups diverting effort from production to lobbying and redistribution (Anderson and Boettke, 1993; Olson, 1993; Wintrobe, 1998: 228–232; Grossman, 1998; Solnick, 1998). However, economic models of rent-seeking and corruption typically explain only poor performance and slow growth, not collapse. Exceptionally, Murphy *et al.*, (1993) showed that an economy that starts in a "good" equilibrium of high output and low rent-seeking may

[5]The causes of post-war slowdown in Soviet growth are still debated. Explanations, not all mutually exclusive, include diminishing technical progress (Bergson, 1978), lack of scope for capital-labor substitution (Weitzman, 1970), and Easterly and Fischer, 1995), the end of post-war recovery (Harrison, 1998a), falling plan tension (Schroeder, 1985), and rising misallocation (Allen, 2001).

slide to a "bad" equilibrium in which rent-seekers take over and output collapses; this happens if there is an adverse supply shock and property rights are poorly defended.[6] If so, it is not clear why an apparently stable regime of the Soviet type should suddenly have ceased to defend state property with such catastrophic results.

An explanation that is similar in the sense that it may make a contribution but is hardly complete in itself is the evident *loss of fiscal control* after 1985 (CIA Directorate of Intelligence, 1988; Ofer, 1989). In the second half of the 1980s, the budget deficit widened and almost all of it was monetised, leading to an increase in the monetary overhang (Kim, 1999). In consequence, the real gains to worker households from the nominal wage rises of the period resulted mainly in lengthening queues. A significant problem was accumulating. But since the authorities largely created this problem for themselves, it is not clear why they did so. In any case, the implosion of the real economy did not begin until 1990.

Some explanations for transitional recession have been based on *restrictive fiscal and monetary policies*, but these do not apply to Russia. Jan Winiecki and János Kornai have attributed output decline in central and east European transitions to demand restriction (Winiecki, 1993; Kornai, 1994). In the Russian case, when real output fell 40 per cent below the previous peak despite persistent budget deficits, monetary growth, and price inflation, the binding constraint was supply, not demand (World Bank, 1992: 13). Also placing the accent on budgetary policy, Olivier Blanchard has offered a model of output decline involving a supply shock that begins with the removal of subsidies from the state sector (Blanchard, 1997: 26–35). While it is perfectly plausible that removing subsidies adversely effected output, it is difficult to apply to the Russian collapse, which began in 1990 well before subsidies were first withdrawn (Kim, 2002).

A separate aspect of Blanchard's analysis that has been widely accepted as a factor in Soviet economic collapse is the sudden *loss of inter-industry coordination*; Blanchard ascribes it to a "CMEA shock" but there was a similar shock associated with the national independence struggles of the former Soviet republics that pulled apart the Soviet trade area. In fact, the sectoral and regional patterns of decline provide little support for this idea. If trade disintegration were the key factor, one would expect output decline to have been greater in the smaller, more trade-dependent states than in

[6]Solnick (1998) studies Soviet state collapse as a slide into a "bad" equilibrium.

the relatively self-sufficient Russian republic, but the evidence for it is very weak. [7] In Russia itself one would expect that output decline was greater in the industry sectors with more highly fabricated final products, but the evidence is weaker still.[8]

Beforehand, many practitioners of Sovietological economics and political science, including the present writer, were sceptical of prophecies of imminent Soviet economic collapse. After the event they tended to place much weight on *the Gorbachev factor*: the Soviet economy was murdered by caprice, not run down by a deterministic trend.[9] Vladimir Kontorovich has written: "We tend to confer the mantle of inevitability on accomplished facts, and arguing that what happened did not have to happen is likely to be dismissed as inventing excuses for the losing side. But the collapse of the Soviet system was the unintended result of a small number of disastrous decisions by a few individuals" (Kontorovich, 1993: 44; Brown, 1997; Dallin, 1992; Ellman and Kontorovich, 1992; 1998; Treml and Ellman, 1993; Khanin, 1991; 1992; Becker, 1994; Schroeder, 1995).

Acceptance of a certain role for particular individuals and policies, however plausible, does not reveal what would have happened without them. At one extreme Kontorovich has asserted: "Had Andropov lived longer or had Gorbachev turned out to be less self-assured, the Soviet system might still be with us". More modestly, Alexander Dallin considered that, "had Gorbachev and his associates *not* come to power, the Soviet Union would have hobbled along, and might have continued to muddle through without overt instability" (Kontorovich, 2001: 676; Dallin, 1992: 296). Archie Brown concluded only: "There was nothing inevitable either about *the timing of the end* of the Soviet state or about *the way in which*,

[7]The change in real GDP per head in various years compared with 1990 can be regressed on the log of 1990 population across 15 former Soviet republics; data are from Maddison (2001: 340–1). The estimated slope coefficient for each year after 1990, while always positive as suggested by the Blanchard hypothesis, fails a 10 per cent significance test over every interval except that which extends to 1992.

[8]The change in real output compared with 1990 across 17 industry branches in the Russian Federation (data from Goskomstat Rossii (2000: 302)), can be regressed on a dummy variable of "fabricatedness" that I set to 0 for extractive sectors (oil, gas, and coal mining, ferrous and nonferrous metallurgy, the timber and paper industry, and the industry for construction materials), 1 for processing sectors (electricity generation, oil refining, chemicals and petrochemicals, and six branches of the light and food industries), and 2 for engineering and metalworking. Again the estimated slope coefficient for each year after 1990, while consistently negative as expected, fails a 10 per cent significance test over every interval for which figures are available.

[9]*The Gorbachev Factor* is the title of the well-known book by Brown (1997).

under Gorbachev's leadership, the system was transformed" (Brown, 1997: 317, with added emphasis). Additionally, these views do not explain what features of the Soviet system made it so vulnerable to the unintended consequences of the actions of a few and why these consequences, if unintended, were not reversed.

Why did the Soviet command economy collapse? Are command economies intrinsically unstable? I will argue that stability of a command system is conditional; I will seek to identify some general conditions that may demarcate the command economy's "good" and "bad" states, and some particular circumstances that may have pushed Soviet institutions from one to the other. Part I considers the nature of economic coercion in general and the Soviet command system in particular, and identifies some costs of coercion. Part II sets the command system in an historical context of twentieth-century trends in these costs. Part III defines the players in the command system as self-interested producers and a dictator and sets out a game between the players. Part IV explores the long-run properties of the game. Part V shows that trends in variables and the actions of the players may cause a command system to collapse in various ways. In part VI, this model leads us to a narrative of Soviet economic decline and collapse that is both logical and consistent with known facts. Part VII considers the inevitability, reversibility, and welfare implications of the collapse and suggests some limitations of the approach followed. Part VIII concludes.

I

In this section, I consider the nature of economic coercion in general and the Soviet command system in particular, and identify some costs of coercion. I take coercion as the core relationship of a command economy. The dictator uses it to prevent the population from working for anyone else; he becomes a monopolist of capital and a monopsonist of labor. I distinguish economic coercion from repression as follows: economic coercion directs the labor of producers and the choices of consumers and punishes disloyalty to the dictator's economic interests, whereas repression punishes political disloyalty (Wintrobe, 1990). Starting from this, I will develop a model that allows coercion to be set at different levels, analyses the costs and benefits, and considers the conditions for a solution in which output is high.

Coercion brings a return that diminishes. Holland Hunter argued that developing economies may gain from a degree of tension in economic planning (Hunter, 1961). With moderate tension, a command economy mobilises resources and grows as a result. As tension rises, returns to coercion diminish because of growing disproportions and errors. Eventually, violence rises to the point where growth declines. Therefore, coercion also has an optimum. In this, it is like repression: according to Ronald Wintrobe, political loyalty to a dictator increases in repression up to a point as repression reduces the value of "disloyal" investments but beyond that point, loyalty declines again because of the rising probability that even behaviour that is loyal by intention will be repressed (Wintrobe, 1990).

Effective coercion requires a willingness to comply, and securing compliance can be costly. Mills and Rockoff studied the regulation of food supplies in wartime Britain and the USA (Mills and Rockoff, 1987). They found that compliance was positively associated with the resources invested in coercion. The British enforced food regulations more strictly, detected small infractions more frequently and prosecuted them. As a result there was less free-riding, and food restrictions were widely accepted as equitable. In the United States, monitoring and enforcement were starved of resources and fell below the level that would have secured consent. With rule-breaking endemic and unpunished, the system decayed. Machiavelli would have agreed that, given the will to sustain the costs, coercion can build consent and does not crowd it out. He considered "why all armed prophets have conquered, and unarmed prophets have come to grief". He proposed that "the populace is by nature fickle; it is easy to persuade them of something, but difficult to confirm them in that persuasion. Therefore one must urgently arrange matters so that when they no longer believe they can be made to believe by force" (Machiavelli, 1961).

In short, coercion brings a return that varies with its intensity. There is a level that is "just right", but coercion may also be too much and too little. Coercion mobilises economic resources, just as repression mobilises political assets. Its effectiveness relies on the resources invested in monitoring and incentives.

Soviet coercion arose in the context of a command system: production took place in state-owned enterprises regulated by a compulsory plan. By exercising coercion, the dictator established a near monopoly over capital and monopsony over labor, and was able to mobilise output. He returned part of the social product to producers for their subsistence on a line written into early Soviet constitutions but first recommended by St Paul: "he that does not work, neither shall he eat". Being far-sighted, the dictator aimed to maximise the long-run surplus or rent that he retained after covering his costs and this led him to invest a considerable part of his current surplus in defending and developing the territory under his control (Olson, 1993). Other costs were the resources he invested in monitoring and incentives. By varying the latter the dictator controlled the degree of coercion.

Think of the level of output as dependent on producers' effort in a deterministic way. Producers worked for the dictator, who gave them access to a basic income and extra incentives conditional upon the monitoring system. For the latter producers had to supply effort and this supply was what they controlled. The dictator could observe their output if he paid for monitoring. He could not observe producers' effort directly but while output depended observably on effort, this did not matter since he could observe output.

In the Soviet command system, who were the producers and who was the dictator? The Soviet command system comprised several multi-level, parallel hierarchies with overlapping spheres of responsibility, and this creates significant problems of definition. In theorising, I will define the dictator and the producer by their roles: producers controlled effort, while the dictator controlled coercion. The dictator was first and foremost an individual — Stalin, Khrushchev and so on. The dictator was an economic principal who stood above the law and whose word was law. Thus planning under dictatorship was not "law-governed"; instead, "the plan *is* the law". Nonetheless, the dictator necessarily delegated some powers to a favoured subset of the party's Politburo; for the 1930s Eugenia Belova and Paul Gregory call this group "Team Stalin".[10] In all periods, the dictator also required a wider but still numerically small group of economic coordinators and monitors: the central planning staff of Gosplan. But so far as has been learnt from the archives, these acted as the dictator's loyal agents, implementing his decisions and reporting truthfully to him on the outcomes as they saw them (Gregory, 2001; Belova and Gregory, 2002). Thus, the dictator gathered around him a number of individuals bound to his regime by a shared and encompassing interest.

Some scholars have argued that the Soviet economy's principal was not the individual dictator but a collective entity, the *nomenklatura* or party-approved list of names available for appointment to privileged state posts (Voslensky, 1984; Gershenson and Grossman, 2000; Lazarev, 2001). In this

[10]The "fractional" Politburo revealed in 1956 by Khrushchev (1976: 76–77), originated in 1926–27 and was formalised in 1937 (Khlevniuk, 1996: 48, 237–8); on its post-war forms see Gorlizki (2001). On "Team Stalin", see Belova and Gregory (2002).

view the Soviet leader was just the self-interested, sometimes badly behaved agent of the nomenklatura. Historical research on the nomenklatura is still in its infancy, but early results suggest strongly that it was far too large, too poorly defined and factionalised in real life to be thought of as a single agent.[11]

Just as the dictator's hierarchy extended downward from above, a hierarchy of producers rose from the factory workers to managers and officials of fundholding ministries; what the latter shared was a common interest not in fulfilling the dictator's orders as such, but in tilting the ratio of the incentives they gained from the dictator relative to their effort (Harrison, 1998; Markevich, 2000). At the apex of the system, a few of the most important ministers acted both to represent the interests of producers and at the same time as insiders of the dictatorship. The resultant risk of divisions in the Politburo preoccupied Soviet leaders (Belova and Gregory, 2002). This is because the dictator's agents had an incentive to cheat on him when the value of an asset they could steal exceeded the value of their share in the long-run rent from that asset under the dictator.[12] The dictator had to counter the incentive to cheat by forcing his subordinates to cooperate. Thus Stalin imprisoned his prime minister Molotov's wife and also invited him to frequent meetings not because he valued his company but to keep him under surveillance (Ellman, 2001: 142).

How was the Soviet command economy monitored? Monitoring made producers account for outputs and inputs and verified its distribution between the dictator's and the producers' uses. Thus, it prevented producers from stealing from the dictator. Unhindered, producers would steal inputs and products and consume them directly or trade them illicitly, and this would dissipate the dictator's rent. The possibility of stealing rents arose because the dictator had to delegate some control rights over the assets of state-owned enterprises to the managers and workers who were his agents.

Monitoring was costly, however. The dictator could not check information and enforce decisions without so-called transmission belts: party structures and party-dominated mass organisations at every level of the

---

[11]Khlevniuk (2001); and, Lewin (2001). According to Lewin, there were separate nomenklaturas at every level and regional division of the party and state. In 1946, that of the party central committee alone comprised more than 40,000 positions; summing all the lower level and regional nomenklaturas might generate a million posts.

[12]For illustration, see the analysis of "hostile takeovers" in the unofficial secondary market for motor vehicles by Gregory and Lazarev (2002).

apparatus and in every workplace. The dictator's planners could not gather products without security guards, transport police, market inspectors, enterprise and ministry accountants, ministry and Gosplan sectors of material balances, Gosbank records, and finance ministry auditors. In fact, planners could not even *count* products particularly well; they had to aggregate them at plan prices, and the definition of real output was subject to inflationary bargaining between producers and planners (Harrison, 1998). Thus, the dictator had to choose: monitor and pay monitoring costs, or don't monitor but let producers steal some output.

Unlike output, monitored imperfectly at some finite cost, effort could be monitored only at a cost that was prohibitive. We know this from evidence of systematic labor-hoarding by enterprises, combined with the fact that officials responded to suspected labor-hoarding not by increasing monitoring but by revising incentives (Dearden *et al.*, 1990). Effort matters when intrinsic productivities vary and the ratio of output to effort is randomised. The literature on ratcheting starts from such premises (Weitzman, 1980). However, for present purposes the greater difficulty of monitoring effort is not important: the essential properties of the model are established when effort generates output deterministically.

How were incentives designed in the Soviet command system? The dictator offered rewards and punishments. These were "artificial" in that they depended on the dictator's discretion, not market automatism. The incentives were necessary because, without them, producers would always prefer low effort. Soviet leaders learnt this the hard way from the command economy's formative years: in 1929 and 1930 they witnessed a vicious circle of wage equalisation, rising coercion, and declining productivity.[13] After that, the dictator rewarded producers when output was high and punished them when output was low. Rewards were additional income in cash and kind. Punishments were firing and forced labor. Firing, although an ineffective threat against ordinary workers under conditions of a general labor shortage, powerfully threatened managers and officials who stood to lose higher pay, privileges, and promotion prospects.

Both rewards and penalties were costly to the dictator. To pay a reward the dictator had to transfer part of his rent to the producer. In enforcing a penalty, he incurred both direct and collateral costs. *Direct costs* were those

[13]Kuromiya (1988) and Davies (1989; 1996). This phase ended with Stalin's attack on wage levelling which he called "petty-bourgeois egalitarianism", followed by the rehabilitation of payment by results and large skill and seniority premia.

of enforcement, which alone could be very large: for example, at its height the Gulag employed hundreds of thousands of guards to detain, transport, and supervise forced labor. To cover such costs labor camps were treated as self-financing units where prisoners and other "special settlers" paid for their own accommodation, subsistence, and detention out of reduced consumption (Khlevniuk, 1993). However, while the dictator could shift the direct costs onto the victims, he could not do this with the *collateral costs* that arose because firing and forced labor reallocated workers to jobs of lower intrinsic productivity so that output was lost. This loss fell unavoidably on the dictator.

Historically, command economies have always used both rewards and punishments. At first sight it may not be clear why a dictator should offer rewards if punishments would work as well. Call an incentive that works "efficient". Rewards that were efficient and resulted in the behaviour that the dictator desired had to be paid, whereas penalties that were efficient did not have to be enforced. Thus, efficient penalties were cheaper than efficient rewards. This suggests that an all-powerful dictator should set penalties at whatever level is required to make them efficient subject to rewards set at zero. For example, we could make our students always submit essays on time, even for zero credit, by punishing late submission with death.

In practice, the dictator did not have the absolute discretion that this required. Suppose there is a maximum penalty for shirking that cannot be exceeded. If the maximum penalty is less than the efficient level, then rewards for not shirking are required as well. One upper limit is set by the potential income of which producers can be deprived; for this reason positive inducements were significant in Soviet labor camps (Karklins, 1989). Another upper limit is set by inherited social norms: if society expects the punishment to fit the crime, unlimited penalties for shirking may not be enforceable. Specifically, if the ratio of output to effort is partly stochastic and effort cannot be observed so incentives are attached to output, then low effort may be unjustly punished; unlimited punishments may lead to unlimited injustice.

For incentives to be efficient, the promise of incentives must be credible. Incentives that, when implemented, are seen to impose large costs on those who offer them may lose credibility for that reason. However, the credibility of rewards and punishments was tested with different frequency. The credibility of rewards that were large enough to be efficient was tested continually. In contrast that of punishments was only tested when they failed. If punishments were mostly efficient, their credibility was tested at correspondingly infrequent intervals.

II

Long-run trends in the dictator's costs provide a motivation for the model to be developed below. Here, I suggest that a variety of influences, some domestic and some global, worked to make Soviet penalties progressively more costly, to increase the rewards required by the population to invest effort in production, and to raise the costs of monitoring the level and uses of output.

First, the *collateral costs of the penal system* are likely to have risen through time. In the 1930s and 1940s, Soviet leaders experimented with very harsh penalties to combat shirking (Filtzer, 1986; Manning, 1993). Minor production failures were treated as "wrecking" by "enemies of the people". Most voluntary turnover and minor timekeeping infractions by ordinary workers were eventually criminalised regardless of individual circumstances and intentions. Those punished were commonly sentenced to forced labor in establishments subject to harsh financial constraints and physical conditions. Stalin died, and the penal system was relaxed in two stages (Kontorovich, 1986: 21). Initially, Khrushchev dismantled the system of large-scale forced labor and repealed the harsh laws criminalising petty indiscipline. Then, Brezhnev increased managers' job security and greatly reduced the threat of dismissal for senior officials accused of poor performance.

An economic interpretation is as follows. Soviet human capital accumulation steadily widened the gap between the average productivities of free and forced labor. Skilled labor also became more productive relative to raw untrained labor. Thus firing and forced labor wrote off human capital of increasing value and pushed the productivity of each demoted manager or imprisoned worker down a gradient of increasing slope. Over time this steadily raised the dictator's collateral cost of reliance on extreme penalties to induce effort.

Dependence on extreme penalties also carried other kinds of collateral costs with a rising trend. For example, the rules governing the treatment of industrial absentees and quitters under Stalin were so harsh that in the late 1940s the dictator's hitherto loyal agents began to refuse to enforce them (Filtzer, 2007: 158–200). This weakened the credibility of the penal system and raised the spectre of having to intensify monitoring of the monitors in order to restore it.

The growing postwar difficulties of Stalin's penal system were not just a domestic issue. There was a global context. After World War II, the victors acclaimed the outcome as a victory over fascism and exploitation. In the Cold War, the superpowers increasingly competed over civil and 'human' rights. As state after state signed up to conventions that guaranteed ever higher standards of treatment of citizens by government, the Soviet Union paid a rising price for its penal system in losses of international reputation and commercial opportunities abroad.

Second, the *real rewards required to motivate effort* in the Soviet economy clearly rose through the postwar period, and probably rose faster than productivity. For example, of nearly 3,000 Brezhnev-era emigrants surveyed by Paul R. Gregory, three quarters reported the impression that average productivity was falling (although it was not); of these, three fifths listed inadequate incentives as the main cause of productivity problems (Gregory, 1987). Several possible reasons offer themselves. One is that more educated consumers might become less easily satisfied; Joseph Schumpeter (1954: 145–55) suggested long ago that the accumulation of human capital stimulates the development of elements in society that are more interested in criticising than accumulating.[14] Another is that the Soviet economy may have overinvested in human capital.[15] Diminishing returns to rising human capital might make it difficult to maintain rewards with the result that those endowed with it might become less likely to apply it to production.

Again, any trend in the efficient reward was not a purely domestic matter. From the 1970s onwards, the Soviet lag in income per head behind the United States, already large, was slowly widening, and the lag behind western Europe, although somewhat smaller, was widening more rapidly (Maddison, 2001). An increasingly educated Soviet workforce was becoming more accurately informed about this gap. Comparisons with the West that were increasingly invidious may have contributed to Soviet producers' psychological devaluation of the real rewards available from their own economy.

[14]Schumpeter's argument was made in relation to the intellectual under capitalism, but seems potentially applicable to any society that promotes education. Isaac Deutscher (1967) shared this belief, predicting that Soviet educational policies would eventually undermine the basis of bureaucratic rule.

[15]The economic literature that analyses Soviet growth in terms of a CES aggregate production function suggests that the Soviet economy overinvested in capital generally (for example, Easterly and Fischer (1995)). David Granick (1987: 202–34), has suggested that the Soviet economy overinvested in human capital specifically.

Third, it is likely that the *costs of monitoring production* were rising too. Soviet post-war product and process innovations and structural changes promoted this trend. Product innovation brought ceaseless change in the range of products and variation of product attributes. These increased planning costs, made output less measurable, and impeded the monitoring of productivity. Product innovations assigned from above by powerful user ministries could be monitored directly, but producers channeled much innovation from below into shifting the ratio of reward to effort in their favour in ways that planners could not detect (Berliner, 1976: 375–80; Harrison, 1998b). As the quality and variety of products grew in economic significance, this strategy proved ever more rewarding. And this is to consider only the 'productive' sphere of the economy. In the growing sphere of non-military services, which raised its share of Soviet employment from one sixth before the war to one third in the 1980s, real output was essentially unmeasured.

Process innovations associated with the transition from mass to flexible production may have raised monitoring costs by fostering producers' control over effort and information. Soviet industry had moved from craft production to mass production before and during World War II not only because of its production cost advantages but also to cut monitoring costs (Harrison and Simonov, 2000: 237–8). If mass production began to lose its production cost advantage after the war, flexible production threatened planners with a return to the high monitoring costs previously associated with artisan control of production. One could interpret the Soviet post-war failure to engage with flexible production as a decision to forego its growing cost advantages so as to avoid the higher monitoring costs that would come with it.

Is there direct evidence of rising monitoring costs? There was no obvious upward trend in the costs officially reported in budget outlays on planning and administration. The proportion officially engaged in "administration" remained remarkably constant over many decades at approximately two per cent of Soviet public-sector employment (Goskomstat SSSR, 1987: 410). This was despite fears expressed by apparently qualified observers that the returns to growth were being eaten up by bureaucracy. For example, in the 1960s a prominent systems analyst predicted that "at the present rate of development, by 1980 the entire adult population of the USSR will be engaged in planning" (V.M. Glushkov, cited by Ellman, 1969: 288).

However, important monitoring costs normally go unreported because of the hidden regulatory burden on producers. As recent experience in UK higher education indicates, hidden burdens may substantially exceed those reported in the regulator's budget (Harrison and Lockwood, 2001). One hidden burden on the Soviet economy was maintenance of the communist party's organisations in every workplace and office. These played an essential role in verifying the information and decisions of managers and officials. Producers were obliged to cover the overheads and direct costs of party facilities and activists' time. Outlays on party maintenance, if we knew them, would be a revealing proxy for the trend in Soviet monitoring costs. We do not know outlays, only membership. Membership was exclusive and controlled on the basis of personal recommendations and a probation period, carried with it an obligation to take part in the work of the party, and was subject to periodic "cleansing" (Wintrobe, 1998: 219–20). Overall, party membership rose steadily in proportion to the adult population from less than one per cent in the early 1920s to 3 per cent in the mid-1930s, five per cent in the 1950s, eight per cent in the 1970s, and nine per cent in the late 1980s.[16] Multiplied by average hours of party work and an imputed wage these figures would suggest a similarly rising share of GDP. Thus, the upward trend of party membership suggests that the monitoring of production was driven by growing needs and carried rising costs.

III

In this section, I define the players in the command system as self-interested producers and a dictator and I set out a game between the players. The game shows that high coercion can produce a stable state of high output: instability is not intrinsic to a command economy. It reveals other possible equilibria including one of reduced coercion and output — a state of collapse. It defines the conditions under which the economy may slip from a high state to a low state. It shows that as a result everyone may suffer a loss of income.

The dictator maximises a payoff made up by the value of rents *less* his costs and losses, $Z - M - R - S$, while the producer maximises income received in wages and bonuses and appropriated through theft, *less* costs

[16] Gershenson and Grossman (2000: 19). Taking into account numbers of pensionable age, nine per cent of the adult Soviet population in the late 1980s was approximately 15 per cent of the working population. Gershenson and Grossman analyse rising party membership as the outcome of a shift by the nomenklatura between strategies of repression and cooption. Lazarev (2001), analyses the same as the outcome of a Ponzi-type pyramid scheme operated by the nomenklatura.

of effort and punishments, $Y + S + R - E - P$, defining each symbol as follows:

E	subjective cost of effort	$(E = e, 0; e > 0)$
F	value of effort: the extra output produced by effort	$(F = f, 0; f > 0)$
M	monitoring costs	$(M = m, 0; m > 0)$
P	direct cost of punishments	$(P = p, 0; p > 0)$
Q	collateral cost of punishments	$(Q = q, 0; q > 0)$
R	value or cost of rewards	$(R = r, 0; r > 0)$
S	value or cost of stolen rents	$(S = s, 0; s > 0)$
X	value of output	$(X = x + f, x, x - q)$
Y	value of producer incomes	$(Y = y; y > 0)$
Z	value of rents	$\begin{pmatrix} Z = X - Y \\ \quad = z + f, z, z - q \end{pmatrix}$

The dictator and producer make choices in sequence, and both players are fully informed about each other's choices at each stage. At stage I, the dictator sets coercion high or low by deciding whether or not to monitor output. When coercion is high the dictator monitors output; this prevents stealing and also enables him to promise to reward high output and penalise low output. When coercion is low the dictator is indifferent to both effort and stealing and does not monitor output: coercion is low but not zero, and the dictator still extracts a rent, but part is reallocated to producers through theft. Monitoring is costly and, when implemented, is a charge on the dictator's rent:

$$\left.\begin{array}{r} M = m \\ M = 0 \end{array}\right\} \text{ when output is } \left\{\begin{array}{l} \text{monitored} \\ \text{not monitored.} \end{array}\right. \tag{1}$$

Monitoring output efficiently eliminates stealing. Thus, although stealing is done by producers, the dictator decides whether producers can steal:

$$\left.\begin{array}{r} S = 0 \\ S = s \end{array}\right\} \text{ when } \left\{\begin{array}{l} M = m \\ M = 0. \end{array}\right. \tag{2}$$

At stage II, the producer observes the dictator's choice and responds in that light. Depending on the promised rewards and punishments she sets effort high or low. When effort is low, the value of output is positive and the producer cost of effort is zero. When effort is high and has a positive cost, output is raised by the value of effort. Thus, controlling for the allocation of producers among employments of different intrinsic productivities,

$$\left.\begin{aligned} X &= x + f \\ X &= x \end{aligned}\right\} \text{ when } \begin{cases} E = e \\ E = 0. \end{cases} \tag{3}$$

At stage III, provided output has been monitored, the dictator observes the results of producers' choices made at stage II, and makes a further choice: whether or not to honor his stage I promises: to reward high output and punish low output. In the case when the dictator only makes promises that he will keep,

$$\left.\begin{aligned} P &= 0 \\ P &= p \\ P &= 0 \end{aligned}\right\} \text{ and } \left\{\begin{aligned} R &= r \\ R &= 0 \\ R &= 0 \end{aligned}\right\} \text{ when } \begin{cases} M = m \text{ and } \begin{cases} E = e \\ E = 0 \end{cases} \\ M = 0. \end{cases} \tag{4}$$

However, if the dictator defaults on incentives $P = R = 0$ in all outcomes.

When the producer is punished for shirking there is a collateral loss of output since firing and forced labor reallocate workers towards employments of lower intrinsic productivity. Output is high when effort is high, low when effort is low and low output is unpunished, and lower still when low output is punished. Equation 3 was written for a given allocation of producers among employments of different intrinsic productivities, which is no longer the case when shirking is punished. In the latter case, setting $X = x$ for calibration when effort is low and unpunished:

$$\left.\begin{aligned} X &= x + f \\ X &= x \\ X &= x - q \end{aligned}\right\} \text{ when } \begin{cases} E = e \\ E = 0 \text{ and } \begin{cases} P = 0 \\ P = p. \end{cases} \end{cases} \tag{4a}$$

Finally, the dictator's rent, the excess of output over wages, is decided as follows, setting $Z = z$ for calibration when effort is low and unpunished:

$$\left.\begin{aligned} Z = z + f &= x - y + f \\ Z = z &= x - y \\ Z = z - q &= x - y - q \end{aligned}\right\} \text{ when } \begin{cases} E = e \\ E = 0 \text{ and } \begin{cases} P = 0 \\ P = p. \end{cases} \end{cases} \tag{5}$$

This framework is based on two simplifications with the aim of avoiding an unduly complex parameterisation. First, the choices facing the dictator and producer are binary rather than continuously variable. Second, I assume that the dictator and producer value each unit of current income the same; however, they will discount the future differently. The simplicity of the

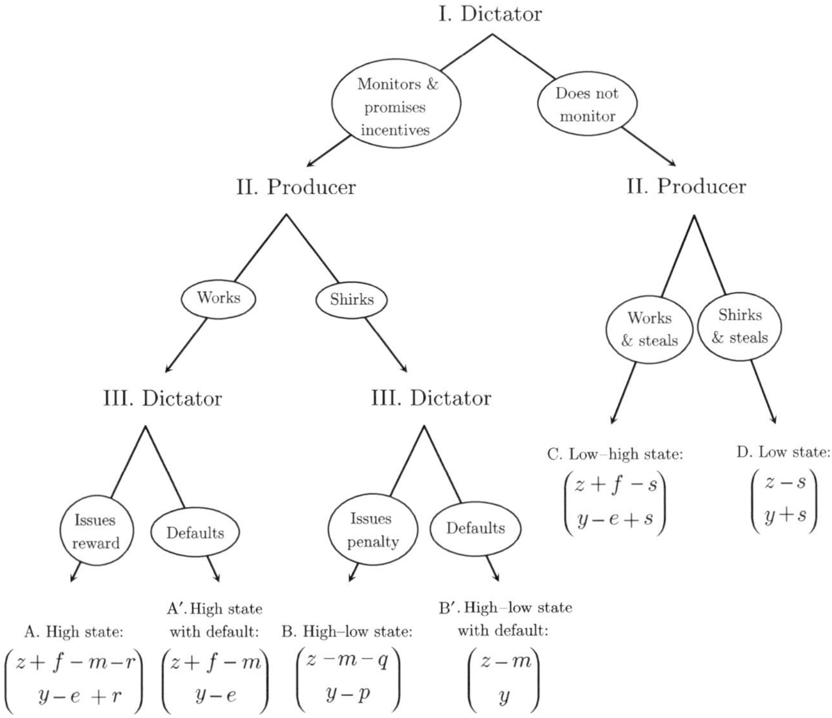

Fig. 2. Players and Payoffs.

model that results is limiting. Designed to illustrate the conditions of a sudden collapse, it cannot be expected to shed much light on the continuous small variations of effort and monitoring that are historically observed in Soviet history before the collapse.

Figure 2 sets out the players, choices, and payoffs. Call outcome A on the far left combining high coercion with high effort a "high" state and outcome D on the far right combining low coercion with low effort a "low" state. In between are some mixed states, outcome B which is "high-low" (with high coercion and low effort), and C which is "low-high" (low coercion and high effort). Finally, the high-coercion states include outcomes A′ and B′ in which the dictator defaults on incentive promises. But in the low-coercion states the dictator promises nothing and cannot default.

When played once, the game has the following features. First, effort is costly so low effort is unambiguously the producer's best response when high effort is unrewarded and shirking unpunished. This means two cases:

when the dictator does not monitor output, and when the dictator monitors output but is expected to default on promises. In each case the producer loses from high effort; the payoff to her from D is superior to that from C, and B' is superior to A'.

The dictator may try to induce high effort by means of high coercion. He needs enough room within the economy's parameters that both players may gain something from high output. He has to make it possible for the producer to prefer outcome A over B, offering rewards and punishments that are credible so that her payoff becomes higher with high effort. He has to make the expected reward for effort *plus* the penalty for shirking together big enough to exceed the producer's subjective cost of effort. Having done that he, the dictator, must still prefer outcome A over D: his return to monitoring must exceed his costs of monitoring and rewards, the return being the value of the extra effort induced *plus* the output secured from theft. In short, high coercion and high effort will pay both parties when the dictator can offer a credible reward such that $f + s - m > r > e - p$.

High coercion may still be the dictator's best strategy even if he cannot make high effort pay and he will prefer the high-coercion, low-effort outcome B to D. The condition for this is $s > m + q$. This is because monitoring does at least protect his rents from theft. The dictator's downside is the associated monitoring costs and collateral costs of penalties.

A defect of the one-shot game is that it does not allow for consequences that follow the current game or the forward-looking behaviour that anticipates them. Specifically, it fails to show why the dictator should ever keep a promise to reward or punish. This is because the dictator pays for incentives. Whether producers work hard in the expectation of rewards or shirk despite the threat of punishment the dictator always raises his short-run payoff by defaulting; for example, he prefers outcome A' to A and B' to B so that A and B are ruled out. Knowing this the producer will not find the dictator's promises credible and will always choose low effort, so that A' is ruled out too.

In fact, the only outcomes available are B' and D. The dictator can choose between these two. Which is better depends on the relative costs of monitoring and not monitoring. The cost of monitoring is m. The cost of not monitoring is s, the output that the producer will steal. Thus, when $m < s$ the dictator should choose B' and maintain high coercion regardless of the producer's effort choice. But if $m > s$ he will abandon monitoring and choose D.

Although the one-shot game is very restricted it does identify some possible equilibria, some conditions required to sustain monitoring in general and a high-state outcome specifically, and some different ways in which the high state might break down. The dictator will always monitor the producer provided $m < s$. A high state is the equilibrium only when incentives are credible and a reward can be set such that $f + s - m > r > e - p$. Increases in the producer's effort cost or the dictator's monitoring cost, and decreases in the value of effort or penalties and the scope for theft can disrupt this condition. So can anything that undermines the credibility of the dictator's promises. Provided the dictator's residual benefit from monitoring still exceeds his higher costs and $s > m$ the game may now end in high coercion followed by low effort, a condition closer to Poland in the 1980s than Russia in any period. Beyond a point, however, the dictator may prefer to switch to a strategy of low coercion, the economy slumping from a high state to a low state.

As a former Soviet official once told William Keegan, "We used to work in a centrally controlled system where they told you what to produce. Now they've stopped telling us what to produce, so we don't produce anything".[17]

IV

In this section, I explore the long-run properties of the game set out above. The long run matters because the dictator's gain from keeping incentive promises only becomes apparent when the game is repeated. A repeated game offers a far-sighted dictator the possibility of a stream of rewards from high effort, but he can only secure this stream by delivering a matching stream of rewards and penalties. Promises have to be credible to be efficient, and the prospect of repetition can make his promises credible. The repeated game shows the conditions under which the dictator's best choice is to sustain a high-state equilibrium; this means it also shows conditions under which the dictator may let it collapse. The repeated game additionally shows conditions under which the producer can force the dictator's hand and precipitate a collapse by staging a slowdown or strike.

[17] *The Observer*, 18 October, 1998.

At stage I, the dictator's regime choice is conditioned by the relative costs of monitoring and not monitoring. The cost of monitoring to the dictator is m. The cost of not monitoring is now at least s, the output that the producer will steal. Take the case when $m < s$; this means the dictator's best choice at stage I is to maintain high coercion regardless of the return he expects from the producer's effort choice at stage II. Then, at stage III the dictator chooses between keeping his word and cheating on incentives; under what circumstances should a forward-looking dictator pay up? When effort is high and $m < s$ paying a reward is the dictator's best long-run strategy provided he can set r such that $\delta_D \cdot f > r > e - p$ [for proof see the appendix, Proposition 1], where δ_D denotes the dictator's discount factor and $0 < \delta < 1$. As before, he must be able to fix the producer's net return from compliance higher than her cost of effort. This gains him a discounted stream of revenue from the high effort that is induced. If the stream is large enough both players gain from the dictator's honoring his promises, and the high state of outcome A will be the unique equilibrium. This is more likely the more far-sighted the dictator and the larger his discount factor. If the condition is not met the dictator will cancel rewards, but continue to monitor output, and the high-low outcome B', will ensue.

However, when $m > s$ monitoring is unprofitable unless the dictator's payoff is bolstered by a return from the producer's high effort. His regime choice at stage I will then be conditioned by the efficiency of incentives. Faced with the producer's effort choice at stage II, the dictator must pay up or default at stage III; if he defaults, he implicitly also chooses to move to a low-coercion regime at stage I of the following game. This is because if there is no point in paying incentives and $m > s$ there is no point in paying monitoring costs either. Under what circumstances will the dictator prefer to pay up? When $m > s$ paying a reward and continuing to monitor is the dictator's best strategy provided he can set r such that $\delta_D \cdot (f + s - m) > r > e - p$ [Proposition 2]. Again the dictator must be able to fix the producer's reward higher than the cost of effort less the penalty that the producer avoids by working hard. This gains him a discounted stream of revenue from the producers' high effort that is induced, plus the rents secured from theft, net of his monitoring costs. Again, the larger the stream and the more far-sighted the dictator, the more likely that all players will gain from the dictator's honoring his promises, making a high state the unique long-run equilibrium. If not, the dictator will cancel rewards and cease to monitor output, and the outcome will be a low state.

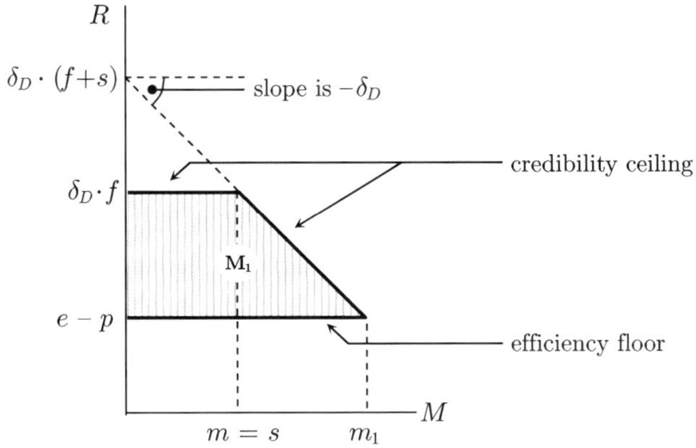

Fig. 3. The Efficiency and Credibility Frontiers.

Figure 3 maps Propositions 1 and 2 onto a space defined by rewards on the vertical axis and monitoring costs on the horizontal axis. The result is a shaded envelope M_1 that marks out the feasible space for a high-state equilibrium: for any $m < m_1$ it is possible to set r such that a high-state equilibrium is feasible.

M_1 is bounded by two frontiers. The lower surface is an *efficiency floor*: below it, incentives are not efficient. Thus, a producer who receives a reward less than $e - p$ will prefer to shirk, so the efficiency floor is a horizontal line at $r = e - p$.

The upper surface of M_1 is a *credibility ceiling*: above it, incentives are not credible. The credibility ceiling is kinked at $m = s$. From Proposition 1, over the range $0 < m < s$ a dictator who has to pay a reward greater than $\delta_D \cdot f$ will prefer to cheat, but without switching regime, so the ceiling is horizontal at $r = \delta_D \cdot f$. From Proposition 2, however, a dictator who has to pay a reward in excess of $\delta_D \cdot (f + s - m)$ would prefer not only to cheat but also to abandon monitoring, so at $m = s$ the credibility ceiling turns down along a line with vertical intercept at $r = \delta_D \cdot (f + s)$ and slope of $-\delta_D$.

Outside the envelope the high state will collapse; it will collapse into a high-low state to the left of $m = s$, and into a low state to the right. The intersection of the credibility and efficiency frontiers defines the maximum of m_1 to which the dictator's monitoring costs may rise before he can be induced to abandon monitoring as well as cheat. I will call the case when $m > m_1$ an m_1 violation.

What happens when a producer with incomplete information decides to test the credibility of the official punishment for low effort? The rationale for asking this question is as follows. To avoid public accountability the dictator keeps his costs and revenues secret and may also disguise his valuation of them. But this creates a credibility problem. Incentive promises are credible when meeting them, however costly, is seen to be in the dictator's interest. The credibility of efficient rewards is tested continually. But the credibility of a punishment is only tested when it fails and in a well-functioning command system this may not have happened for some time. In an atmosphere of official secretiveness and dissimulation doubts may accumulate in society as to whether the authorities remain prepared to implement penalties that are rarely enforced and potentially costly. In this situation the efficiency and credibility frontiers are redefined. We should ask what level of rewards will be efficient when producers do not expect to be punished; and what level of rewards is credible when, before the dictator can pay them again, he must first bear the cost of restoring the credibility of punishments. In short, when is it that producers who do not expect to be punished will be proved wrong? Clearly the question can only be applied to producers acting in partial ignorance.

Reconsider the producer's problem. The efficiency floor in Figure 3 showed the level of reward below which producers will choose to shirk even when they expect to be punished for it. But if producers would prefer to shirk when they expect to be punished, i.e. if in Figure 2 they would prefer outcome B to A, they would like outcome B' (shirk with impunity) still better, and outcome D (steal as well as shirk) best of all. If the producer could choose these she would reject a reward at the efficiency floor in Figure 3. The producer's problem is how to bring these outcomes within her choice set when the dictator has the first move. Under certain conditions, there is a solution and it is to stage a slowdown or strike.

In Figure 4, these conditions are mapped onto the same space as Figure 3. When punishments have lost credibility there is a new efficiency floor that is higher than before and steps upward at $m = s$; below it, the reward offered will not be enough to put off a strike. A new credibility ceiling, horizontal at first as in Figure 3, turns down at $m = s$. Above it, the reward is so high that the dictator could not be expected to gain by punishing shirkers enough to restore high effort. Within the frontiers a shaded envelope M_2 marks out the feasible space for a high-state equilibrium when punishments have lost credibility: only for $m < m_2$

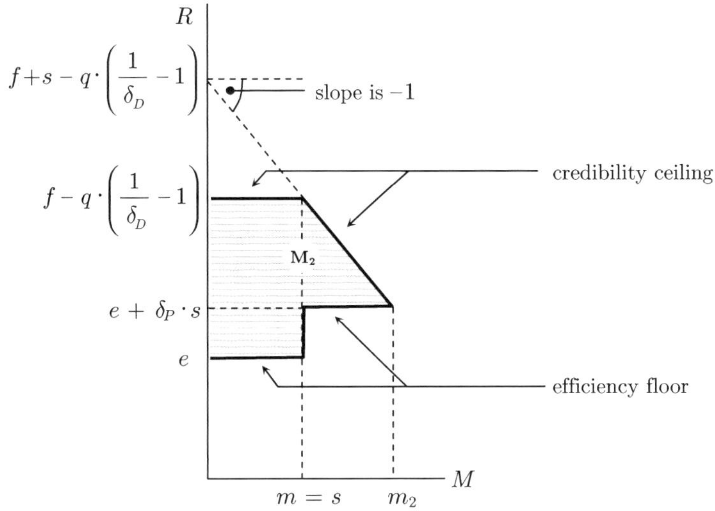

Fig. 4. When Producers Expect not to be Punished.

is it possible to set r such that a high-state equilibrium is feasible. When $m > m_2$ I will call it an m_2 violation.

The frontiers are found as follows. First, as for Fig. 3, the producer's options are limited by the dictator's regime choice, and the latter are again conditioned by the relative costs of monitoring and not monitoring. When $m < s$ the dictator cannot be induced to abandon monitoring, but he can be induced to condone a strike if the cost of punishing it is high enough. To maintain a high-state equilibrium and avoid a strike when the producer does not expect to be punished it is no longer enough that $r > e - p$; the efficiency floor lifts to $r = e$. As for the credibility ceiling, it moves to $r = f - q \cdot \left(\frac{1}{\delta_D} - 1\right)$; a credible reward must be low enough that, in combination with the collateral cost of imposing penalties, it does not exhaust the dictator's gain from restoring high output. In short, the producer's best strategy is to strike and the dictator's best response is to cancel penalties but maintain monitoring, i.e. choose outcome B', if he cannot set r such that $f - q \cdot \left(\frac{1}{\delta_D} - 1\right) > r > e$ [Proposition 3].

Where will the new credibility ceiling lie in relation to the old one? The new vertical intersect may be either lower or higher. If q is small, specifically if $q < \delta_D \cdot f$, the credibility ceiling lifts and the dictator's problem is actually made easier. The fact that the stoppage temporarily relieves him of paying a costly reward relaxes his long-run budget constraint.

The dictator's strategy is to punish the strike, then reinstate rewards at a level that may be higher than before; Hungary in the 1960s or Poland in the 1970s may have resembled this outcome. If q is large, however, the cost of punishing the strike causes his credibility ceiling to descend at the same time as the efficiency floor rises, compressing his options from both sides.

When in contrast $m > s$ the dictator will abandon monitoring unless his payoff is supplemented from the producer's high effort. Faced with a strike, his regime choice is conditioned by the efficiency of incentives. In this case the producer's best strategy is to strike and the dictator's best response is to cancel penalties and abandon monitoring, i.e. choose outcome D, if he cannot set r such that $f + s - m - q \cdot \left(\frac{1}{\delta_D} - 1\right) > r > e + \delta_P \cdot s$ [Proposition 4], where δ_P denotes the producer's discount factor. The logic of this is that when the producer chooses to strike she not only does not expect to be punished but also takes into account her potential gain from forcing the dictator into a regime change. As a result the efficiency floor rises to $r = e + \delta_P \cdot s$; it is no longer enough that $r > e$. At the same time the credibility ceiling turns down along the line of $r = f + s - m - q \cdot \left(\frac{1}{\delta_D} - 1\right)$: for it to be worth restoring a high-state equilibrium in face of a strike the efficient reward also must be low enough that, in combination with the dictator's immediate cost of imposing penalties and flow of monitoring costs, it does not exhaust the subsequent gain from restoring high output. Thus, when there is an m_2 violation a strike can force even a long-sighted dictator who is also faced with rising costs of monitoring and penalties, or a falling value of effort, from the high state to the low state.

Proposition 4 has two further notable features. One is the general effect of the rents that may be stolen when monitoring is lifted. As s increases the incentive for producers to strike rises, but the incentive of the dictator to resist rises by more. Thus the spread of corruption opportunities, controlling for other factors, does not threaten the stability of the high state but enhances it.

Another feature is the effect of the players' different discount factors. Without a capital market to bring about convergence, the Soviet dictator discounted the future by less than producers. In the Stalin era the gap was wide enough that the dictator would impose famine on the rural population if necessary to safeguard accumulation targets.[18] After Stalin there were no more famines, and the gap between the dictator's and

[18]Wheatcroft and Davies (1999) and Ellman (2000b) show that famine processes were also partly attributable to unintended consequences of policy.

producers' discount factors perhaps diminished. However, forced saving remains the best explanation of observed consumer behaviours through the postwar period (Kim, 1997; 1999). Proposition 4 suggests that under some circumstances an excess of δ_D over δ_P helped to sustain the command system: the dictator, focused on the long term, remained willing to invest in incentives while producers, struggling to survive from day to day, remained willing to comply.

V

In this section, I show that trends in variables and the actions of the players may cause a command system to collapse in various ways. Figures 3 and 4 show that a collapse could be triggered by two types of violation. The possibility of an m_1 violation is shown in Figure 3: beyond a certain point the dictator would choose to cheat on rewards and abandon monitoring, and m_1 shows the maximum level of monitoring costs that the dictator could sustain before taking this option. The other possibility, shown in Figure 4, is an m_2 violation: the dictator's surrender could be forced when he could no longer afford the collateral costs of an efficient penalty for shirking, and m_2 shows the maximum level to which monitoring costs could rise before producers could force him to run up a white flag.

These are mapped onto the same space in Figure 5. In each diagram the vertically shaded area M_1 marks out the feasible space for a high state equilibrium when both rewards and penalties are credible. The horizontally shaded area M_2 marks out the same when producers no longer expect to be punished. As monitoring costs rise, which trigger point will the command economy encounter first? The two diagrams suggest the likely relationship between m_2 and m_1. In Figure 5.1 the collateral costs of penalties to the dictator are relatively small, and M_2 is so large that $m_1 < m_2$. In Figure 5.2 penalty costs are so large that M_2 is completely enclosed by M_1; this sufficiently determines $m_2 < m_1$. A comparison of the two figures suggests that m_2 remained less than m_1 for a wide range of intermediate values.[19] Therefore, taking into account the likely upward trend of punishment costs, I think of Figure 5.1 as a special case and other cases where $m_2 < m_1$, including Figure 5.2, as more general, at least for the late Soviet period.

[19] To be precise, $m_1 < m_2$ requires that $q < e - \frac{1}{1-\delta_D} \cdot (p - \delta_D \cdot \delta_P \cdot s)$; bear in mind that probably $0 < \delta_P < \delta_D < 1$ while $\frac{1}{1-\delta_D} > 1$ and was also potentially a large number, so the condition is relatively restrictive.

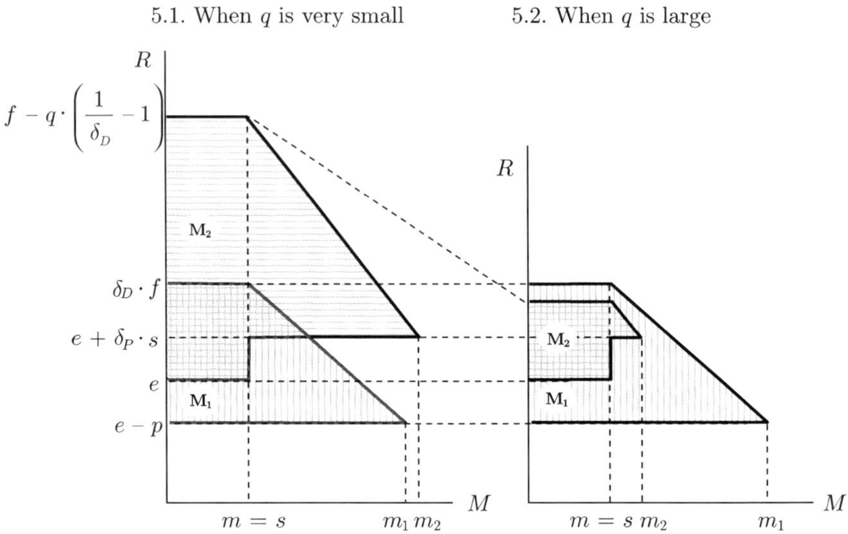

Fig. 5. Triggers for Regime Insolvency.

When $m_2 < m_1$ it did not follow that the dictator was certain to be brought down by a strike as soon as monitoring costs passed the m_2 level. Rather, one could think of the range above m_2 and below m_1 as a zone in which the dictator gained from continued operation of the command system but could not afford to defend it if challenged. In this range, therefore, the stability of the command system could be upheld by a strategem but compromised if the strategem was exposed.

The first element in this strategem was concealment of the dictator's costs. Secretiveness was a double-edged weapon; it created doubt where the the dictator could otherwise have gained credibility from revelation, but it also served to obscure potentially dangerous truths. The policy of concealing the true maintenance costs of the monitoring and penal systems that was pursued from Stalin's time onward ensured that producers could not know at what point the dictator might self-interestedly choose to condone a strike or fail to defend the regime.

The second element was the dictator's reputation for brutality and intransigence, which combined elements of truth and bluff. It was in the dictator's interest not to be perceived as a rational actor who chooses freely and self-interestedly among available options. Thus Stalin encouraged the belief that he did not count the cost of casualties or the harsh penalisation of shirkers and deserters whether in war or in economic mobilisation, and he supported this belief through his actions. In this way he acquired the reputation of a dictator who would never condone a go-slow or strike, who would always choose high coercion, punish efficiently under all circumstances, and defend his regime regardless of the producer's effort choice and the cost of changing it. He presented himself not as a free agent but as an instrument of a deterministic History that marches ever onwards and precludes past decisions from ever being reversed. He bequeathed this reputation to his successors.

Believing in this reputation and subject to secrecy, the producer should always be satisfied with any reward above the efficiency floor of M_1, and would never strike with the hope of forcing the dictator to concede monitoring. The dictator's strategem would permit him to sustain a level of monitoring that would be instantly compromised if the bluff were called.

VI

The framework outlined above enables us to distinguish those narratives of Soviet collapse that are logical and consistent with known facts from those that are not. First, was the dictator's reputation at issue?

The record shows that every Soviet leader until Gorbachev attached grave importance to the reputation of the regime. Each took every opportunity to uphold the permanence of the monitoring mechanism. Most famously, in 1937, Stalin said that mass terror would actually intensify as Soviet society marched nearer to communism. Two decades later Khrushchev withdrew this specific commitment but reasserted "the unshakeable unity of our party, its cohesiveness around the Central Committee, its resolute will to accomplish the great task of building communism [...] We are absolutely certain that our party, armed with the historical resolutions of the 20th congress, will lead the Soviet people along the Leninist path to new successes, to new victories" (Khrushchev, 1976: 80–1). Hindsight suggests that such claims protested too much, because the only thing that guaranteed them was an unbending will. If monitoring was relaxed once, the claim that it could never be relaxed would no longer be believed. The dictator's reputation was fragile: a stumble could weaken it and a major slip could destroy it.[20]

[20]Fragile reputation might contribute to "coercion fatigue": in the end, every dictatorship fails (Kalin Nikolov, personal communication, 23 March 1999). Fragile reputation also reinforces Machiavelli's dictum that the most dangerous moment for a regime is when it begins to reform itself.

In the post-war years the process of socialist 'economic reform' undermined this reputation. Declining Soviet postwar growth with rising costs of both monitoring and not monitoring set the scene for a complex cycle of systemic reforms in the Soviet Union and eastern Europe.[21] These reforms were driven over more than two decades by the need to improve allocation while reducing monitoring costs. This was expressed as a search for a socialist economic mechanism that would regulate itself, realigning incentives so that plans and producers could coexist with greater harmony than under traditional command arrangements. If reforms succeeded, the dictator could efficiently delegate control rights to managers without continual costly monitoring. The final attempt at such a rearrangement was Gorbachev's industrial reform of 1987.

The goal of improved allocation was an important dimension of economic reform. However, empirical studies of economic reforms suggest that allocation outcomes were often negative. Efficiency was worsened as new incentives provided producers with new avenues for rent-seeking. Hours fell and production discipline was relaxed. Controls were reimposed to correct the consequences. Meanwhile the original problems had not been solved, so a cycle of reforms and counterreforms resulted. Thus reforms failed to stem the rising tide of regime costs.

In the cycle of reforms and counterreforms, moves towards either too much or too little monitoring prompted reversals. Did the opposing shocks cancel out, with counterreforms enabling the dictator exactly to recapture the power devolved in each reform phase? There is evidence of net slippage. Although the market was not strengthened, the plan was gradually weakened. Post-Stalin regimes in the USSR and eastern Europe increasingly tolerated sideline economic activities, which sometimes reallocated resources more efficiently but at the same time undermined state ownership rights; James R. Millar described this as Brezhnev's "Little Deal" with the Soviet Union's urban population (Millar, 1985). According to Mancur Olson, state-owned enterprises became "more nearly insider lobbies or organised special interests than productive enterprises" (Olson, 1995: 32).

In short, each reform cycle weakened the dictator's reputation. This reputation could not be rebuilt by counterreforms because it relied on an unbroken history that was being destroyed. Declining reputation and rising

[21]This account rests on Schroeder (1972; 1979; 1982); Hanson (1983); Bornstein (1985); Brus (1986a; 1986b); Kornai (1986); and, Kontorovich (1988).

costs associated with monitoring the changing production system together suggest a story in which the feasible space for a high-state outcome was gradually squeezed. As in figure 5.2 the dictator could profitably sustain monitoring costs beyond the level of m_2 on the strength of his reputation alone. As his reputation shrank, the risk grew that he might suddenly find a given level of monitoring costs that was formerly sustainable, less than m_1 though above m_2, to lie outside his feasible space.

Figure 5 illustrated two triggers for command insolvency. In either case the dictator's best response was to renege on promised incentives, but in the case of an m_2 violation the dictator was forced to abandon monitoring when no longer able to afford an efficient penalty; faced with an m_1 violation the dictator made an unforced decision to abandon monitoring when no longer able to afford an efficient reward. Which of these matches better the sequence of events that led to the collapse of the Soviet command system? The historical record tends to support a narrative of forced surrender.

A few observations from the final years of the Soviet economy are indicative of growing incentive problems, but also suggest that before Gorbachev the difficulties were capable of being put right. Under Brezhnev, economic plans became less demanding while productivity growth declined (Schroeder, 1985). Rewards were perceived to have become unsatisfactory, and this was widely believed to contribute to productivity problems (Gregory, 1987). Failures of work discipline became increasingly commonplace and were unpunished. But a policy shift in 1983 under Andropov, continued by Chernenko, heightened monitoring and the penalisation of labor violations. There was also a brief reduction in repressed inflation. This led to improved effort for a few years (Kontorovich, 1986; Kim, 1999).

In short, incentives that had gradually become less efficient were temporarily made more efficient again. Not wishing to increase rewards substantially, the dictator was not afraid to raise penalties and incentive credibility was maintained. Still there is a lot that we do not know about this episode, including the scale of additional monitoring costs and collateral productivity losses. Besides, the variation of both effort and incentives through these years, although noticeable in time series, were relatively small compared with what was to follow.

Under Gorbachev producers lost their fear of penalisation and unprecedentedly went on strike for higher pay. Equally without precedent, their withdrawal of effort was not only not punished but was rewarded with

Table 1. Growing Soviet Financial Disequilibrium, 1985 to 1989.

	State budget deficit, % of revenues plus deficit[a]	Average nominal public sector earnings, % change over previous year[b]	Monetary overhang, % of money stock[c]	Market-clearing retail prices, % of prevailing prices[d]	Real GDP per head, % change over previous year[e]
1985	4.6	2.9	16	119	0.0
1986	11.4	2.9	17	120	3.1
1987	13.1	3.7	19	123	0.3
1988	19.2	8.3	21	127	1.3
1989	18.6	9.4	23	130	0.7

Sources:
[a]Kim, 'Causes', Table 3.
[b]Computed from TsSU SSSR (1986:397), and Goskomstat SSSR (1989: 76).
[c]Kim, 'Income', p. 662.
[d]Computed as $\frac{1}{1-MO}$ where MO is the percentage monetary overhang.
[e]Computed from Maddison (2001: 278).

concessions: the leader reneged on the repressive response to worker unrest required by Soviet tradition.[22]

Oddly enough the situation on the eve of this decisive event, shown in Table 1, although disturbing, still looked far from catastrophic. Wage inflation, well controlled in 1985 and 1986, rose modestly to 8 per cent in 1988. With price controls still in place, and little productivity growth, the result was rising budget subsidies and an increase in retail shortages and queues. Accordingly the monetary overhang also increased: if prices needed to rise by 19 per cent in 1985 to clear the retail market, by 1988 the required price increase had risen to 27 per cent. Real output, however, continued to grow slowly and had not yet attained its Soviet-era peak.

In principle this pattern could be interpreted in two ways. One implication could be that the Soviet economy was experiencing an m_1 violation in which the rewards offered to producers ceased to be credible or efficient. But the fact that output continued to rise through 1989 indicates that producers did not treat it as such. It was not until the credibility of punishments was tested in the strike wave in the summer of 1989 that the collapse began.

[22]On the social history of the 1989 strikes, see Siegelbaum and Walkowitz (1995). On traditional responses, see Kozlov (1999), and for the treatment meted out to strikers in Novocherkassk in 1962, Baron (2001).

What happened in 1989 is better described as an m_2 violation caused by a collapse of the dictator's reputation that led both producer and dictator to recalculate their options. Consider figure 5.2: suppose that monitoring costs, less than m_1 but higher than m_2 and rising, were squeezing the economy into the diminishing space between the efficiency floor and the credibility ceiling of M_1. The combination of rewards and monitoring costs already lay outside M_2 but at first producers didn't strike because of the dictator's reputation for regime intransigence and brutal penalisation of shirking. Thus far the monitoring system remained profitable for the dictator and the costs of monitoring remained affordable. Then a new leader resolved to experiment with something new and unprecedented involving more rules and citizens' rights, wider choices, less monitoring, and less brutal punishment than before. At a certain point producers woke up: the move Stalin had played as if irreversibly in 1929, the year of the Great Breakthrough, was being replayed! The regime could change and would respond to pressure! Producers recalculated their options and set about testing the credibility of the penal system.

The unpunished strike wave of 1989 provided the confirmation they sought. The minimum reward required to keep producers at high effort rose instantly from the efficiency floor of M_1 to that of M_2. In the relevant range of monitoring costs the minimum efficient reward was above the M_2 credibility ceiling: an m_2 violation. The decisive sequence was, therefore, a producer strike followed a failure to punish, an inflationary implosion of real output, and the dismantling of the monitoring regime.

VII

In this section, I consider the inevitability, reversibility, and welfare implications of the collapse and I suggest some limitations of the approach followed. In the narrative of Soviet collapse that has emerged, both deterministic trends and willful acts play intrinsic roles. The trends, all adverse, were the rising costs to the dictator of punishments, rewards, and monitoring. These alone would one day have squeezed the life out of the command system. Eventually monitoring would become a loss-making activity and would be closed down.

The adverse trends were partly contingent upon the successes of the command system. I have suggested that the costs of punishments, rewards, and monitoring were bound to rise with human capital accumulation, the limits to further exploitation of mass production, and the rise of services, together with the Soviet engagement in superpower rivalry. To this extent a deterministic process was at work that would end one day in the command system's collapse. However, it does not mean that command systems are unstable under all circumstances or may collapse at any time.

The collapse was conditioned not only by deterministic trends but also by willful acts that had unintended consequences. Players made mistakes that reflected biased expectations. These arose because only one outcome of the game was normally experienced. An inexperienced dictator inheriting a high-coercion, high-effort equilibrium might underestimate how much its maintenance relied on his reputation. He might also undervalue the rents that would be lost to shirking and theft without monitoring. These might make the dictator cancel penalties or stop monitoring too easily. Beforehand Gorbachev believed that with less monitoring producers would reduce rent-seeking and increase effort: scaling down plan assignments in his 1987 industrial reform, he expected output to rise as effort was shifted into profit-seeking, market-oriented activities, but the unintended consequence was that effort and output fell. Another unintended consequence of his policies was the widespread public discrediting of basic Soviet institutions (Ellman and Kontorovich, 1998; Solnick, 1998).

Producers also probably suffered from biased expectations. Influenced by the command system's reputation for permanence, they might have colluded in a high-coercion, high-effort outcome for years beyond the point where they could have mounted a strike that would have forced the monitoring regime into insolvency. But producers might also go on strike too easily if they wrongly equated a relaxation of monitoring with limitless possibilities for securing income without effort.

If monitoring was abandoned because of mistakes arising from misperceptions, then the misperceptions should have been corrected by experience. Thus a recent survey showed 48 per cent of Russians in favour of a return to state planning and distribution; 58 per cent believed it would have been better if the country had stayed as it was before 1985.[23] Of course this too could reflect misperceptions: the evidence is that most people, including most Russians, are very poor judges of the trend in their own incomes and often tend to skew their perceptions downward.[24]

[23] *The Economist*, 18 December, 1999.

[24] In a large sample of Russian households used by Graham and Pettinato (2001), nearly three quarters of those whose incomes doubled or more between 1995 and 1998 assessed their income change negatively, whereas three quarters of those who lost income made accurate assessments.

When misperceptions were corrected, why were mistaken policy choices not reversed? If the Soviet command system was made to collapse in error, why has it not been restored? One reason is that the dictator lost reputation. To restore the command system today would require the Russian government to act much tougher — to monitor more thoroughly and increase the size of punishments — compared with the old days, just to restore this reputation. A dictator might be willing to pay for this investment if the command system could still generate enough rents. But perhaps in the last years the regime was kept running on reputation alone; if reputation were lost there was not enough revenue to restore it. Thus, though the system may have been destroyed in error, deterministic trends have prevented the error from being reversed.

The framework developed in this article throws only limited light, however, on what happened after collapse and on the accompanying redistribution of income and welfare.

The coercion-effort model interprets the Soviet collapse as the outcome of a struggle between a dictator and a producer over the distribution of income. To maintain high output, costs were incurred. These costs fell on both players: the producer bore the cost of effort, and the dictator bore the costs of coercion. Incomes exceeded payoffs when output was high, since part of the incomes of both players merely compensated them for the costs of the high state. High effort became unsustainable when high output was no longer enough to cover the combined coercion costs of the dictator and effort costs of producers. When monitoring was abandoned effort and incomes fell. In the process, incomes fell by more than welfare. The dictator's income was certain to fall. Producers' true incomes were likely to fall, and their reported incomes would certainly fall if, as seems likely, their new revenues from theft were unreported. In short, it is readily shown how a command economy of the Soviet type may look stable, in fact be stable, then be induced to collapse as a result of adverse trends and a dictator's decisions that leave everyone with a lower income than before.

To analyse the actual redistribution of producer incomes among nomenklatura capitalists, new Russians, and the impoverished rest would require extensions of the model to understand the process of power conversion.[25]

[25]'Power conversion' is the term used by Mawdsley and White (2000) for the private capitalisation of party networks and enterprises that accompanied the collapse of the command system.

Instead of a dictator and a producer, we should differentiate producers by their abilities to grab existing assets and create new ones, distinguish those elements of the dictatorship that were capable of capitalising their political associations as the endgame approached, and understand the strategy of those other elements that sought to retain political power by other means (Lazarev, 2001). To explain a general collapse of welfare would require further extensions to analyse producers' competitive asset-grabbing as a negative-sum game, or a general increase in transaction costs and loss of inter-industry coordination, giving rise to further unintended consequences (Blanchard, 1997; Solnick, 1998).

VIII

A simple model of the costs and benefits to a dictator and a producer of maintaining a command system has supported a consistent interpretation of the trends and events leading up to Soviet economic collapse.

The Soviet economy was stable until it collapsed. The Soviet economy grew along a stable trend for most of the twentieth century. Production and allocation were inefficient, and welfare outcomes fell short of what may have been feasible under different arrangements. Its economic and international successes may have conditioned those adverse trends in regime costs that ensured that one day the command system would become unstable. This does not mean that the command system was unstable by its very nature or could have collapsed at any time.

A simplified model of the command system suggests that *command economies could secure stable high output under specific historical circumstances.* Stability meant that the perceived benefits to the players in the command economy from combining high levels of effort and coercion outweighed the costs. Stability was conditional: the dictator had to be able offer a reward for effort that conformed to limits imposed by considerations of efficiency and credibility.

The stability of high output in a command system rested on the dictator's credibility and reputation. Under all circumstances high output relied on the credibility of the rewards and punishments promised by the dictator for high and low effort. Under specific circumstances, that is, when the collateral cost to the dictator of implementing punishments became particularly large, high output could also become reliant on the dictator's reputation as a intransigent and brutal leader.

The stability of the command system could be damaged by mistake. To the extent that a high-output equilibrium relied on the dictator's reputation, policy errors that undermined that reputation undermined the equilibrium. In particular, command economies may have been undermined by socialist economic reforms. These were designed to reduce monitoring costs but probably failed to do so. The cycle of reforms and counterreforms harmed regime credibility in the process.

The command system could follow two routes to insolvency. When adverse trends had sufficiently eroded the feasible space for a command economy, the high-output equilibrium could collapse in different ways involving the dictator's unforced or forced surrender. *The Soviet case took the form of a forced surrender*: in 1989, faced with a strike movement and no longer able to afford an efficient penalty, the dictator was compelled to abandon the monitoring regime.

References

Official Papers

CIA Directorate of Intelligence USSR (1988). Sharply Higher Budget Deficits Threaten Perestroyka (SOV 88–10043U), Washington, DC.
Goskomstat Rossii (2000). Rossiiskii statisticheskii ezhegodnik, Moscow.
Goskomstat SSSR (1987). Narodnoe khoziaistvo SSSR za 70 let. Iubileinyi statisticheskii sbornik, Moscow.
Goskomstat SSSR (1990). Narodnoe khoziaistvo SSSR v 1989 g. Statisticheskii ezhegodnik, Moscow.
TsSU SSSR (1986). Narodnoe khoziaistvo SSSR v 1986 g. Statisticheskii ezhegodnik, Moscow.
World Bank (1992). Russian Economic Reform: Crossing the Threshold of Structural Change, Washington, DC.

Publications

Allen, RC (2001). The Rise and Decline of the Soviet Economy. *Canadian Journal of Economics/Revue Canadien d'Ēconomique,* 34(4): 859–81.
Anderson, GM and Boettke, PJ (1993). Soviet Venality: A Rent-seeking Model of the Communist State. *Public Choice*, 93(1–2): 37–53.
Bahry, D (1993). Society Transformed? Rethinking the Social Roots of Perestroika. *Slavic Review*, 52(3): 512–54.
Bahry, D and Silver, BD (1990). Soviet Citizen Participation on the Eve of Democratization. *American Political Science Review*, 84(3): 821–47.
Baron, SH (2001). *Bloody Saturday in the Soviet Union: Novocherkassk, 1962.* Stanford, CA: Stanford University Press.
Becker, AC (1994). Intelligence Fiasco or Reasoned Accounting? CIA Estimates of Soviet GNP. *Post-Soviet Affairs*, 10(4): 291–329.
Belova, E and Gregory, PR (2002). Dictators, Loyal and Opportunistic Agents: The Soviet Archives on Creating the Soviet Economic System. *Public Choice*, 113(3–4): 265–286.

Bergson, A (1978). *Productivity and the Social System: The USSR and the West. Cambridge*. MA: Harvard University Press.

Bergson, A (1989). *Planning and Performance in Socialist Economies: The USSR and Eastern Europe*. Boston, MA: Unwin Hyman.

Berliner, JS (1976). *The Innovation Decision in Soviet Industry*. Cambridge, MA: MIT Press.

Birman, I (1980). The Financial Crisis in the USSR. *Soviet Studies*, 32(1): 84–105.

Blanchard, O (1997). The Economics of Post-communist Transition. Oxford: Oxford University Press.

Blanchflower, DG and Freeman, RB (1997). The Attitudinal Legacy of Communist Labor Relations. *Industrial and Labor Relations Review*, 50(3): 438–59.

Bornstein, M (1985). Improving the Soviet Economic Mechanism. *Soviet Studies*, 37(1): 1–30.

Brown, A (1997). *The Gorbachev Factor*. Oxford: Oxford University Press.

Brus, W (1986a). 1956 to 1965: In Search of Balanced Development. In: Kaser M (ed.) *The Economic History of Eastern Europe 1919–75, Volume 3: Institutional Change Within a Planned Economy*. Oxford: Oxford University Press, pp. 71–138.

Brus, W (1986b). 1966 to 1975: Normalization and Conflict. In: Kaser M (ed.) *The Economic History of Eastern Europe 1919–75, Volume 3: Institutional Change Within a Planned Economy*. Oxford: Oxford University Press, pp. 139–249.

Brzeski, A (1999). The end of communist economics. In: Edwards L (ed.) *The Collapse of Communism*. Stanford, CA: Hover Press, pp. 119–39.

Churchward, LG (1975). *Contemporary Soviet Government* (2nd edn.), New York: Routledge & Kegan Paul.

Dallin, A (1992). Causes of the Collapse of the USSR. *Post-Soviet Affairs*, 8(4): 279–302.

Davies, RW (1989). *The Industrialisation of Soviet Russia, Volume 3: The Soviet Economy in Turmoil, 1929–1930*. Volume 3. Basingstoke: Palgrave Macmillan.

Davies, RW (1996). *The Industrialisation of Soviet Russia, Volume 4: Crisis and Progress in the Soviet Economy, 1931–1933*. Volume 4. Basingstoke: Palgrave Macmillan.

Dearden, J, Ickes, BW and Samuelson, L (1990). To Innovate or Not to Innovate: Incentives and Innovation in Hierarchies. *American Economic Review*, 80(5): 1105–24.

Deutscher, I (1967). *The Unfinished Revolution: Russia, 1917–1967*. Oxford: Oxford University Press.

Easterly, W and Fischer, S (1995). The Soviet Economic Decline. *World Bank Economic Review*, 9: 341–71.

Ellman, M (1969). Economic reform in the Soviet Union. *Political and Economic Planning, PLANNING Broadsheet 509*, pp. 283–371.

Ellman, M (2000a). The Social Costs and Consequences of the Transformation Process. *Economic Survey of Europe*, 2–3: 125–40.

Ellman, M (2000b). The 1947 Soviet Famine and the Entitlement Approach to Famines. *Cambridge Journal of Economics*, 24: 603–30.

Ellman, M (2001). The Road from Il'ich to Il'ich: The Life and Times of Anastas Ivanovich Mikoian. *Slavic Review*, 60(1): 140–150.

Ellman, M and Kontorovich, V (eds.) (1992). *The Disintegration of the Soviet Economic System*. New York: Routledge.

Ellman, M and Kontorovich, V (eds.) (1998). *The Destruction of the Soviet Economic System: An Insiders' History*. New York: Routledge.

Filtzer, D (1986). *Soviet Workers and Stalinist Industrialization: The Formation of Modern Soviet Production Relations, 1928–1941*. New York: M E Sharpe Inc.

Filtzer, D (2007). *Soviet Workers and Late Stalinism: Labour and the Restoration of the Stalinist System after World War II*. Cambridge: Cambridge University Press.

Finifter, AW and Mickiewicz, E (1992). Redefining the Political System of the USSR: Mass Support for Political Change. *American Political Science Review*, 86(4): 857–74.

Fleron, FJ (1996). Post-Soviet Political Culture in Russia: An Assessment of Recent Empirical Investigations. *Europe-Asia Studies*, 48(2): 225–60.

Gershenson, D and Grossman, HI (2000). Cooption and repression in the Soviet Union. *International Monetary Fund Working Paper WP/00/201*, Washington, DC: IMF.

Gibson, JL (1993). Perceived Political Freedom in the Soviet Union. *Journal of Politics*, 55(4): 936–74.

Gorlizki, Y (2001). Stalin's Cabinet: The Politburo and Decision-Making in the Post-War Years. *Europe-Asia Studies*, 53(2): 291–312.

Graham, C and Pettinato, S (2001). Frustrated Achievers: Winners, Losers, and Subjective Well-Being in New Market Economies. *The Brookings Institution, Center on Social and Economic Dynamics, Working Paper Series no. 21*, Washington, DC: The Brookings Institution.

Granick, D (1987). *Job Rights in the Soviet Union: Their Consequences*. Cambridge: Cambridge University Press.

Gregory, PR (1987). Productivity, Slack, and Time Theft in the Soviet Economy. In: Millar JR (ed.) *Politics, Work, and Daily Life in the USSR: A Survey of Former Soviet Citizens*. Cambridge, pp. 241–75.

Gregory, PR (2001). The Dictator's Orders. In: Gregory PR (ed.) *Behind the Façade of Stalin's Command Economy*. Stanford, CA: Hoover Press, pp. 11–33.

Gregory, PR and Lazarev, V (2002). The Wheels of a Command Economy: Allocating Soviet Vehicles. *Economic History Review*, 55(2): 324–48.

Grossman, G (1998). Subverted Sovereignty: Historic Role of the Soviet Underground. In: Cohen SS, Schwartz A and Zysman J (eds.). *The Tunnel at the End of the Light: Privatization, Business Networks, and Economic Transformation in Russia*. University of California, International and Area Studies Research Series no. 100, Berkeley, CA, pp. 24–50.

Hanson, P (1983). Success Indicators Revisited: The July 1979 Decree on Planning and Management. *Soviet Studies*, 35(1): 1–13.

Harrison, M (1998a). Trends in Soviet Labour Productivity, 1928–1985: War, Postwar Recovery, and Slowdown. *European Review of Economic History*, 2(2): 171–200.

Harrison, M (1998b). Prices, Planners, and Producers: An Agency Problem in Soviet Industry, 1928–1950. *Journal of Economic History*, 58(4): 1032–62.

Harrison, M (2002). Economic Growth and Slowdown. In: Bacon E and Sandle M (eds.) *Brezhnev Reconsidered*. London and Basingstoke: Palgrave, pp. 38–67.

Harrison, M and Lockwood, B (2001). What Price Teaching Quality? *Royal Economic Society Newsletter*, 113: 3–5.

Harrison, M and Simonov, NS (2000). Voenpriemka: Prices, Costs, and Quality in Defence Industry. In: Harrison M and Barber JD (eds.) *The Soviet Defence Industry Complex from Stalin to Khrushchev*. Basingstoke: Palgrave Macmillan, pp. 223–45.

Hunter, H (1961). Optimum Tautness in Development Planning. *Economic Development and Cultural Change*, 9(4): 561–72.

Karklins, R (1989). The Organisation of Power in Soviet Labour Camps. *Europe-Asia Studies,* 41(2): 276–97.

Khanin, GI (1991). Novyi Etap Krisiza. *Kommunist,* 12: 71–83.

Khanin, GI (1992). Economic Growth in the 1980s. In: Ellman M and Kontorovich V (eds.) *The Disintegration of the Soviet Economic System*. New York: Routledge, pp. 73–85.

Khlevniuk, OV (1993). Prinuditel'nyi trud v ekonomike SSSR. 1929–1941 gody. *Svobodnaia mysl'*, 13: 73–84.

Khlevniuk, OV (1996). Politbiuro. Mekhanizmy politicheskoi vlasti v 1930–e gody. Moscow.

Khlevniuk, OV (2001). *Sovetskie regional'nye rukovoditeli ot Stalina do Brezhneva: Politizatsiia "nomenklatury"*. Conference on 'Serving the State: Administrative Practices in the Soviet Union', École des Hautes Études en Sciences Sociales, Paris.

Khrushchev, NS (1976). *The Secret Speech Delivered to the Closed Session of the Twentieth Congress of the Communist Party of the Soviet Union*. Introduced by Z.A. and R.A. Medvedev, Nottingham.

Kim, B-Y (1997). Soviet Household Saving Function. *Economics of Planning,* 30: 181–203.

Kim, B-Y (1999). Income, Savings, and Monetary Overhang of Soviet Households. *Journal of Comparative Economics*, 27(4): 644–68.

Kim, B-Y (2002). Causes of Repressed Inflation 1 in the Soviet Consumer Market, 1965–1989: Retail Price Subsidies, the Siphoning Effect, and the Budget Deficit. *Economic History Review*, 55(1): 105–127.

Kontorovich, V (1986). Discipline and Growth in the Soviet Economy. *Problems of Communism,* 34(6): 18–31.

Kontorovich, V (1988). Lessons of the 1965 Soviet Economic Reform. *Soviet Studies*, 40(2): 308–316.

Kontorovich, V (1993). The Economic Fallacy. *The National Interest*, 31: 35–45.
Kontorovich, V (2001). Economists, Soviet Growth Slowdown and the Collapse. *Europe-Asia Studies*, 53(5): 675–95.
Kornai, J (1986). The Hungarian Reform Process: Visions, Hopes, and Reality. *Journal of Economic Literature*, 24(4): 1687–1737.
Kornai, J (1994). Transformational Recession: The Main Causes. *Journal of Comparative Economics*, 19(1): 39–63.
Kozlov, VA (1999). *Massovye besporiadki v SSSR pri Khrushcheve i Brezhneve. 1953-nachalo 1980-kh gg.* Novosibirsk: Sibirskii khronograf.
Kuromiya, H (1988). *Stalin's Industrial Revolution: Politics and Workers, 1928–1932.* Cambridge: Cambridge University Press.
Lane, D (1976). *The Socialist Industrial State: Towards a Political Sociology of State Socialism.* Colorado: Westview Press.
Lazarev, V (2001). *Evolution of the Soviet Elite and its Post-Communist Transformation.* University of Warwick, Department of Economics, PERSA Working Paper No. 11.
Lewin, M (2001). *The Soviet nomenklatura.* Conference on 'Serving the State: Administrative Practices in the Soviet Union', École des Hautes Études en Sciences Sociales, Paris.
Machiavelli, N (1961). *The Prince.* Harmondsworth: Penguin.
Maddison, A (1995). *Monitoring the World Economy, 1820–1992.* Paris: OECD.
Maddison, A (2001). The World Economy: A Millenial Perspective. Paris: OECD.
Malia, M (1999). The Highest Stage of Socialism. In: Edwards L (ed.) *The Collapse of Communism.* Stanford, CA: Hoover Press, pp. 71–92.
Manning, RT (1993). The Soviet economic crisis of 1936–1940 and the Great Purges. In: Getty JA and Manning RT (eds.) *Stalinist Terror: New Perspectives.* Cambridge: Cambridge University Press, pp. 116–41.
Markevich, AM (2000). Otraslevye narkomaty i glavki v sovetskoi ekonomike 30-ykh gg. (na primere NKTP i GUMPa). Russian Academy of Sciences Institute of Russian History.
Mawdsley, E and White, S (2000). *The Soviet Elite from Lenin to Gorbachev: The Central Committee and its Members, 1917–1991.* Oxford: Oxford University Press.
McNeill, T (1998). Soviet studies and the collapse of the USSR: In: Cox M (ed.) *Rethinking the Soviet Collapse: Sovietology, the Death of Communism and the New Russia.* London and New York: Pinter, pp. 51–72.
Millar, JR (1985). The Little Deal: Brezhnev's Contribution to Acquisitive Socialism. *Slavic Review*, 44(4): 694–706.
Millar, JR (1987). An Overview. In: Cracraft J (ed.) *The Soviet Union Today: An Interpretive Guide* (2nd edn.). Chicago, IL: University of Chicago Press, pp. 177–90.
Millar, JR and Clayton, E (1987). Quality of life: Subjective measures of relative satisfaction. In: Millar JR, (ed.) *Politics, Work, and Daily Life in the USSR: A Survey of Former Soviet Citizens.* Cambridge: Cambridge University Press, pp. 31–60.

Mills, G and Rockoff, H (1987). Compliance with Price Controls in the United States and the United Kingdom During World War II. *Journal of Economic History,* 47(1): 191–213.

Murphy, KM, Shleifer, A and Vyshny, RW (1993). Why is Rent-Seeking so Costly to Growth? *American Economic Review Papers and Proceedings,* 83(2): 409–14.

Ofer, G (1987). Soviet Economic Growth: 1928–1985. *Journal of Economic Literature,* 25(4): 1767–1833.

Ofer, G (1989). Budget Deficits, Market Disequilibrium and Soviet Economic Reforms. *Soviet Economy,* 5(2): 107–61.

Olson, M (1993). Dictatorship, Democracy, and Development. *American Political Science Review*, 87(3): 567–76.

Olson, M (1995). The Devolution of Power in Post-Communist Societies. In: Skidelsky R (ed.) *Russia's Stormy Path to Reform.* The Social Market Foundation, Centre for Transition Economies Paper no. 1, pp. 9–42.

Pipes, R (1999). The Fall of the Soviet Union. In: Edwards, L (ed.) *The Collapse of Communism.* Stanford, CA: Hoover Press, pp. 35–49.

Pryce-Jones, D (1995). *The War That Never Was: The Fall of the Soviet Empire, 1985–1991.* London: Weidenfeld and Nicolson.

Rutland, P (1998). Sovietology: Who Got it Right, and Who got it Wrong? And Why? In: Cox, M (ed.) *Rethinking the Soviet Collapse: Sovietology, the Death of Communism and the New Russia*, London and New York: Pinter, pp. 32–50.

Schroeder, GE (1972). The "Reform" of the Supply System in Soviet Industry. *Soviet Studies*, 24(1): 97–119.

Schroeder, GE (1979). The Soviet Economy on a Treadmill of "Reforms". In: U.S. Congress, Joint Economic Committee, Soviet Economy in a Time of Change, Volume 1. Washington, DC, pp. 312–40.

Schroeder, GE (1982). Soviet Economic "Reform" Decrees: More steps on the Treadmill. In: U.S. Congress, Joint Economic Committee, Soviet Economy in the 1980s: Problems and Prospects, Part 1. Washington, DC, pp. 65–88.

Schroeder, GE (1985). The Slowdown in Soviet Industry, 1976–1982. *Soviet Economy*, 1(1): 42–74.

Schroeder, GE (1995). Reflections on Economic Sovietology. *Post-Soviet Affairs,* 11(3): 197–234.

Schumpeter, JA (1954). *Capitalism, Socialism and Democracy* (4th edn.). New York: Routledge.

Siegelbaum, LH and Walkowitz, DJ (1995). *Workers of the Donbass Speak: Survival and Identity in the New Ukraine, 1989–1992.* Albany, NY: SUNY Press.

Silver, BD (1987). Political Beliefs of the Soviet Citizen: Sources of Support for Regime Norms. In: Millar, JR (ed.) *Politics, Work, and Daily Life in the USSR: A Survey of Former Soviet Citizens.* Cambridge: Cambridge University Press, pp. 100–41.

Solnick, SL (1998). *Stealing the State: Control and Collapse in Soviet Institutions.* Cambridge, MA: Harvard University Press.
Ticktin, HH (1998). Soviet studies and the collapse of the USSR: In Defence of Marxism. In: Cox M (ed.) *Rethinking the Soviet Collapse: Sovietology, the Death of Communism and the New Russia.* London and New York: Pinter, pp. 73–94.
Treml, VG and Ellman, M (1993). Debate: Why Did the Soviet Economic System Collapse? *Radio Free Europe/Radio Liberty Research Report,* 2(23): 53–8.
Voslensky, M (1984). Nomenklatura: The Soviet Ruling Class. Garden City, NY: Doubleday.
Weitzman, ML (1970). Soviet Postwar Growth and Capital-Labor Substitutability. *American Economic Review,* 60(4): 676–92.
Weitzman, ML (1980). The "Ratchet Principle" and Performance Incentives. *Bell Journal of Economics,* 11: 302–8.
Wheatcroft, SG and Davies, RW (1999). The Soviet famine of 1932–33 and The Crisis in Agriculture. British Association for Slavonic and East European Studies Annual Conference, Cambridge.
White, S (1990). *Gorbachev in Power.* Cambridge: Cambridge University Press.
Winiecki, J (1993). Post-Soviet-Type Economies in Transition. Aldershot: Averbury Publishers.
Wintrobe, R (1990). The Tinpot and the Totalitarian: An Economic Theory of Dictatorship. *American Political Science Review,* 84(3): 849–72.
Wintrobe, R (1998). *The Political Economy of Dictatorship.* Cambridge: Cambridge University Press.

Appendix

The purpose of this appendix is to prove the conditions under which the high state may be maintained in the long run assuming complete information (propositions 1 and 2) and those under which a producer who does not expect to be punished may precipitate a forced surrender by going on strike. For lower-case symbols and definitions see section 2.1 and equations 1 to 5. The dictator's and producer's discount factors are denoted by δ_D and δ_P respectively. Bold upper-case roman symbols (**A**, **B**, **C**) denote long-run payoffs as defined below.

Proposition 1. *When effort is high and $m < s$ paying a reward is the dictator's best strategy provided he can set r such that $\delta_D \cdot f > r > e - p$.*

Proof. Under high coercion with incentives that are credible the producer's long-run payoffs to high and low effort are $\frac{1}{1-\delta_P} \cdot (y - e + r)$ and $\frac{1}{1-\delta_P} \cdot (y - p)$; call these **A** and **B** respectively. Producers' best response is high effort provided **A** > **B**, that is $r > e - p$.

Faced with high effort the dictator must choose between **C** that denotes $\frac{1}{1-\delta_D} \cdot (z + f - m - r)$ if he pays the reward promised and two alternative payoffs: $\mathbf{D} = z + f - m + \frac{\delta_D}{1-\delta_D} \cdot (z - m)$ if he cheats on rewards now but continues to monitor in the future, the difference arising because if he cheats then in future games the producer will shirk; or $\mathbf{G} = z+f-m+\frac{\delta_D}{1-\delta_D}\cdot(z-s)$ if he cheats now, then chooses not to monitor in the future, the difference arising because in the absence of monitoring the producer will steal as well as shirk.

When $m < s$ a high state will be the outcome provided both $r > e - p$ and $\mathbf{C} > \mathbf{D}$, that is $\delta_D \cdot f > r$. □

Proposition 2. *When $m > s$ paying a reward and continuing to monitor is the dictator's best strategy provided he can set r such that $\delta_D\cdot(f+s-m) > r > e - p$.*

Proof. From the proof of Proposition 1, when $m > s$, $\mathbf{D} < \mathbf{G}$, so a high state will be the outcome only if both $r > e - p$ and $\mathbf{C} > \mathbf{G}$, that is $\delta_D \cdot (f + s - m) > r$. □

Proposition 3. *Even if $r > e - p$ and rewards are credible, when $m < s$ the producer's best strategy is to strike and the dictator's best response is to cancel penalties, but maintain monitoring, if he cannot set r such that $f - q \cdot \left(\frac{1}{\delta_D} - 1\right) > r > e$.*

Proof. Faced with a strike under high coercion, the dictator must choose between $\mathbf{H} = z - m - q + \frac{\delta_D}{1-\delta_D} \cdot (z + f - m - r)$ from imposing penalties that, while costly to him, efficiently restore high effort in future games, and two alternative payoffs: $\mathbf{J} = \frac{1}{1-\delta_D} \cdot (z - m)$ if he cheats now on penalties and accepts low effort but maintains monitoring in the future, the difference arising because producers will then continue to shirk; and $\mathbf{K} = z-m+\frac{\delta_D}{1-\delta_D}\cdot(z-s)$ if he not only cheats now on penalties but also chooses low coercion in future games accepting that producers will then steal as well as shirk. Thus when $m < s$ the dictator will maintain the high state provided $\mathbf{H} > \mathbf{J}$, that is $f - q \cdot \left(\frac{1}{\delta_D} - 1\right) > r$. However, when $m > s$, $\mathbf{J} < \mathbf{K}$ and the dictator will maintain the high state only if $\mathbf{H} > \mathbf{K}$, that is $f + s - m - q \cdot \left(\frac{1}{\delta_D} - 1\right) > r$.

Under high coercion when $r > e - p$ and rewards are credible the producer must choose between high effort that yields $\mathbf{A} = \frac{1}{1-\delta_P} \cdot (y - e + r)$ and two alternative payoffs from low effort: a strike that is unpunished yields $\mathbf{L} = \frac{1}{1-\delta_P} \cdot y$, and an unpunished strike followed by a shift to low coercion yields $\mathbf{N} = y + \frac{\delta_P}{1-\delta_P} \cdot (y + s)$. Thus under high coercion the producer's best strategy is to strike if either of **L** and **N** is superior to **A**, so long as the dictator will also accept low effort without penalty.

When $m < s$ the dictator will prefer to accept low effort without penalty if he cannot set r such that $f - q \cdot \left(\frac{1}{\delta_D} - 1\right) > r$ and the producer will get **L**. If she is to be induced to reject the latter and prefer **A** the dictator must be able to set $r > e$; it is not enough that $r > e - p$. □

Proposition 4. *When $m > s$ the producer's best strategy is to strike and the dictator's best response is to cancel penalties and abandon monitoring if he cannot set r such that $f + s - m - q \cdot \left(\frac{1}{\delta_D} - 1\right) > r > e + \delta_P \cdot s$.*

Proof. From the proof of proposition 3, when $m > s$ the dictator will prefer to accept low effort without penalty if he cannot set r such that $f + s - m - q \cdot \left(\frac{1}{\delta_D} - 1\right) > r$ and the producer will get **N**. If she is to reject the latter and prefer **A** the dictator must be able to set $r > e + \delta_P \cdot s$; it is not enough that $r > e$. □

Index